W9-CLB-646

Lecture Notes in Mathematics

Edited by A. Dold and B. Eckmann

412

Classification of Algebraic Varieties and Compact Complex Manifolds

Edited by H. Popp

Springer-Verlag
Berlin · Heidelberg · New York 1974

Prof. Dr. Herbert Popp
Lehrstuhl für Mathematik VI
Universität Mannheim (WH)
68 Mannheim/BRD
Schloß

Library of Congress Cataloging in Publication Data

Popp, Herbert.
Classification of algebraic varieties and compact complex manifolds.

(Lecture notes in mathematics, 412)
1. Algebraic varieties--Addresses, essays, lectures.
2. Complex manifolds--Addresses, essays, lectures.
I. Title. II. Series: Lecture notes in mathematics (Berlin), 412.
QA3.L28 no. 412 [QA564] 510'.8s [516'.353]
74-16463

AMS Subject Classifications (1970): 13F15, 13J05, 14C99, 14D05, 14D20, 14D99, 14F05, 14H10, 14H25, 14J05, 14J10, 14J15, 14J25, 14K10, 14K15, 14K30, 14M15, 32C15, 32C45, 32G05, 32G13, 32J05, 32J99, 55F05

ISBN 3-540-06951-8 Springer-Verlag Berlin · Heidelberg · New York
ISBN 0-387-06951-8 Springer-Verlag New York · Heidelberg · Berlin

Offsetdruck: Julius Beltz, Hemsbach/Bergstr.

Vorwort

Die in diesem Band enthaltenen Beiträge von W.D. Geyer, H. Grauert, B. Moishezon, Y. Namikawa, H. Popp und K. Ueno ergeben in ihrer Gesamtheit eine Darstellung des jetzigen Standes der Klassifikationstheorie algebraischer Mannigfaltigkeiten und kompakter komplexer Mannigfaltigkeiten mit Ausnahme der Griffiths'schen Theorie der Periodenabbildungen.

Die Beiträge von W. Barth und A. Van de Ven, F. Hirzebruch, J. Lipman, J.P. Murre, F. Oort, W. Schmid und T. Shioda und N. Mitani behandeln Fragen, die in engem Zusammenhang mit der Klassifikationstheorie stehen.

Über alle Arbeiten wurde auf der Mannheimer Arbeitstagung von den jeweiligen Verfassern referiert.

Die Stiftung Volkswagenwerk hat in grosszügiger Weise diese Arbeitstagung finanziell unterstützt; die Verwaltung der Universität Mannheim hat einen guten Ablauf derselben ermöglicht. Bei beiden Institutionen darf ich mich sehr bedanken.

H. Popp

Inhaltsverzeichnis

W. Barth and A. Van de Ven — On the geometry in codimension 2 of Grassmann manifolds 1

W.D. Geyer — Invarianten binärer Formen 36

H. Grauert — Deformation kompakter komplexer Räume 70

F. Hirzebruch — Kurven auf den Hilbertschen Modulflächen und Klassenzahlrelationen 75

J. Lipman — Picard schemes of formal schemes; Application to rings with discrete divisor class group 94

B. Moishezon — Modifications of complex varieties and the Chow Lemma 133

J.P. Murre — Some results on cubic threefolds 140

Y. Namikawa — Studies on degeneration 165

F. Oort — Hyperelliptic curves over number fields 211

H. Popp — Modulräume algebraischer Mannigfaltigkeiten 219

W. Schmid — Abbildungen in arithmetische Quotienten hermitesch symmetrischer Räume 243

T. Shioda and N. Mitani — Singular abelian surfaces and binary quadratic forms 259

K. Ueno — Introduction to classification theory of algebraic varieties and compact complex spaces 288

Adressen der Autoren 333

Theorem 1. *Let α be an algebraic 2-vector bundle on $G(n,k)$, $n-k \geqslant 2$. If for each $m \geqslant 0$ there exists a bundle α_m on $G(n+m,k)$ with $i_m^*(\alpha_m) \cong \alpha$, then either α is decomposable, or $k=1$ and $\alpha \cong \omega(h)$ for a suitable $h \in \mathbb{Z}$.*

This is a generalization of Theorem I in [1]. It partially confirms a conjecture of Hartshorne's, stating that every vector bundle (of arbitrary rank) on $G(n,k)$, which is unlimitedly extendable in the sense of Theorem 1, is in fact homogeneous.

There also exists a generalization of Theorem II in [1], but we shall not consider it here. We shall however prove generalizations of Theorems III and IV.

In order to pass from bundles to submanifolds on $G(n,k)$, we shall prove the following "generalized Lefschetz theorem".

Theorem 2. *Let $V \subset G(n,k)$ be a closed algebraic submanifold of pure codimension 2. Then, if $n-k \geqslant 6$, restriction induces isomorphisms*

$$H^i(G(n,k), \mathbb{Z}) \longrightarrow H^i(V, \mathbb{Z})$$

for $i = 0,1,2$.

Using this fact one obtains from Theorem 1 :

Theorem 3. *Let $n-k \geqslant 6$ and let $V \subset G(n,k)$ be an algebraic submanifold of codimension 2. If for each m there exists a submanifold $V(m) \subset G(n+m,k)$, intersecting $i_m(G(n,k))$ transversally along V , then either V is a complete intersection of two hypersurfaces, or $k=1$, and for a suitable $h \in \mathbb{Z}$ there is a section of $\omega(h)$, vanishing exactly and transversally on V .*

The submanifold V represents a homology class $[V]$ of codimension 4 on $G(n,k)$. The integer homology group of $G(n,k)$ in this dimension is isomorphic to $\mathbb{Z} \oplus \mathbb{Z}$. Taking as a base for this group the Schubert classes of type $(n-k-2, n-k, \ldots, n-k)$ and $(n-k-1, n-k-1, n-k, \ldots, n-k)$ respectively (see Chapter 2 for details) and expressing $[V]$ in this base, we can attach to V two integers c and c' . The pair (c,c') will be called the bidegree of $[V]$ and also of V .

Theorem 4. *Let $V \subset G(n,k)$ be an algebraic submanifold of pure codimension 2 and of bidegree (c,c') . Then, if $n-k \geqslant 6$ and if furthermore*

$$c < \tfrac{1}{4}(n-k+6)$$

$$c' < \tfrac{1}{2}\sqrt{n-k-2}-1 ,$$

we have that either V is a complete intersection (and hence *c = c'), or k=1 and there exists for a suitable $h \in \mathbb{Z}$ a section of $\omega_{n,k}(h)$ vanishing exactly and transversally on V (and hence $c = h(h+1)$, $c' = c+1$).*

Theorems 3 and 4 contribute to the solution of the smoothing problem for algebraic cycles on Grassmannians, which problem is treated in [3] .

2. Preliminaries.

We shall use the following notation : if X is a (k+1)-dimensional linear subspace of $\mathbb{C}^{n+1}$, then the point in G(n,k) , representing X , will be denoted by $x \in G(n,k)$.

Furthermore, we shall make use of the flag manifold F(n,k,k-1) , i.e. the manifold of pairs (X,Y) , $X \supset Y$, where X and Y are linear subspaces in $\mathbb{C}^{n+1}$ of dimension k and k-1 respectively. There is a canonical diagram :

$$\begin{array}{ccc} F(n,k,k-1) & \xrightarrow{\;q\;} & G(n,k-1) \\ {\scriptstyle p}\big\downarrow & & \\ G(n,k) & & \end{array} \qquad (2.1)$$

where p and q are the projections of fibre bundles with

typical fibre $\mathbb{P}_n$ and $\mathbb{P}_{n-k}$ respectively.

Also, we shall have to deal with several types of Schubert cycles on $G(n,k)$ (see for example [2], p 66).

Let

$$0 \leqslant a_o \leqslant a_1 \leqslant \dots \leqslant a_k \leqslant n-k$$

be a sequence of integers, and let

$$L^{(o)} \subset L^{(1)} \subset \dots \subset L^{(k)} \subset \mathbb{C}^{n+1}$$

be a flag of linear spaces in $\mathbb{C}^{n+1}$, such that $L^{(i)}$ is of dimension a_i+i+1. Then the Schubert cycle on $G(n,k)$, corresponding to this flag, is the algebraic variety

$$\left\{ x \in G(n,k) \text{ , } \dim(X \cap L_i) \geqslant i+1 \text{ , } 0 \leqslant i \leqslant k \right\}$$

This cycle has dimension $a_o + a_1 + \dots + a_k$, and it is called a Schubert cycle of type $(a_o, a_1, \dots, a_k)$.

The special types of Schubert cycles we shall need are the following :

a) type $(n-k-1, n-k, \dots, n-k)$. The cycles of this type will be denoted by $H(L^{(o)}, \dots, L^{(k)})$, or simply by H. They are hypersurfaces on $G(n,k)$, and their homology class is dual to a generator of $H^2(G(n,k), \mathbb{Z})$.

b) type $(n-k-2, n-k, \dots, n-k)$ and type $(n-k-1, n-k-1, n-k, \dots, n-k)$. The cycles of this type will be denoted by $C(L^{(o)}, \dots, L^{(k)})$ and $C'(L^{(o)}, \dots, L^{(k)})$ or simply C and C' respectively. The duals of their homology classes form a base of $H^4(G(n,k), \mathbb{Z})$; it is this base that was referred to in the Introduction.

c) type $(0, \dots, 0, n-k)$. The cycles of this type will be denoted by S_L (or S), where $L = L^{(k-1)}$. This notation is justified by the fact that by definition S_L is the set of all $x \in G(n,k)$, such that $X \supset L^{(k-1)}$. All these Schubert cycles are isomorphic to $\mathbb{P}_{n-k}$, and they are exactly the sets $pq^{-1}(y)$, $y \in G(n,k-1)$.

d) type $(0, \dots, 0, 1, 1)$, to be denoted by $P(L, \bar{L})$, where $L = L^{(k-2)}$ and $\bar{L} = L^{(k)}$. These cycles are isomorphic to $\mathbb{P}_2$.

e) type $(0, \dots, 0, 1)$, to be denoted by $E(L, \bar{L})$, where $L = L^{(k-1)}$ and $\bar{L} = L^{(k)}$. We call these cycles the lines on $G(n,k)$, for they are the lines in the cycles S_L , and also the true lines on $G(n,k)$ if $G(n,k)$ is projectively imbedded by means of $\mathcal{O}_{G(n,k)}(1)$. Clearly they represent a generator of $H_2(G(n,k), \mathbb{Z})$.

We shall use the following simple geometric "connecting lemma's".

<u>Lemma 2.2.</u> <u>Be given two Schubert cycles S_o and S of type $(0, \ldots, 0, n-k)$ on $G(n,k)$. Then there exists a chain $S_1, \ldots, S_k$ of Schubert cycles of the same type, with $S_{i-1} \cap S_i \neq \emptyset$ for $i = 1, \ldots, k$, and with $S_k = S$.</u>

<u>Proof.</u> Let $S = S_L$ and $S_o = S_{L_o}$. Choose L_i with $\dim L_i = k$, such that

$$\dim L_i \cap L \geqslant i , \quad L_i \supset L_{i-1} \cap L ,$$

$$\dim L_i \cap L_o \geqslant k-i , \; L_i \cap L_o \subset L_{i-1} .$$

Then the S_i , with $S_i = S_{L_i}$, will have the required properties.

<u>Lemma 2.3.</u> <u>Let $E \subset G(n,k)$ be a line, intersecting some Schubert cycle S of type $(0, \ldots, 0, n-k)$. Then there is always a cycle P of type $(0, \ldots, 0, 1, 1)$, containing both E and some line in S .</u>

<u>Proof.</u> Let $E = E(N,N')$, and let $S = S_L$. By assumption, $L \cup N \subset X \subset N'$ for some $x \in E \cap S$. Then $L \cap N$ contains a linear space M of dimension $k-1$. The cycle $P = P(M,N')$

contains both E and the line $E(L,N')$.

Frequently we shall use the same notation for a Schubert cycle, the homology class represented by this cycle, and the cohomology class dual to this class.

The following intersection properties are as well-known as they are easy to verify.

$$H^2 = C + C'$$

$$HE = C'P = 1$$

$$CP = C'S = 0 \qquad (2.4)$$

CS_L is represented by the $(n-k-2)$-dimensional linear subspaces of S_L .

A closed algebraic subvariety $V \subset G(n,k)$ of codimension 2 represents a class in $H^4(G(n,k), \mathbb{Z})$ of the form $cC + c'C'$, with non-negative integers c, c' . We call the ordered pair (c, c') the <u>bidegree</u> of V .

If α is a 2-vector bundle on $G(n,k)$, we may consider its first Chern class as an integer $c_1(\alpha)$, and its second Chern class as a pair of integers $(c_2(\alpha), c_2'(\alpha))$. From (2.4) it follows that

$$c_2(\alpha|P) = c_2'(\alpha)$$

$$c_2(\alpha|S) = c_2(\alpha) .$$

We shall call the bundle α <u>normalized</u> if $c_1(\alpha)$ is either 0 or -1. For a given α, there is a unique integer h, such that $\alpha(h)$ is normalized. We call $\alpha_{norm} = \alpha(h)$ the <u>normalization</u> of α.

We recall that the bundle space of the <u>universal subbundle</u> $\omega(n,k)$ on $G(n,k)$ is the variety

$$\{(g,c) \in G(n,k) \times \mathbb{C}^{n+1} , c \in G\} ,$$

which projects canonically on $G(n,k)$.

In case $k = 1$, the bundle $\omega(n,1)$ can also be described in the following way, using diagram (2.1). On $G(n,k-1) \simeq \mathbb{P}_n$, we can identify $\mathbb{C}^{n+1}$ with the dual of $\Gamma(\mathcal{O}_{\mathbb{P}}(1))$, and $X \subset \mathbb{C}^{n+1}$ with the dual of $\Gamma(p^{-1}(x), \mathcal{O}_{\mathbb{P}}(1))$. Then $\omega(n,1)$ is dual to the bundle $p_*q^*\mathcal{O}_{\mathbb{P}}(1)$, and also isomorphic to $(p_*q^*\mathcal{O}_{\mathbb{P}}(1))(-1)$. As to the Chern classes of $\omega(n,1)(h)$, for all $h \in \mathbb{Z}$ they are given by

$$\text{(2.5)} \qquad \begin{aligned} c_1(\omega(n,1)(h)) &= 2h - 1 \\ c_2(\omega(n,1)(h)) &= (h^2 - h,\ h^2 - h + 1) \end{aligned}$$

3. A generalized Lefschetz theorem for Grassmannians.

For projective spaces the following generalized Lefschetz theorem holds.

Theorem 3.1. Let V be a subvariety of $\mathbb{P}_n$ of (pure) dimension d. Then restriction induces isomorphisms in cohomology

$$H^i(\mathbb{P}_n, G) \xrightarrow{\sim} H^i(V, G) ,$$

a) with $G = \mathbb{Z}$, and for $0 \leqslant i \leqslant 2d-n$ if V is non-singular ([5] , Theorem) ;

b) with $G = \mathbb{C}$, and for $0 \leqslant i \leqslant 2d-n$ if V carries a (possibly non-reduced) structure sheaf $\mathcal{O}_V$ which makes $(V, \mathcal{O}_V)$ locally a complete intersection ([7] , Theorem 2.3.3) ;

c) with $G = \mathbb{Z}$, if $n \geqslant 6$ and for $i = 0, 1, 2$ under the same conditions as in b) ([7] , Theorem 4.9) .

We shall need the following partial generalization of c) , the proof of which was communicated to us by A. Ogus.

Theorem 3.2. Let V be a subvariety of $\mathbb{P}_n$, $n \geqslant 6$, defined by an ideal sheaf J , which is locally generated by two functions. Then restriction induces isomorphisms

$$H^i(\mathbb{P}_n, \mathbb{Z}) \longrightarrow H^i(V, \mathbb{Z})$$

for $i = 0, 1, 2$.

<u>Proof.</u> From the assumption it follows that $V = A \cup B'$, where A is a divisor and where B' has codimension 2 everywhere, and where no component of B' is contained in A . Then $J \subset J_A \simeq \mathcal{O}(-g)$, where g is the degree of A . Using the embedding $J(g) \subset J_A(g) \simeq \mathcal{O}_{\mathbb{P}}$, we can consider $J(g)$ as an ideal sheaf in $\mathcal{O}_{\mathbb{P}}$. The sheaf $J(g)$ is locally isomorphic to J , and therefore also $J(g)$ is locally generated by two functions. The support B of the quotient sheaf $\mathcal{O}_{\mathbb{P}}/J(g)$ coincides with B' outside A , and the space $(B, \mathcal{O}_{\mathbb{P}}/J(g))$ is locally a complete intersection of codimension 2 . If we apply Theorem 3.1 , part c) to the space B , we find that restriction induces isomorphisms

$$H^i(\mathbb{P}_n, \mathbb{Z}) \longrightarrow H^i(B, \mathbb{Z}), \qquad i = 0, 1, 2 .$$

The same holds if we replace B by the divisor A . Furthermore, by [7], Theorem 2.3.1 , we have that $H^1(A \cap B, \mathbb{Z}) = 0$.
From the Mayer-Vietoris sequence

$$\longrightarrow H^0(A, \mathbb{Z}) \oplus H^0(B, \mathbb{Z}) \longrightarrow H^0(A \cap B, \mathbb{Z}) \longrightarrow$$

$$\longrightarrow H^1(V, \mathbb{Z}) \longrightarrow H^1(A, \mathbb{Z}) \oplus H^1(B, \mathbb{Z}) \longrightarrow H^1(A \cap B, \mathbb{Z}) \longrightarrow$$

$$\longrightarrow H^2(V,\mathbb{Z}) \longrightarrow H^2(A,\mathbb{Z}) \oplus H^2(B,\mathbb{Z}) \longrightarrow H^2(A \cap B,\mathbb{Z}) \rightarrow$$

we find first of all that $H^1(V,\mathbb{Z}) = 0$, since $A \cap B$ is connected by the connectedness theorem. Furthermore, by combining part of this sequence with the sequence

$$0 \rightarrow H^2(\mathbb{P}_n,\mathbb{Z}) \rightarrow H^2(\mathbb{P}_n,\mathbb{Z}) \oplus H^2(\mathbb{P}_n,\mathbb{Z}) \longrightarrow H^2(\mathbb{P}_n,\mathbb{Z}) \text{---} 0$$

where the second arrow sends c to $(c,-c)$ and where the third arrow maps (c,d) in $c+d$, we obtain a commutative diagram

$$\begin{array}{ccccccc} 0 \rightarrow & H^2(\mathbb{P}_n,\mathbb{Z}) & \rightarrow & H^2(\mathbb{P}_n,\mathbb{Z}) \oplus H^2(\mathbb{P}_n,\mathbb{Z}) & \rightarrow & H^2(\mathbb{P}_n,\mathbb{Z}) & \rightarrow 0 \\ & \downarrow & & \downarrow\wr & & \downarrow & \\ 0 \rightarrow & H^2(V,\mathbb{Z}) & \rightarrow & H^2(A,\mathbb{Z}) \oplus H^2(B,\mathbb{Z}) & \rightarrow & H^2(A \cap B,\mathbb{Z}) & , \end{array}$$

with vertical arrows denoting restriction homomorphisms. Since $A \cap B$ has positive dimension, the last vertical arrow is injective. This implies that the first vertical arrow gives an isomorphism.

We now are ready to prove the main result of this chapter.

<u>Theorem 3.3. Let V be a non-singular subvariety in $G(n,k)$ of (pure) codimension 2 . Then for $i = 0, 1, 2$ restriction induces an isomorphism</u>

$$H^i(G(n,k),\mathbb{Z}) \longrightarrow H^i(V,\mathbb{Z}) ,$$

<u>provided that $n-k \geqslant 6$.</u>

<u>Proof.</u> We use diagram (2.1).

Let $W = p^{-1}(V)$. Then we claim: if restriction induces isomorphisms

$$H^i(F, \mathbb{Z}) \longrightarrow H^i(W, \mathbb{Z}), \qquad 0 \leq i \leq i_o ,$$

then restriction also induces isomorphisms

$$H^i(G(n,k), \mathbb{Z}) \longrightarrow H^i(V, \mathbb{Z}), \qquad 0 \leq i \leq i_o .$$

To prove this, we use the following facts ([9] , Lemma II.1). Let $u \in H^2(F, \mathbb{Z})$ restrict to a generator of $H^2(p^{-1}(x), \mathbb{Z})$, $x \in G(n,k)$. Then every element $a \in H^i(F, \mathbb{Z})$ can be written in a unique way as

$$p^*(a_o) + p^*(a_1)u + \ldots + p^*(a_m)u^m$$

where $a_j \in H^{i-2j}(G(n,k), \mathbb{Z})$ and $m = \min([\tfrac{1}{2}i], k)$.

A similar fact holds for the elements of $H^i(W, \mathbb{Z})$, provided that we replace $G(n,k)$ by V and p by the restriction p_W of p to W. Using this facts, both the injectivity and the surjectivity of the restriction homomorphisms

$$H^i(G(n,k), \mathbb{Z}) \longrightarrow H^i(V, \mathbb{Z})$$

can be verified immediately.

To prove the statement of the theorem for F and W,

we distinguish between three cases :

i) $\mathrm{cod}_{G(n,k-1)} q(W) = 2$;

ii) $\mathrm{cod}_{G(n,k-1)} q(W) = 1$;

iii) $q(W) = G(n,k-1)$.

Since W is of codimension 2 in F , these three cases exhaust all possibilities.

In fact, case i) cannot occur. For in this case we would have $q^{-1}q(W) = W$, and by Lemma 2.2 , V would be all of $G(n,k)$, a contradiction.

In the case ii) , $q(W)$ is a hypersurface, and hence a hyperplane section for a suitable embedding of $G(n,k-1)$ in a projective space. Thus from Lefschetz theorem we derive that restriction induces isomorphisms

$$H^{\ell}(G(n,k-1), \mathbb{Z}) \longrightarrow H^{\ell}(q(W), \mathbb{Z}) , \quad 0 \leq \ell \leq k(n-k)-2 .$$

Furthermore, each fibre $q_W^{-1}(x)$, $x \in q(W)$ of $q_W = q|W$ is either a hypersurface in $q^{-1}(x)$ or equal to $q^{-1}(x)$. Therefore, restriction induces again isomorphisms

$$H^m(q^{-1}(x), \mathbb{Z}) \longrightarrow H^m(q_W^{-1}(x), \mathbb{Z}) , \quad 0 \leq m \leq n-k-2 .$$

Since a fibre of q_W has a neighbourhood on W , which can be retracted into that fibre (this follows for example from Lojasiewics results on the triangulability of algebraic sets, see [6]), there are isomorphisms for the direct images of the constant sheaves $\mathbb{Z}_F$ and $\mathbb{Z}_W$:

$$R^m q_*(\mathbb{Z}_F)|q(W) \longrightarrow R^m(q_W)_* \mathbb{Z}_W , \qquad 0 \leq m \leq n-k-2 .$$

Using Leray's spectral sequence for both q and q_W, we obtain a commutative diagram

$$\begin{array}{ccccc} & & H^\ell(G(n,k-1), R^m q_* \mathbb{Z}_F) & \Longrightarrow & H^{\ell+m}(F,\mathbb{Z}) \\ & \nearrow\sim & \downarrow & & \downarrow \\ H^\ell(q(W), R^m q_* \mathbb{Z}_F) & & & & \\ & \searrow\sim & & & \\ & & H^\ell(q(W), R^m(q_W)_* \mathbb{Z}_W) & \Longrightarrow & H^{\ell+m}(W,\mathbb{Z}) \end{array}$$

It is well known that the first spectral sequence degenerates on the E_2-level. Consequently, for the second spectral sequence $d_2^{\ell,m} = 0$ for $0 \leq \ell \leq k(n-k)-2$, $0 \leq m \leq n-k-2$. It follows that restriction induces isomorphisms

$$H^i(F,\mathbb{Z}) \longrightarrow H^i(W,\mathbb{Z})$$

for $0 \leq i \leq n-k-2$, in particular for $i = 0, 1, 2$, since $n-k \geq 6$.

Finally, in the case iii) we use again Leray's spectral sequence to obtain a commutative diagram

$$\begin{array}{ccc} H^{\ell}(G(n,k-1), R^m q_* \mathbb{Z}_F) & \Longrightarrow & H^{\ell+m}(F, \mathbb{Z}) \\ \downarrow & & \downarrow \\ H^{\ell}(q(W), R^m(q_W)_* \mathbb{Z}_W) & \Longrightarrow & H^{\ell+m}(W, \mathbb{Z}) \end{array}$$

It will be sufficient to show that the restriction $q|W = q_W$ induces an isomorphism of sheaves

$$R^m q_* \mathbb{Z}_F \longrightarrow R^m((q_W)_* \mathbb{Z}_W)$$

for $i = 0, 1, 2$. Again this may be done fibre-wise. But for $x \in G(n,k-1)$ the fibre $q_W^{-1}(x)$ is a subvariety in $q^{-1}(x)$, defined by an ideal sheaf locally generated by two functions. Thus our claim is a consequence of Theorem 3.2 (note that $\dim q^{-1}(x) = n-k \geqslant 6$).

4. A class of 2-vector bundles on Grassmannians.

In this chapter we want to study those 2-vector bundles α on $G(n,k)$, which have the property that all restrictions $\alpha|S_L$ are decomposable. Among these bundles are, apart from the decomposable bundles on $G(n,k)$, the bundles $\omega(n,1)(h)$ on $G(n,1)$. For if we look at diagram (2.1), then in this case L is a point in $G(n,k-1) = \mathbb{P}_n$, and if J is a hyperplane of $\mathbb{P}_n$, not containing L, then (g,L) and $(g, qp^{-1}(g) \cap J)$,

$g \in S_L$, are two regular sections of $\mathrm{Proj}(\omega(n,1)(h))$ which never meet. In fact, the two forementioned types exhaust already all possibilities.

<u>Theorem 4.1.</u> <u>Let α be a 2-vector bundle on $G(n,k)$, $n-k \geqslant 2$, such that the restriction of α to all cycles S_L is decomposable. Then either α is itself decomposable, or $k = 1$ and $\alpha \simeq \omega(n,1)(h)$ for a suitable $h \in \mathbb{Z}$.</u>

<u>Proof.</u> By assumption, the cycles S_L have dimension at least 2 . Since the Chern classes of $\alpha | S_L$ are independent of L , for all S_L we have :

$$\alpha | S_L \simeq \mathcal{O}_{\mathbb{P}_{n-k}}(a_1) \oplus \mathcal{O}_{\mathbb{P}_{n-k}}(a_2) ,$$

with a_1, a_2 independent of L . We may of course assume that $a_1 = 0$ and $a_2 = -a \leqslant 0$. As in [1] , we can apply [8] , p 40 to the proper map q and the $\mathcal{O}_{G(n,k-1)}$-flat sheaf $p^*(\alpha)$. We find that $q_* p^*(\alpha)$ is locally free of rank 1 if $a \neq 0$, and locally free of rank 2 if $a = 0$. Furthermore, $q^* q_* p^*(\alpha)$ is a 1-subbundle of $p^*(\alpha)$ in the first case, and isomorphic to $p^*(\alpha)$ in the second. From now on we consider both cases separately.

If $a \neq 0$, we put $q_* p^*(\alpha) = \mathcal{O}_{G(n,k-1)}(\ell)$. Restricting to the fibres of p, we immediately find that $\ell \leq 0$. If $\ell = 0$, the bundles $q_* p^*(\alpha)$ and $q^* q_* p^*(\alpha)$ are trivial, and the direct image under p of this last bundle gives a trivial 1-subbundle of α itself. This implies that α is decomposable. If $\ell < 0$, we find in the same way as in [10] (p. 247) that the fibres of p must have dimension 1, i.e. $k = 1$. We fix a Schubert cycle $P(M,M')$, which we can also denote by $P(M')$, for in this case M is empty. Since every line in $P(M')$ can be obtained as the intersection of $P(M')$ with a cycle S_L, the restriction $\alpha \mid P(M')$ is uniform. Restricting everything to $P(M')$, we obtain from diagram (2.1) the following one :

$$\begin{array}{ccc} F' & \xrightarrow{q'} & G(2,0) \\ {\scriptstyle p'}\downarrow & & \\ P(M') & & \end{array}$$

Here $G(2,0) \subset G(n,0)$ is given by M', furthermore $F' = q^{-1}(G(2,0)) = F(2,1,0) \subset F(n,1,0)$, and p' and q' are the restrictions of p and q respectively. But now we have obtained exactly the situation described in [10] on p. 247. Thus we find

$$\mathcal{O}_{G(n,0)}(\ell)|G(2,0) = (q')_*(p')^*\alpha = \mathcal{O}_{G(2,0)}(-1) ,$$

and hence $\ell = -1$. Now, since the quotient bundle

$p^*(\alpha)/\, q^*(\mathcal{O}_{G(n,0)}(-1))$ is isomorphic to some line bundle

$p^*(\mathcal{O}_{G(n,1)}(h)) \otimes q^*\mathcal{O}_{G(n,0)}(1)$, it follows from the exact sequence

$0 \longrightarrow q^*(\mathcal{O}_{G(n,0)}(-1)) \longrightarrow p^*(\alpha) \longrightarrow p^*(\alpha)/\, q^*(\mathcal{O}_{G(n,0)}(-1)) \rightarrow 0$

that

$$\alpha \simeq p_*p^*(\alpha) \simeq \mathcal{O}_{G(n,1)}(h) \otimes p_*(\mathcal{O}_{G(n,0)}(1)) \simeq$$
$$\simeq \omega(n,1)^*(h) \simeq \omega(n,1)(h+1) ,$$

which had to be proved.

If $a = 0$, since $p^*(\alpha)|p^{-1}(x)$ is trivial for every point $x \in G(n,k)$, so are $q^*(\alpha')|p^{-1}(x)$ and $\alpha'|qp^{-1}(x)$, where $\alpha' = q_*p^*(\alpha)$. Since every line $E \subset G(n,k-1)$ is contained in some set $qp^{-1}(x)$, the restriction $\alpha'|E$ is trivial for all lines E in $G(n,k-1)$. So, if L' is a (k-2)-dimensional linear subspace of $\mathbb{C}^{n+1}$, then by [10] the restriction $\alpha'|S_{L'}$ is trivial. Consequently we can apply induction with respect to k , starting with the known case $k = 1$, provided that we show : if α' is trivial, then so is α . But this is clear : if α' is trivial, then $q^*(\alpha') = p^*(\alpha)$ is trivial, and $\alpha = p_* p^*(\alpha)$ is trivial too.

5. A review of earlier results.

In this chapter we recall in a word or two the results of [1].

Let α be a 2-vector bundle on $\mathbb{P}_1$. By a theorem of Grothendieck α is decomposable, i.e.

$$\alpha \simeq \mathcal{O}_{\mathbb{P}}(a_1) \oplus \mathcal{O}_{\mathbb{P}}(a_2) \quad ,$$

with $a_1, a_2 \in \mathbb{Z}$ uniquely determined up to permutation.

We put

$$b(\alpha) = \begin{cases} \frac{1}{2}\,|a_1 - a_2| & \text{if } c_1(\alpha) \text{ is even} \\ \frac{1}{2}\,|a_1 - a_2| - \frac{1}{2} & \text{if } c_1(\alpha) \text{ is odd .} \end{cases}$$

If α is a 2-bundle on $\mathbb{P}_n$, we put

$$b(\alpha) = \min_{L \text{ line in } \mathbb{P}_n} b(\alpha | L)$$

$$B(\alpha) = \max_{L \text{ line in } \mathbb{P}_n} b(\alpha | L)$$

If $b(\alpha) = B(\alpha)$, i.e. when α is uniform, then α is either decomposable or projectively equivalent to the tangent bundle of $\mathbb{P}_2$ (this fact was already used in Chapter 4).

Concerning the numbers $b(\alpha)$ and $B(\alpha)$ - which in a sense measure the complexity of α - we have ([1], Proposition 3.3) :

Proposition 5.1. For each 2-bundle α on $\mathbb{P}_2$ the inequality

$$B(\alpha) \leqslant c_2(\alpha_{norm}) + (b(\alpha) + 1)^2$$

holds, where α_{norm} is the normalization of α.

Furthermore, from the Propositions 3.1 and 3.2 in [1] we can conclude

Theorem 5.2. Let α be a 2-vector bundle on $\mathbb{P}_n$ with

$$B(\alpha) < \tfrac{1}{4}(n - 2) .$$

Then α is decomposable.

The following result slightly improves upon Theorem 4.6 of [1] .

Theorem 5.3. Let α be a 2-bundle on $\mathbb{P}_n$, $n \geqslant 3$. If α has a section vanishing transversally and precisely on a submanifold V of degree g , with $n > 4g - 6$, then α is decomposable (and hence V is a complete intersection).

Proof. Let s be the section. By [1], (4.9) we have $4 \leqslant c_1(\alpha) \leqslant g + 1$.

If $L \subset \mathbb{P}_n$ is a line not contained in V, then $s|L$ is a section in $\alpha|L$ with at most g zeros (counted with multiplicity). Let $\alpha|L \simeq \mathcal{O}_{\mathbb{P}_1}(a_1) \oplus \mathcal{O}_{\mathbb{P}_1}(a_2)$. Then either both integers a_1 and a_2 are non-negative - which implies

$$|a_1 - a_2| \leqslant a_1 + a_2 = c_1(\alpha) \leqslant g + 1 ,$$

and therefore $b(\alpha|L) \leqslant \frac{1}{2}(g+1)$ - or one of them, say a_1, is negative. In this case the section $s|L$ is contained in the subbundle isomorphic to $\mathcal{O}_{\mathbb{P}_1}(a_2)$, and $a_2 \leqslant g$. Then $a_1 \geqslant 4 - g$, and $|a_1 - a_2| \leqslant 2g - 4$, which implies $b(\alpha|L) \leqslant g - 2$. In both cases it follows that $b(\alpha|L) \leqslant g - 2$, since we may assume $g \geqslant 3$ without any loss of generality.

If L is a line contained in V, then we observe that $\alpha|V$ is isomorphic to the normal bundle of V in $\mathbb{P}_n$, and therefore generated by its global sections. As in the first case above (a_1 and a_2 both non-negative) we obtain $b(\alpha|L) \leqslant \frac{1}{2}(g+1)$.

We conclude that in any case $B(\alpha) \leqslant g - 2$, and our claim follows from <u>Theorem 5.3</u>.

Another consequence of Theorem 5.3 ([1] , Theorem 3.8) is

Theorem 5.4. Let α be a 2-bundle on $\mathbb{P}_n$. Assume that for each standard embedding $\mathbb{P}_n \subset \mathbb{P}_m$ there exists a bundle α_m on $\mathbb{P}_m$ with $\alpha_m | \mathbb{P}_n \simeq \alpha$. Then α is decomposable.

6. Proof of the main results.

Theorem 1 from the Introduction follows easily from Theorems 5.4 and 4.1 . In fact, the standard embedding of $G(n,k)$ in $G(n+m,k)$ maps every cycle S_L $(\simeq \mathbb{P}_{n-k}) \subset G(n,k)$ into a cycle S_L' $(\simeq \mathbb{P}_{n+m-k}) \subset G(n+m,k)$. So if α is a 2-bundle on $G(n,k)$, which can be extended to $G(n+m,k)$ for all $m \geqslant 0$, then by Theorem 5.4 the restricted bundles $\alpha | S_L$ must all be decomposable. By Theorem 4.1 therefore α is either itself decomposable, or $\alpha \simeq \omega(n,1)(h)$ for a suitable $h \in \mathbb{Z}$.

For the proof of the other theorems stated in the Introduction we need some preliminaries.

Let V be a submanifold of pure codimension 2 in $G = G(n,k)$, with $n-k \geqslant 6$. Then we know from Chapter 3 that V is connected, that $H^1(V,\mathcal{O}_V) = 0$ and that restriction induces an isomorphism $H^2(G,\mathbb{Z}) \longrightarrow H^2(V,\mathbb{Z})$. Since also $H^2(G,\mathcal{O}_G) = 0$, it follows that $H^1(G,\mathcal{O}_G^*)$ is mapped isomorphically onto $H^1(V,\mathcal{O}_V^*)$. In particular, there is a line bundle $\mathcal{O}_G(\ell)$, restricting to $\overset{2}{\Lambda}\,\nu_{V/G}$ on V.

<u>Proposition 6.1.</u> <u>There is a 2-bundle $\alpha = \alpha_V$ on G with the following properties :</u>

i) <u>$\Gamma(\alpha)$ contains a section vanishing transversally and precisely on V ;</u>

ii) <u>$c_1(\alpha) = \ell$, where $\mathcal{O}_V(\ell) = \overset{2}{\Lambda}\,\nu_{V/G}$;</u>

iii) <u>$c_2(\alpha) = (c,c')$, the bidegree of V.</u>

<u>Proof.</u> Because of the remark preceding this proposition, and since $H^1(G,\mathcal{O}_G(h)) = H^2(G,\mathcal{O}_G(h)) = 0$ for all $h \in \mathbb{Z}$, the proposition can be proved by the methods of the first part of Chapter 4 in [1].

<u>Remark.</u> For a more formal proof of this result, see [4], Proposition 6.1.

If $S_L \subset G$ is a Schubert cycle in general position with respect to V, then V and S_L intersect transversally and $V \cap S_L$ is non-singular of codimension 2 in $S_L \simeq \mathbb{P}_{n-k}$ (Appendix, Lemma 1). If (c,c') is the bidegree of V, then the degree of $V \cap S_L$ is c, because of (2.4). Thus we obtain :

Proposition 6.2. Assume that in the situation of Proposition 6.1 the variety V has bidegree (c,c'), and that

$$n - k > 4c - 6 .$$

Then the restriction $\alpha_V|S_L$ is decomposable for every cycle S_L, in general position with respect to V.

To apply Theorem 4.1, however, we need that *all* restrictions $\alpha_V|S_L$ are decomposable. To arrive at this conclusion, we first prove

Proposition 6.3. Assume $c \neq 0$. Let S_L be a Schubert cycle intersecting another cycle $S^o = S_{L_o}$ on which α is decomposable. Then

$$B(\alpha|S_L) \leqslant (c_2'(\alpha) + 1)^2 .$$

Proof. Let $\alpha|S^o \simeq \mathcal{O}_{S^o}(a_1) + \mathcal{O}_{S^o}(a_2)$. Then $a_1 a_2 = c_2(\alpha|S^o) = c_2(\alpha)$, because of (2.4). In particular we have that $1 \leqslant a_i \leqslant c_2(\alpha)$ for $i = 1,2$.

Hence

$$(6.4) \qquad B(\alpha|S^o) \leqslant \tfrac{1}{2}|a_1 - a_2| \leqslant \tfrac{1}{2}(c_2(\alpha) - 1) .$$

Furthermore, from the definition of α_{norm} it follows that

(6.5) $$(B(\alpha|S^o))^2 \leqslant -c_2(\alpha_{norm}) .$$

Now let $E \subset S_L$ be a line intersecting S^o, and let $P = P(M,M') \simeq \mathbb{P}_2$ be a Schubert cycle, containing E and a line $E^o \subset S^o$. Such a P exists by Lemma 2.3. From Proposition 5.1 and the inequalities (6.4) and (6.5) we obtain :

$$\begin{aligned} b(\alpha|E) &\leqslant c_2((\alpha|P)_{norm}) + (b(\alpha|E^o) + 1)^2 \leqslant \\ &\leqslant c_2'(\alpha_{norm}) - c_2(\alpha_{norm}) + c_2(\alpha) \leqslant \\ &= c_2'(\alpha) \end{aligned},$$

since $c_2(\alpha|P) = c_2'$ in view of (2.4), and $c_2'(\alpha_{norm}) - c_2(\alpha_{norm}) = c_2'(\alpha) - c_2(\alpha)$. To estimate $b(\alpha|E')$ for an arbitrary line $E' \subset S_L$, we may assume to have chosen E such that it intersects E' too. Then we can apply Proposition 5.1 once more, to get

$$\begin{aligned} b(\alpha|E') &\leqslant c_2(\alpha_{norm}) + (c_2'(\alpha) + 1)^2 \leqslant \\ &\leqslant (c_2'(\alpha) + 1)^2 \end{aligned},$$

for $c_2(\alpha_{norm})$ is always negative.

As a consequence, we obtain

<u>Theorem 6.6.</u> <u>Let $V \subset G(n,k)$, $n-k \geqslant 6$, be a submanifold of pure codimension 2 and bidegree (c,c'). If $c \neq 0$ and</u>

$$\underline{n - k > \max \{ 4c - 6, 4(c' + 1)^2 + 2 \}} ,$$

<u>then all restrictions $\alpha | S_L$ are decomposable.</u>

<u>Proof.</u> If $c \neq 0$, by Proposition 6.2 $\alpha_V | S_L$ is decomposable for general S_L. By Proposition 6.3 and Theorem 5.2 $\alpha_V | S_L$ is decomposable for every cycle S_L intersecting some cycle already having this property. Thus Lemma 2.3 can be applied.

Theorem 6.6 and Theorem 4.1 together immediately yield Theorem 4 in Chapter 1 for the case that $c \neq 0$. The case $c = 0$ is covered by the Appendix. Since Theorem 3 is a trivial consequence of Theorem 4, we now have proved all of the results stated in the Introduction.

7. Some comments.

In this paper we have proved theorems about algebraic submanifolds in Grassmannians by intersecting them with the many projective spaces, contained in the Grassmann varieties, and then using the corresponding facts for submanifolds of projective spaces. Though of course more direct methods are to be preferred, our approach nevertheless illustrates how easily results about vector bundles on projective spaces can be used to obtain the same type of results for vector bundles on Grassmannians.

It is amusing to compare the conclusions of this paper with the content of the recent paper [3] by Hartshorne, Rees, and Thomas. For example, their Theorem 3 reads in our language :

The bidegree (c, c') of any 2-codimensional algebraic submanifold in $G(5,1)$ satisfies either $c - c' \equiv 0 \bmod 4$ or $c - c' \equiv -1 \bmod 4$. The bidegree of a 2-codimensional algebraic submanifold in $G(6,2)$ satisfies $c - c' \equiv 0 \bmod 4$.

Using standard embeddings one finds that the first statement holds for all $G(n,1)$, $n \geqslant 5$, and the second for all $G(n,k)$, with $k \geqslant 2$ and $n-k \geqslant 4$.

The common theme of both papers seems to be that - at least for n large enough - the bidegrees of 2-codimensional algebraic submanifolds concentrate on the diagonal $c = c'$, with the exception of the case $k = 1$, for which also the line $c' = c - 1$ is allowed.

The advantage of the topological method of [3] is, that the results hold without bounds for c and c'. On the other hand, our paper shows that the equations $c - c' = 0$ or $c - c' = -1$ are actually true, and not only mod 4 . We also get more information about the nature of the subvarieties which exist. For it is not obvious - and at least on $G(3,1)$ not true - that a subvariety of bidegree $(c\ ,c\ )$ is a complete intersection (a classical example is obtained by intersecting $G(3,1)$ - which is a quadric in $\mathbb{P}_5$ - with the image of a generic Segre embedding of $\mathbb{P}_1 \times \mathbb{P}_2$ in $\mathbb{P}_5$).

Appendix.

There are in Chapter 6 a few places where we have used a couple of results which, properly speaking, belong to the elementary projective geometry of Grassmann manifolds. Since we do not know of a suitable reference for them, we shall prove these results in this appendix. But we shall do just that and nothing more ; we shall not seek any generality, nor shall we be fastidious as far as our proofs are concerned.

Lemma 1 . *Let V be a submanifold of codimension 2 and bidegree (c,c') in $G(n,k)$, with $1 \leqslant k \leqslant n-2$. If the Schubert* cycle S_L in $G(n,k)$ is chosen in general position, then V and L intersect transversally along $V \cap S_L$, this intersection being a submanifold of codimension 2 and degree c in $S_L \simeq \mathbb{P}_{n-k}$. In particular, $V \cap S_L$ is empty for general S_L if and only if $c = 0$.

Proof. As in the proof of Theorem 3.3, let $W = p^{-1}(V) \subset F$. If z is the point in $G(n,k-1)$, corresponding to the linear subspace $Z \subset \mathbb{C}^{n+1}$, then $p|q^{-1}(z)$ maps $W_z = W \cap q^{-1}(z)$ isomorphically onto $V \cap S_Z$. When $q(W)$ is all of $G(n,k-1)$, then $q|W$ is regular along the general fibre W_z by Bertini's theorem. This means that V and S_Z intersect transversally along $V \cap S_Z$. By (2.4), the smooth intersection $V \cap S_Z$ has degree c in $S_Z \simeq \mathbb{P}_{n-k}$. Finally, $q(W)$ is a proper subset of $G(n,k-1)$ if and only if $c = 0$.

Lemma 2. Let V be a submanifold of codimension 2 and bidegree $(0,c')$ in $G(n,k)$, with $n-k \geq 6$. Then $k = 1$ and V is a Schubert cycle C'. In particular, $c' = 1$.

Proof. To start with, we consider the case $k = 1$. Since for every submanifold both c and c' are always non-negative, it is sufficient to prove that every irreducible submanifold V is a Schubert cycle C'. Again we use diagram (2.1), putting $p^{-1}(V) = W$ and $W_z = W \cap q^{-1}(z)$. By the argument used in the proof of Theorem 3.3, we find that $q(W) \subset G(n,k-1) \simeq \mathbb{P}_n$ is an irreducible hypersurface. Now for each point $z \in q(W)$, the variety $q(p^{-1}(p(W_z)))$ has codimension 1 in $\mathbb{P}_n$, and it is contained in $q(W)$. Since $q(W)$ is irreducible, we find $q(W) = q(p^{-1}(p(W_z)))$ for all $z \in q(W)$, i.e. $q(W)$ is a cone over all of its points. It follows that $q(W)$ is a hyperplane H in $\mathbb{P}_n$, and V coincides with the Schubert cycle

$$C' = \left\{ x \in G(n,1) \; ; \; X \subset H \right\} .$$

In the case $k \geq 2$ we again can restrict ourselves to the case that V is irreducible. We consider the diagram

$$\begin{array}{ccc} F(n,k,k-2) & \xrightarrow{g} & G(n,k-2) \\ f \downarrow & & \\ G(n,k) & & \end{array}$$

which is built in the same way as (2.1). We set $U = f^{-1}(V)$. If the variety $U_y = U \cap g^{-1}(y)$ has codimension 2 in $g^{-1}(y)$, then it represents a class of bidegree $(0,c')$ on $g^{-1}(y) \simeq G(n-k+1,1)$. From the case $k = 1$ it follows that

$c' = 1$, and U_y is a Schubert cycle C' . Let $\alpha = \alpha_V$ be the 2-bundle provided by Proposition 6.1 . Since $\alpha | g^{-1}(y)$ is uniquely determined by U_y (this follows from the construction of α_V , see [1] , Chapter 4) , we have that

$\alpha | g^{-1}(y) \simeq \omega(n-k+1,1)(1)$ for all $y \in G(n,k-2) \setminus T$, where T is the subvariety of points $y \in G(n,k-2)$, with $\operatorname{codim}_{g^{-1}(y)} U_y \leq 1$. T has codimension 2 at least.

From the diagram

$$\begin{array}{ccc} F(n,k-1,k-2) & \longrightarrow & G(n,k-2) \\ \downarrow & & \\ G(n,k-1) & & \end{array}$$

we deduce, as in the proof of Theorem 3.3 , that the set R of points $z \in G(n,k-1)$ with $y \in T$ for all $Y \subset Z$ has codimension at least 3 in $G(n,k-1)$. Each line $E(Z,Z')$ with $z \notin R$ is contained in some set $f(g^{-1}(y))$, $y \notin T$, so

$\alpha | E \simeq \mathcal{O}_{\mathbb{P}_1} \oplus \mathcal{O}_{\mathbb{P}_1}(1)$, and hence $\alpha | S_Z$ is uniform for all $z \notin R$. Since $c_2(\alpha | S_Z) = 0$, it follows from [10] that $\alpha | S_Z \simeq \mathcal{O}_{\mathbb{P}_{n-k}} \oplus \mathcal{O}_{\mathbb{P}_{n-k}}(1)$. Making use of the diagram

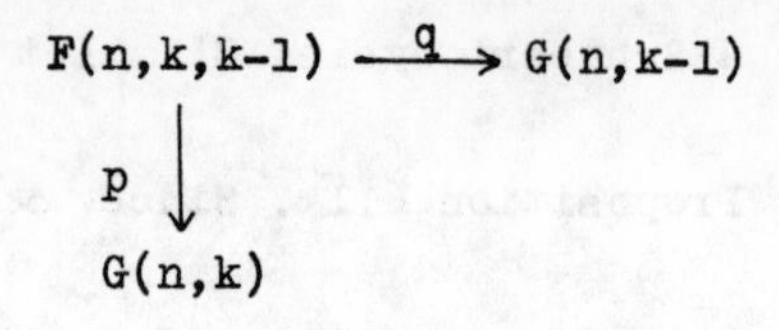

we can conclude as in [1] that $q^* q_* p^* \alpha(-1)$ is a 1-subbundle of $p^*(\alpha(-1))$ outside $q^{-1}(R)$. Since $q^{-1}(R)$ has codimension at least 3, $p^*(\alpha(-1)$ has a 1-subbundle all over $F(n,k,k-1)$. Consequently, $\alpha | S_Z$ is decomposable for all Schubert cycles S_Z, and hence α is itself decomposable by Theorem 4.1. This contradiction proves that in the case of $k \geqslant 2$ there is no V of bidegree $(0,c')$.

References.

1. W. Barth and A. Van de Ven. A decomposability criterion for algebraic 2-bundles on projective spaces. To appear in Inv. Math.
2. S.S. Chern. Complex manifolds without potential theory. Van Nostrand, Princeton N.J. (1967).
3. R. Hartshorne, E. Rees and E. Thomas. Non-smoothing of algebraic cycles on Grassmann varieties. To appear.
4. R. Hartshorne. Varieties of small codimension in projective space. To appear.

5. M.E. Larsen. On the topology of complex projective manifolds. Inv. Math. 19 , 251-260 (1973).

6. S. Lojasiewicz. Triangulation of semi-analytic sets. Ann. Scuola Norm. Pisa 18 , 449-474 (1964).

7. A. Ogus. On the formal neighborhood of a subvariety of projective space. To appear.

8. O. Riemenschneider. Über die Anwendung algebraischer Methoden in der Deformationstheorie komplexer Räume. Math. Ann. 187, 40-55 (1970).

9. A. Van de Ven. Over de homologiestructuur van enige typen vezelruimten. Van Gorcum, Assen (1957).

10. A. Van de Ven. On uniform vector bundles. Math. Ann. 195, 245-248 (1972).

INVARIANTEN BINÄRER FORMEN

von

W.D. Geyer (Erlangen)

Die Arbeit zerfällt in 2 Kapitel, im ersten werden hyperelliptische Kurven untersucht zur Motivation für das Interesse an der klassischen Invariantentheorie. Im II. Kapitel wird dann die Modulmannigfaltigkeit hyperelliptischer Kurven vom Geschlecht g explizit für Charakteristik $\neq 2$ konstruiert. Diese Mannigfaltigkeit ist wesentlich der Quotient des Raumes der binären Formen vom Grad $n = 2g + 2$ nach einer Operation der Gruppe PGL_2. In Primzahlcharakteristik ist diese Quotientenbildung (vgl. Mumford [12]) mit Komplikationen verbunden, da die Gruppe nicht voll reduzibel operiert. Inzwischen sind diese Schwierigkeiten nach Vorarbeit von Nagata [14] von Seshadri [16], vgl. auch [17], behoben, doch scheint eine explizite Konstruktion, die auch auf das Reduktionsverhalten modulo p eingeht, von unabhängigem Interesse. Für $g = 2$ wurde eine ganz explizite Konstruktion von Igusa [11] gegeben, aus der man entnimmt, daß das Reduktionsverhalten bei sehr kleinen Primzahlen (analog zu neu auftretenden Automorphismengruppen) etwas exzentrisch ist. Interessanterweise treten diese Schwierigkeiten nicht auf, wenn man die Invarianten binärer Formen in irrationaler Darstellung, d.h. als Funktionen der Wurzeln betrachtet. Der durch die PGL_2-Operation auf den Wurzeln entstehende Quotient wird in II A behandelt, hier ist der Quotient über $\mathbb{Z}$ definiert. Um zum klassischen Invariantenring und zur Modulmannigfaltigkeit hyperelliptischer Kurven zu kommen, wird dann in II B noch ein Quotient nach der symmetrischen Gruppe S_n gebildet. Dies liefert in jeder Charakteristik einen Quotienten, das Reduktionsverhalten ist nur für $p \leq n$ nicht übersichtlich.

Die Arbeit fußt auf den Hilbertschen Arbeiten zur Invariantentheorie, insbesondere auch auf dem in Math.Ann 33(1889) publizierten Endlichkeitsbeweis.

I. Hyperelliptische Kurven

Sei k ein algebraisch abgeschlossener Körper, char $k \neq 2$, sei Γ eine hyperelliptische Kurve über k vom Geschlecht $g > 1$ mit dem Funktionenkörper K, so gibt es eine von den Differentialen 1. Gattung induzierte kanonische Überlagerung $h : \Gamma \to \mathbb{P}_1$ vom Grad 2, die eine Erzeugung $K = k(x,y)$ mit $y^2 = f(x)$ liefert, in der $f(x) = \prod(x-\lambda_i)$ ein quadratfreies Polynom vom Grad $m = 2g + 2$ ist (vgl.[1], Kap. 16, § 7). $k(x)$ ist der einzige rationale Teilkörper vom Index 2 in K, K ist gekennzeichnet als die quadratische Erweiterung von $k(x)$, die genau an den m Nullstellen $\lambda_1, \ldots, \lambda_m$ von f verzweigt.

Jeder Automorphismus von Γ induziert daher einen Automorphismus von $\mathbb{P}_1$, der die Menge $= \{\lambda_1, \ldots, \lambda_m\}$ permutiert, und ist dadurch bis auf die Involution von Γ über $\mathbb{P}_1$ eindeutig bestimmt. Ebenso sieht man, daß zwei Verzweigungsmengen M, M' aus je m Punkten in $\mathbb{P}_1$ genau dann zu isomorphen Kurven führen, wenn sie durch einen Automorphismus $s \in PGL_2$ von $\mathbb{P}_1$ ineinander überführt werden können. Das liefert die folgenden Sätze.

<u>Satz 1:</u> Sei k ein algebraisch abgeschlossener Körper, char $k \neq 2$, so entsprechen die Isomorphieklassen der hyperelliptischen Kurven vom Geschlecht $g > 1$ über k bijektiv den $PGL_2(k)$-Bahnen auf der Menge der $2g + 2$-elementigen Teilmengen von $\mathbb{P}_1(k)$.

<u>Satz 2:</u> Ist k wie im Satz 1, Γ eine zweiblättrige Überlagerung von $\mathbb{P}_1$ mit der Verzweigungsmenge M aus m Punkten. Dann ist $Gal(\Gamma|\mathbb{P}_1) = \{\pm 1\}$ ein (zentraler) Normalteiler in $Aut(\Gamma)$, und die reduzierte Automorphismengruppe $A = \overline{Aut}(\Gamma) = Aut(\Gamma) \; / \; \{\pm 1\}$ ergibt sich als

$$A = \{s \in PGL_2(k) \;\; ; \;\; sM = M\} \leq S_m$$

Beispiel : $g = 2$

Kurven vom Geschlecht 2 sind stets hyperelliptisch, ihre Isomorphieklassen bilden nach Satz 1 zumindest generisch eine 3-dimensionale Mannigfaltigkeit. Die reduzierte Automorphismengruppe A ist jetzt eine Untergruppe von PGL_2, die eine 6-punktige Menge M invariant läßt. Da ein Element $\neq 1$ aus PGL_2 höchstens 2 Fixpunkte in $\mathbb{P}_1$ hat und im übrigen Bahnen gleicher Länge, ist A eine ebenso geartete Untergruppe von S_6. Die maximale 3-Gruppe dieser Art ist zyklisch, etwa $Z_3 = \langle(123)(456)\rangle$, die maximale 2-Gruppe dieser Art ist eine Diedergruppe der Ordnung 8, etwa

$D_8 = \{(1),(1234),(4321),(13)(24),(13)(56),(24)(56),(12)(34)(56),(14)(23)(56)\}$

1). A enthält ein Element s der Ordnung 5 :

1a). char $k \neq 5$:

Wählt man das Koordinatensystem so, daß 0 und ∞ Fixpunkte von s sind und 0 und 1 in M liegen, so wird $s(x) = \xi x$ mit einer primitiven 5-ten Einheitswurzel ξ, es ist

$$M = \{0, 1, \xi, \xi^2, \xi^3, \xi^4\}$$

das mit Zentrum 0 ergänzte Fünfeck der 5-ten Einheitswurzeln, die zugehörige Kurve ist

$$\Gamma : \; y^2 = x^6 - x$$

Wäre A größer als die zyklische Gruppe $Z_5 = \langle s\rangle$, so enthielte A eine s invertierende Involution t, etwa mit den Fixpunkten 0,1. Diese Involution $t(x) = \frac{x}{2x-1}$ permutiert jedoch M nicht in sich, da $x \to 2-x$ dies nicht tut. Also ist $A = Z_5$, die volle Automorphismengruppe ist $Aut(\Gamma) = Z_{10}$ zyklisch, erzeugt von $s(x) = \xi x$, $s(y) = -\xi^3 y$.

1b) char $k = 5$:

Bei geeignetem Koordinatensystem wird $s(x) = x + 1$ und

$$M = \{ 0, 1, 2, 3, 4, \infty \}$$

die projektive Gerade über $\mathbb{F}_5$, die zugehörige Kurve ist

$$\Gamma : \quad y^2 = x^5 - x$$

Die reduzierte Automorphismengruppe ist offenbar $A = PGL_2(5) = S_5$ (mit nichttrivialer Lage in S_6). $Aut(\Gamma)$ hat die Ordnung 240 und ist die größte und als einzige nicht auflösbare Automorphismengruppe bei Geschlecht 2. $Aut(\Gamma)$ enthält die binäre Ikosaedergruppe (als Überlagerung von $A_5 = PSL_2(5)$) vom Index 2.

2) A enthält ein Element s der Ordnung 6 :

Ist char $k = 3$, so existieren solche Elemente in $PGL_2(k)$ nicht; ist char $k \neq 3$ und wählt man das Koordinatensystem von $\mathbb{P}_1$ so, daß 0 und ∞ Fixpunkte von s sind und 1 in M liegt, so wird $s(x) = \zeta x$ mit einer primitiven 6-ten Einheitswurzel ζ, es ist

$$M = \{ 1, \zeta, \zeta^2, \zeta^3, \zeta^4, \zeta^5 \}$$

die Menge der 6-ten Einheitswurzeln und

$$\Gamma : \quad y^2 = x^6 - 1$$

M läßt außer den Potenzen von s auch die Involution $t(x) = x^{-1}$ zu, also eine Diedergruppe $D_{12} = \langle s,t \rangle$, die in S_6 ihr eigener Normalisator ist. Wäre $A \neq D_{12}$, so müßte A ein Element der Ordnung 5 enthalten - was nach 1a) für char $k \neq 5$ unmöglich, nach 1b) für char $k = 5$ aber möglich ist ($PGL_2(5)$ enthält einen 6-Zyklus).

2a) char $k \neq 3, 5$:

Es ist $A = D_{12}$, die volle Automorphismengruppe $\mathrm{Aut}(\Gamma) = \langle s,t\rangle$ mit $s(y) = y$, $t(y) = ix^{-3}y$ ist ein semidirektes Produkt von Z_6 mit Z_4 : $s^6 = t^4 = 1$, $sts = t$, die in [5], 8.5 als Gruppe $\{4,6|2\}$ bezeichnet wird.

2b) char $k = 5$:

Dann ist Γ isomorph zu der in 1b) behandelten Kurve, ein Isomorphismus wird geliefert durch $g(x) = c\,\frac{x+1}{x-1}$ mit $c = \frac{\zeta-1}{\zeta+1} = \sqrt{-1/3}$.

3) A enthält ein Element s der Ordnung 4 :

Dann besteht M aus den Fixpunkten und einer Bahn von s, in geeignetem Koordinatensystem wird $s(x) = ix$ und

$$M = \{ 0, \infty, 1, -1, i, -i \}$$

(das reguläre Oktaeder auf der Riemannschen Zahlenkugel) mit der Kurve

$$\Gamma : \quad y^2 = x^5 - x$$

Nun läßt M außer der Drehung s auch die Involution $t(x) = -i\,\frac{x-i}{x+1}$ zu, also auch die Oktaedergruppe $S_4 = \langle s,t\rangle$ mit $s^4 = t^2 = (st)^3 = 1$. Ist char $k \neq 5$, so muß $A = S_4$ sein wegen 1a), die volle Automorphismengruppe ist dann die binäre Oktaedergruppe (vgl. [5], 6.5) der Ordnung 48: $\mathrm{Aut}(\Gamma) = \langle s,t\rangle$ mit $s^4 = t^2 = (st)^3$, $t^4 = 1$, wobei $s(y) = \eta y$ mit $\eta^2 = i$ und $t(y) = \sqrt{8}(x+i)^{-3}y$ ist. Für char $k = 5$ fällt Γ mit der in 1b) betrachteten Kurve zusammen.

4) A enthält ein Element s der Ordnung 3 :

Dann besteht M aus 2 Bahnen von s.

4a) char $k \neq 3$:

Wählt man die beiden Fixpunkte von s als $0, \infty$, so wird $s(x) = \rho x$ mit einer dritten Einheitswurzel ρ, und mit $1 \in M$ liefert

$$M = \{1, \rho, \rho^2, \alpha, \rho\alpha, \rho^2\alpha\}$$

mit $\alpha \in k^\times - \{1,\rho,\rho^2\}$ eine 1-parametrige Schar von Kurven

$$\Gamma : \qquad y^2 = (x^3 - 1)(x^3 - \lambda)$$

mit $A \geq Z_3$, wobei $\lambda = \alpha^3 \neq 0,1$. Wählt man den Einheitspunkt 1 in der anderen Bahn von M, so entspricht das einem Parameterwechsel $\alpha \to \alpha^{-1}$, d.h. λ und λ^{-1} liefern isomorphe Kurven. Dem Fixpunkt $\lambda = -1$ (und nur diesem Parameterwert) entspricht dabei die in 2) betrachtete Kurve. Das Vertauschen der Fixpunkte von s stört den Parameter nicht, denn M läßt die Involution $t(x) = \alpha x^{-1}$ zu, also ist $A \geq S_3 = \langle s,t \rangle$. Ist $A \neq S_3$, so muß M eine weitere Involution u mit Fixpunkten in M zulassen, die wir als 1 und α wählen können (siehe die zu Beginn aufgezählten Möglichkeiten für die 2-Sylowgruppe von A). Ist dann $u(\rho) = \rho^2$, so bleibt $\alpha = -1$ als einzige Möglichkeit, ist $u(\rho) = \rho\alpha$, so folgt $\alpha = -2 \pm \sqrt{3}$, $u(\rho) = \rho^2\alpha$ ist unmöglich.

Damit ergibt sich: Der Parameter $\mu = \lambda + \lambda^{-1} \neq 2$ entspricht der Isomorphieklasse von Γ. Für $\mu \neq -2, -52$ ist $A = S_3$ und $\mathrm{Aut}(\Gamma) = A \times Z_2 = D_{12} = \langle s,t \rangle$, $s^6 = t^2 = (st)^2 = 1$, wobei $s(y) = -y$, $t(y) = \beta x^{-3} y$ mit $\beta^2 = \lambda$.

Im Ausnahmefall $\mu = -2$ treffen wir die Kurve aus 2) wieder, im Ausnahmefall $\mu = -52$ ist die Kurve Γ isomorph der in 3) behandelten Kurve, eine Isomorphie wird etwa durch $g(x) = c.\frac{x-1}{x-\varepsilon}$ vermittelt, wo $\varepsilon = -2 + \sqrt{3}$ und $c^2 = \varepsilon i$, etwa $c = (\rho-\varepsilon)(\rho-1)^{-1}$. Für char $k = 5$ fallen diese beiden Parameter zusammen.

4b) char $k = 3$:

In einem geeigneten Koordinantensystem ist $s(x) = x + 1$ und

$$M = \{ 0, 1, 2, \alpha, \alpha + 1, \alpha + 2 \}$$

mit $\alpha \in k - \mathbb{F}_3$, wobei α und $-\alpha$ isomorphe Sechstupel M beschreiben. M läßt wie in 4a) eine bahnenvertauschende Involution $t(x) = -x + \alpha$ zu, die Kurvenschar

$$\Gamma : y^2 = (x^3 - x)(x^3 - x - \lambda)$$

mit $\lambda = \alpha^3 - \alpha \neq 0$ hat also eine reduzierte Automorphismengruppe $A \geq S_3 = \langle s,t \rangle$, λ und $-\lambda$ liefern isomorphe Kurven. Ist $A \neq S_3$, so läßt M (mit gleicher Begründung wie in 4a)) die Involution $u(x) = \frac{\alpha x}{2x-a}$ mit den Fixpunkten 0, α zu, was zu $\lambda = \pm i$ führt, diese Kurve ist isomorph der in 3) behandelten Kurve, ein Isomorphismus wird durch $g(x) = \frac{(1-i)x}{x-1}$ vermittelt. Andernfalls ist $\mathrm{Aut}(\Gamma) = D_{12}$ wie in 4a).

5) A enthält eine Involution s mit Fixpunkt in M:

Wähle 0, ∞ als Fixpunkte von s und $1 \in M$, so wird $s(x) = -x$, und

$$M = \{ 0, \infty, 1, -1, \alpha, -\alpha \}$$

mit $\alpha \in k^{\times} - \{1,-1\}$ läßt offenbar außer s auch die auf M fixpunktfreien Involutionen $t(x) = x^{-1}$, $st(x) = -\alpha x^{-1}$ zu. Die 1-parametrige Schar der Kurven

$$\Gamma : y^2 = x(x^2 - 1)(x^2 - \lambda)$$

hat also eine reduzierte Automorphismengruppe $A \geq D_4 = Z_2 \times Z_2$, dabei ist $\lambda = \alpha^2 \neq 0,1$, λ und λ^{-1} liefern isomorphe Kurven. Der Fixpunkt $\lambda = -1$ dieser Parametertransformation führt zu der in 3) betrachteten Kurve. Ist $A \neq D_4$, so muß A eine weitere Involution u mit Fixpunkten in $\{1, -1, \alpha, -\alpha\}$ enthalten. Sind die Fixpunkte ± 1, also $u(x) = x^{-1}$, so folgt $\lambda = -1$; sind die Fixpunkte $1, \alpha$, also $u(x) = \frac{(1+\alpha)x-2\alpha}{2x-(1+\alpha)}$, so folgt $\lambda = 9$ bzw. $1/9$, für diesen Parameterwert ist char $k \neq 3$ und Γ isomorph zu der in 2) betrachteten Kurve, ein Isomorphismus wird durch $g(x) = \frac{x+\sqrt{-3}}{x-\sqrt{-3}}$ geliefert.

Damit ergibt sich: Der Parameter $\mu = \lambda + \lambda^{-1} \neq 2$ entspricht der Isomorphieklasse von Γ. Für $\mu \neq -2,\ 9 + 1/9$ ist $A = D_4$ und $\mathrm{Aut}(\Gamma) = D_8 = \langle s,t \rangle$, $s^4 = t^2 = (st)^2 = 1$, wobei $s(y) = iy$ und $t(y) = \beta x^{-3} y$ mit $\beta^2 = \lambda$ ist. Für $\mu = -2$ bzw. $\mu = 9 + 1/9$ ergeben sich die in 3) bzw. 2) behandelten Kurven, für char $k = 5$ fallen diese Ausnahmeparameter zusammen.

6) A enthält eine in M fixpunktfreie Involution s :

Bei Einführung passender Koordinaten wird $s(x) = -x$ und

$$M = \{ 1, -1, \alpha, -\alpha, \beta, -\beta \}$$

wodurch eine 2-parametrige Schar von Kurven

$$\Gamma : \qquad y^2 = (x^2 - 1)(x^2 - \lambda)(x^2 - \mu)$$

des Geschlechtes 2 mit $A \geq Z_2$ geliefert wird, wobei $\lambda = \alpha^2$, $\mu = \beta^2$ ist, $\lambda, \mu \in k'$ mit $k' = k - \{0,1\}$, $\lambda \neq \mu$. Dabei sind die Parameter (λ,μ) durch die Isomorphieklasse von Γ nicht wohlbestimmt, Vertauschen der Fixpunkte von s führt zu der Substitution $(\lambda,\mu) \to (\lambda^{-1}, \mu^{-1})$, andere Bezeichnung der Ecken des "Oktaeders" M liefert die von $(\lambda,\mu) \to (\mu,\lambda)$ und $(\lambda,\mu) \to (\lambda^{-1}\mu, \lambda^{-1})$ erzeugte Permutationsgruppe S_3, insgesamt operiert also die Gruppe $G = S_3 \times Z_2$ auf dem Parameterraum $P = (k' \times k') - \Delta$, wo Δ die Diagonale sei, sie operiert fixpunktfrei, solange $\lambda^{-1} \neq \mu \neq \lambda^2$ ist.

Ist $A = Z_2$, so gehört zu Γ eine 12-elementige G-Bahn in P, die volle Automorphismengruppe $Aut(\Gamma) = D_4$ ist die Vierergruppe $x \to \pm x$, $y \to \pm y$. Die beiden von der kanonischen Involution σ verschiedenen Involutionen s und $s\sigma$ liefern zweiblättrige Überlagerungen von Γ über elliptischen Kurven

$$E_1 \quad : \quad y^2 = (x_1 - 1)(x_1 - \lambda)(x_1 - \mu)$$

und

$$E_2 \quad : \quad y_1^2 = x_1(x_1 - 1)(x_1 - \lambda)(x_1 - \mu) \quad ,$$

wobei $x_1 = x^2$, $y_1 = xy$ ist. Da die durch Γ induzierte Korrespondenz von E_1 nach E_2 trivial ist, ist die Jacobische Mannigfaltigkeit von Γ isogen zum Produkt $E_1 \times E_2$.

6a) Enthält A außer s eine weitere mit s kommutierende in M fixpunktfreie Involution t, so vertauscht t die Fixpunkte von s, und M muß ein s und t gemeinsames Punktepaar enthalten, etwa ± 1. Dann ist $t(x) = -x^{-1}$, woraus etwa $\beta = \alpha^{-1}$, also $\lambda\mu = 1$ folgt. Die Kurven

$$\Gamma : \quad y^2 = (x^2 - 1)(x^4 - \gamma x^2 + 1)$$

mit $\gamma = \lambda + \lambda^{-1} \neq \pm 2$ lassen dann auch die Involution st zu, die auf M die Fixpunkte ± 1 hat, wurden also bereits in 5) behandelt. Genauer transformiert die Substitution $g(x) = \delta \frac{x-1}{x+1}$ mit $\delta = \frac{\alpha+1}{\alpha-1}$ die hier angegebene Form in die in 5) behandelte Normalform wegen

$$g(M) = \{ 0, \infty, 1, -1, \delta^2, -\delta^2 \} \quad ,$$

die Parametertransformation $\alpha \to \pm\alpha^{\pm 1}$ übersetzt sich zu $\delta \to \pm\delta^{\pm 1}$. Die beiden auf M fixpunktfreien Involutionen s und t sind durch die Involution $v(x) = \frac{x+1}{x-1}$ konjugiert, beim Wechsel von s zu t liefern daher α und $v(\alpha)$ bzw. γ und $w(\gamma) = -2\frac{\gamma-6}{\gamma+2}$ isomorphe Kurven Γ, was dem Parameterwechsel $\delta^2 \to -\delta^{-2}$ entspricht. Für $\delta^4 \neq -1, 9, 1/9$ bzw. $\gamma \neq -6, -1, 14$ wird nach 5) also $A = D_4$

bzw. $\mathrm{Aut}(\Gamma) = D_8$, in diesem Fall entspricht die Isomorphieklasse von Γ zwei 6-elementigen G-Bahnen im Parameterraum P. Vermöge t sind die oben aus s abgeleiteten elliptischen Kurven E_1 und E_2 isomorph, sie haben die Invariante $j = 2^8 \frac{(\gamma+1)^3}{\gamma+2}$, das andere Paar konjugierter Involutionen in D_8 (darunter t) liefert ebenfalls ein Paar isomorpher elliptischer Kurven vom Index 2 unter Γ, die isogen zu den ersten sind, aber i.a. nicht isomorph wegen

$$j = j(\gamma) \neq j(w(\gamma)) = j_1 = 2^4 . \frac{(14-\gamma)^3}{(2+\gamma)^2} \quad \text{für } \gamma \neq 2, -6, -2^{-5} . (17 \pm 3^2 . 5.\sqrt{-7}).$$

(Anmerkung 1: Die Isogenien haben den Grad 2 vermöge Γ oder wegen $\Phi_2(j, j_1) = 0$, wobei

$$\Phi_2(X,Y) = X^3 + Y^3 - X^2Y^2 + 2^4 . 3 . 31XY(X+Y) - 2^4 . 3^4 . 5^3(X^2+Y^2)$$
$$+ 3^3 . 5^3 . 4027\,XY + 2^8 . 3^5 . 5^6(X+Y) - 2^{12} . 3^9 . 5^9$$

das Modularpolynom zweiter Ordnung ist (vgl. Deuring [6], S. 247). Die Parametrisierung der Kurven vom Geschlecht 2 mit $A \geq D_4$ liefert also eine rationale Parametrisierung der Kurve vierten Grades $\phi_2(x,y) = 0$, die im Komplexen die Relation zwischen $j(z)$ und $j(2z)$ darstellt, genauer: γ parametrisiert elliptische Kurven mit gegebenem Zweiteilungspunkt (modulo Automorphismen). Ersetzt man λ durch $-\lambda$, so geht die hier gegebene Parametrisierung in die bei Deuring [6] § 6 vorhandene über.)

(Anmerkung 2: Die Lösungen der Gleichung $j = j_1$ entsprechen elliptischen Kurven mit Meromorphismen vom Grad 2. Nun gibt es in imaginär quadratischen Zahlkörpern Gitterpunktabstände $\sqrt{2}$ bzw. Elementnormen 2 nur bei den Hauptordnungen in $Q(\sqrt{-1})$, $Q(\sqrt{-2})$ und $Q(\sqrt{-7})$, und zwar haben in den ersten Fällen genau ein Hauptideal, im letzten genau zwei Hauptideale die Norm 2. Da die genannten Ordnungen Klassenzahl 1 haben, gehören zu ihnen in Charakteristik 0 je genau eine elliptische Kurve mit Meromorphismen vom Grad 2. Dies sind die elliptischen Kurven zu $\gamma = 2, -6$ bzw. $-2^{-5}(17 \pm 3^2.5\sqrt{-7})$, die die Invarianten $j = j_1 = 2^6 . 3^3$, $2^6 . 5^3$ bzw. $-3^3 . 5^3$ und die Endomorphismenringe $\mathbb{Z}[i]$, $\mathbb{Z}[\sqrt{-2}]$ bzw. $\mathbb{Z}[\frac{1}{2}(1+\sqrt{-7})]$

haben; in Charakteristik p gilt dasselbe, solange die Primzahl p in den genannten Ringen zerfällt, andernfalls wird j supersingulär, und die Endomorphismenringe größer. Der erste Fall $j = j_1 = 12^3$ tritt im Fall 6a) nicht auf, da $\gamma \neq 2$ ist - aber die elliptische Kurve mit $j = 12^3$ kommt dennoch einmal vor, für $\gamma = -5/2$ mit $j_1 = 2^3.3^3.11^3$.)

Im Fall $\gamma = -6$, etwa $\alpha = (1 + \sqrt{2})i$, liegt die in 3) behandelte Kurve vor, sie entspricht einer einzigen 6-elementigen G-Bahn in P, für char $k \neq 5$ ist $A = S_4$ und Γ besitzt 12 isomorphe elliptische Kurven vom Index 2. Im Fall $\gamma = -1$ bzw. 14, also etwa $\alpha = \zeta$ bzw. $\alpha = -2 + \sqrt{3}$, liegen eine 2- und eine 6-elementige G-Bahn in P vor, beide entsprechen der in 2) behandelten Kurve Γ. Ist char $k \neq 5$ (char $k \neq 3$ ohnehin), so ist s für $\gamma = -1$ zentrale, für $\gamma = 14$ nichtzentrale Involution in $A = D_{12}$, Γ besitzt 8 elliptische Kurven vom Index 2, zwei mit $j = 0$ und sechs mit $j = 2^4.3^3.5^3$ (alle 3 Isogenien vom Grad 2 der elliptischen Kurve mit $j = 0$ führen zur gleichen Invariante $j = 54\,000$).

6b) In analoger Weise kann man den verbleibenden Fall behandeln, daß A außer s noch ein Element t der Ordnung 3 enthält. Diese Kurven wurden bereits in 4) diskutiert. Kommutiert s mit t, so ergibt sich die in 6a) erwähnte, $(\lambda,\mu) = (\rho,\rho^2)$ enthaltende 2-elementige G-Bahn. Andernfalls kann t mit $\langle s,t\rangle = S_3$ gewählt werden. Im allgemeinen ist $A = S_3$; da es nur eine Konjugationsklasse von Involutionen in S_3 gibt, entspricht Γ dann einer 12-elementigen G-Bahn in P. Dazu kommen noch die in 6a) schon erwähnten 2 Spezialfälle von 6-elementigen G-Bahnen, bei denen $A = S_4$ bzw. $A = D_{12}$ ist (für char $k \neq 3,5$).

Von den 2 Spezialfällen abgesehen hat Γ dann zwei Tripel isomorpher elliptischer Kurven vom Index 2, die isogen vom Grad 2 sind.

7) $A = 1$: allgemeiner Fall

Die vorstehenden Typen von Automorphismengruppen finden sich für $k = \mathbb{C}$ bei Bolza [2] , allgemein (auch für Charakteristik 2 mit entsprechenden Ergänzungen) bei Igusa [11]. Die hier vorgeführte explizitere Beschreibung liefert zwar in jedem Fall ein genaues Bild, leidet jedoch darunter, daß bei verschiedenen Fällen verschiedene (angepaßte) Parametrisierungen vorgenommen wurden. Eine gute Beschreibung müßte die in Satz 1 genannten Isomorphieklassen der Kurven vom Geschlecht 2, d.h. einen gewissen Quotientenraum nach $PGL_2(k)$ parametrisieren, d.h. aus diesen Mengen eine Mannigfaltigkeit machen, und deren Parameter zur Beschreibung spezieller Kurven benutzen. Bei Bolza [2] werden tatsächlich invariante Parameter zur Beschreibung benutzt, obwohl die klassische Invariantentheorie([4] oder [15]) noch keine eigentliche Modulmannigfaltigkeit konstruierte. Für Geschlecht $g = 2$ und beliebige Charakteristik ist eine solche Konstruktion zuerst von Igusa [11] durchgeführt worden, er zeigte

<u>Satz 3</u>: Sei k ein algebraisch abgeschlossener Körper, char $k \neq 2$. Die Isomorphieklassen der Kurven vom Geschlecht 2 bilden eine 3-dimensionale rationale normale affine Varietät V mit einem einzigen singulären Punkt P, der der Kurve Γ mit $\overline{\mathrm{Aut}(\Gamma)} = Z_5$ entspricht, für char $k = 5$ hat man $\overline{\mathrm{Aut}(\Gamma)} = S_5$.

Damit läßt sich unsere Diskussion so zusammenfassen:

<u>Satz 4</u>: Seien k und V wie in Satz 3. Die Kurven vom Geschlecht 2, die eine elliptische Kurve vom Index 2 besitzen, bilden eine rationale Fläche F in V. Es ist $P \in F$ genau für char $k = 5$. In jedem Fall repräsentiert $F \cup \{P\}$ die Kurven Γ vom Geschlecht 2 mit $\overline{\mathrm{Aut}(\Gamma)} \neq 1$. F ist nicht normal, der singuläre Ort ist eine rationale Kurve C, die die Kurven Γ vom Geschlecht 2 mit $\overline{\mathrm{Aut}(\Gamma)} \geq D_4$

beschreibt. Eine weitere rationale Kurve D auf F beschreibt die Kurven Γ mit $\overline{\mathrm{Aut}}(\Gamma) \geq S_3$. In beiden Fällen steht hier ein Gleichheitszeichen außer bei den Schnittpunkten von C und D. Ist char $k \neq 3,5$, so schneiden sich C und D in 2 Punkten P_1, P_2, die der Kurve Γ mit $\overline{\mathrm{Aut}}(\Gamma) = S_4$ bzw. $\overline{\mathrm{Aut}}(\Gamma) = D_{12}$ entsprechen. Für char $k = 3$ fällt P_2 fort, für char $k = 5$ ist $P_1 = P_2 = P$ mit $\overline{\mathrm{Aut}}(\Gamma) = S_5$.

II. Invariantentheorie

A. Die Invarianten eines n-Tupels

Seien $P_i = (\alpha_i : \beta_i)$ für $i = 1,\dots,n$ Punkte in $\mathbb{P}_1(k)$, später werden dies die $n = 2g+2$ Verzweigungspunkte einer hyperelliptischen Kurve Γ vom Grad 2 über $\mathbb{P}_1$ sein. Wir suchen Funktionen in den P_i, die vom gewählten Koordinatensystem des $\mathbb{P}_1(k)$ unabhängig sind, die also die geometrische Lage der n Punkte zueinander beschreiben. Dazu betrachten wir Polynome $F = F(\underline{\alpha},\underline{\beta})$ aus $k[\alpha_1,\dots,\alpha_n, \beta_1,\dots,\beta_n]$, die für gewisse natürliche Zahlen d_i, $w \geq 0$ folgende Bedingungen erfüllen:

(1) F ist homogen in (α_i,β_i) vom Grad d_i $(i = 1,\dots,n)$

(2) $F(\lambda_1\underline{\alpha} + \lambda_2\underline{\beta}, \lambda_3\underline{\alpha} + \lambda_4\underline{\beta}) = (\lambda_1\lambda_4 - \lambda_2\lambda_3)^w. F(\underline{\alpha},\underline{\beta})$

Die erste Bedingung besagt, daß F bis auf einen Homogenitätsfaktor nur von den P_i abhängt, d.h. eine sinnvolle Form auf dem n-fachen Produkt $\mathbb{P}_1 \times .. \times \mathbb{P}_1$ ist, die zweite Bedingung besagt die Invarianz unter der Gruppe $GL_2(k)$ bis auf einen linearen Charakter dieser Gruppe - der notwendig die Form $(\det)^w$ hat. Die Gleichung (2) ist äquivalent zur Konjunktion folgender Teilgleichungen:

(2a) F ist homogen in $\underline{\alpha}$ und in $\underline{\beta}$, jeweils vom Grad w:

$$F(\lambda\underline{\alpha},\lambda\underline{\beta}) = F(\underline{\alpha}, \lambda\underline{\beta}) = \lambda^w. F(\underline{\alpha},\underline{\beta})$$

(2b) F ist translationsinvariant: $F(\underline{\alpha} + \lambda\underline{\beta}, \underline{\beta}) = F(\underline{\alpha},\underline{\beta})$

(2c) F ist (schief) symmetrisch: $F(\underline{\beta},\underline{\alpha}) = (-1)^w. F(\underline{\alpha},\underline{\beta})$

Denn (2a), (2b) besagen, daß (2) für Dreiecksmatrizen $\begin{pmatrix} \lambda_1 & \lambda_2 \\ 0 & \lambda_4 \end{pmatrix}$ gilt, diese erzeugen mit $\begin{pmatrix} 0 & 1 \\ 1 & 0 \end{pmatrix}$ aber ganz $GL_2(k)$.

Übrigens ist, wie wir auch gleich bei der direkten Berechnung solcher Polynome F sehen werden, die Bedingung (2c) überflüssig: Ist nämlich F unter einer Borelgruppe B von $SL_2(k)$ invariant, wie aus (2a), (2b) folgt, so auch unter ganz $SL_2(k)$, da $SL_2(k)/B$ eine projektive Mannigfaltigkeit ist, die als affine Bilder nur Punkte haben kann.

Der Vergleich von (1) mit (2a) führt, wenn $F \neq 0$, zur Gleichung

$$2w = d_1 + d_2 + \dots + d_n \quad , \tag{3}$$

die wir im folgenden voraussetzen.

Mit der Abkürzung $\underline{d} = (d_1 \dots , d_n)$ bezeichnen wir $R_{\underline{d}}$, genauer

$$R_{\underline{d}}(k) = \{ F \in k[\underline{\alpha},\underline{\beta}] \quad ; \quad (1) \text{ und } (2) \quad \text{gilt} \}$$

als den Modul der Invarianten (zu einem n-Tupel in $\mathbb{P}_1$) vom Gewicht $\underline{d}$ über k.

Beispiele:

(i) Für $i \neq j$ ist $p_{ij} = \alpha_i\beta_j - \alpha_j\beta_i$ eine Invariante vom Gewicht $\underline{d}$ mit $d_i = d_j = 1$, sonst $d_s = 0$.

(ii) Sind F, G Invarianten der Gewichte $\underline{d}$, $\underline{e}$, so ist F.G eine Invariante vom Gewicht $\underline{d} + \underline{e}$ (komponentenweise Addition).

(iii) Ist jedem Paar (i,j) mit $1 \leq i < j \leq n$ eine Zahl $m(i,j) \geq 0$ zugeordnet, so wird

$$X_{\underline{m}} = \prod_{i<j} (\alpha_i\beta_j - \alpha_j\beta_i)^{m(i,j)} = \prod_{i<j} p_{ij}^{m(i,j)}$$

eine Invariante vom Gewicht $\underline{d}$ mit $\quad d_i = \sum_{j=1}^{i-1} m(j,i) + \sum_{j=i+1}^{n} m(i,j) \quad ,$

wofür wir auch kurz $\underline{d} = d(\underline{m})$ schreiben.

Im folgenden empfiehlt es sich, für k einen beliebigen (kommutativen) Ring zuzulassen.

Satz 5: Jede Invariante vom Gewicht $\underline{d}$ ist k-Linearkombination von Produkten $X_{\underline{m}}$ mit $d(\underline{m}) = \underline{d}$.

Beweis:

Wir schließen mit doppelter Induktion nach n und w. Für n = 1 gibt es keine nichtkonstanten Invarianten, für $\underline{d} = \underline{0}$ sind die Invarianten gerade k. Sei also der Satz für kleinere Variablenzahl und kleineres w bewiesen. Sei $F \in R_{\underline{d}}$, so liegt $G' = F(\alpha_1,\ldots,\alpha_{n-1},\alpha_{n-1}, \beta_1,\ldots,\beta_{n-1},\beta_{n-1})$ in $R_{\underline{d}'}$ mit $\underline{d}' = (d_1,\ldots,d_{n-1}+d_n)$. Nach Induktionsvoraussetzung ist G' Linearkombination von Produkten $X_{\underline{m}'}$ mit $d(\underline{m}') = \underline{d}'$. Ersetzt man in irgendwelchen d_n unter den relevanten $d_{n-1} + d_n$ Linearfaktoren von $X_{\underline{m}'}$ jeweils das Paar $(\alpha_{n-1}, \beta_{n-1})$ durch (α_n,β_n) so erhält man ein Produkt $X_{\underline{m}}$ aus $R_{\underline{d}}$, und aus G' wird dann eine Linearkombination G solcher $X_{\underline{m}}$ mit $(F-G)(\alpha_1,\ldots,\alpha_{n-1},\alpha_{n-1},\beta_1,\ldots,\beta_{n-1},\beta_{n-1}) = 0$. Da F - G also für $P_{n-1} = P_n$ verschwindet, ist es durch $\alpha_{n-1}\beta_n - \alpha_n\beta_{n-1}$ teilbar, also

$$F = G + (\alpha_{n-1}\beta_n - \alpha_n\beta_{n-1}).F_o \quad .$$

F_o ist nun eine Invariante mit niedrigerem Gesamtgewicht w als F, nach Induktionsvoraussetzung ist F_o wie gewünscht darstellbar und damit auch F.

Bemerkung: Die $X_{\underline{m}}$ sind nicht linear unabhängig. Wie durch verschiedene Wahl im Beweis des Satzes 5 folgt, gilt z.B. $p_{12}p_{34} - p_{13}p_{24} + p_{14}p_{23} = 0$. Ebenfalls aus dem Beweis folgt, daß alle linearen Relationen zwischen den $X_{\underline{m}}$ aus solchen Relationen folgen.

Folgerung 1: Der graduierte Ring $R = \sum_{\underline{d}} R_{\underline{d}}$ ist endlich erzeugt über k, nämlich von den $\binom{n}{2}$ nichtkonstanten Invarianten kleinsten Grades p_{ij} für $1 \leq i < j \leq n$. Die definierenden Relationen für $R = k[p_{ij}]$ lauten

$$p_{is}p_{jt} - p_{it}p_{js} = p_{ij}p_{st} \qquad (1 \leq i < j < s < t \leq n)$$

<u>Folgerung 2:</u> R ist der Ring der Invarianten bei der in (2) dargestellten Operation der $SL_2(k)$ auf dem Polynomring $k[\underline{\alpha},\underline{\beta}]$. R ist faktoriell. Für jeden Ring k gilt

$$R_{\underline{d}}(k) = R_{\underline{d}}(\mathbb{Z}) \otimes k \quad , \quad R(k) = R(\mathbb{Z}) \otimes k$$

Insbesondere sind alle Invarianten in Charakteristik p Reduktion von solchen in Charakteristik 0.

Beweis:

Die Operation von $SL_2(k)$ auf $k[\underline{\alpha},\underline{\beta}]$ erhält die (α_i,β_i)-Graduierung, also ist R der Invariantenring nach Definition der $R_{\underline{d}}$. Ist F.G invariant unter $SL_2(k)$, so auch die beiden Faktoren F,G ; daher sind Primteiler von Invarianten wieder solche, woraus sich R als faktoriell ergibt. Der Rest folgt aus Satz 5.

<u>Folgerung 3:</u> Verschwinden alle Invarianten (positiven Grades) auf einem n-Tupel $(P_1, \dots, P_n)$, so ist $P_1 = \dots = P_n$, abgesehen von entarteten Punkten $P_i = (0:0)$. Die Umkehrung gilt auch.

Beweis:

$p_{ij} = 0$ ist äquivalent zu $P_i = P_j$, solange P_i, P_j nicht entarten.

<u>Folgerung 4:</u> Sei n > 1. Die 2n-3 Invarianten

$$c_t = \sum_{\substack{i+j=t \\ i<j}} p_{ij} \qquad (t = 3,\dots, 2n-1)$$

sind algebraisch unabhängig über k, der Invariantenring $R = k[p_{ij}]$ ist ganz über $k[c_t]$.

Beweis:

Ist $c_3 = \dots = c_{2n-1} = 0$ für ein n-Tupel $(P_1, \dots, P_n)$ und streicht man in den Gleichungen $c_t = 0$ alle p_{ij} mit Indizes i (oder j), die zu entarteten Punkten P_i (oder P_j) gehören, so erhält man ein Gleichungssystem, das sukzessive die

Gleichheit der nicht entarteten Punkte zeigt, indem sich bei jedem Schritt die betrachtete Gleichung $c_{t+1} = 0$ auf eine einzige Gleichung $p_{ij} = 0$ reduziert, bei nichtentarteten P_1 und P_t etwa zu $p_{1t} = 0$. Damit hat das System $c_3 = \ldots = c_{2n-1} = 0$ dieselben Nullstellen wie das System $p_{ij} = 0$, woraus folgt, daß die p_{ij} ganz über $k[c_t]$ sind (Satz von Hilbert-Zariski, siehe [10] § 3). Da $R[\alpha_1,\alpha_2,\beta_1]$ offenbar den gleichen Quotientenkörper wie $k[\underline{\alpha},\underline{\beta}]$ hat, hat R den Transzendenzgrad 2n-3 über k und die c_t müssen algebraisch unabhängig über k sein.

<u>Bemerkung</u>: Ist char k = 0 oder $\geq$ n, so kann man statt c_t in Folgerung 4 auch

$$c'_t = \sum_{\substack{i+j=t \\ i<j}} (j-i)p_{ij}$$

wählen (vgl. Hilbert [10] § 6). Diese Linearkombinationen der p_{ij} ergeben sich im wesentlichen als Koeffizienten der Funktionaldeterminante $A_X B_Y - A_Y B_X$ der Formen $A = \sum_{i=1}^{n} \alpha_i X^{n-i} Y^{i-1}$ und $B = \sum_{j=1}^{n} \beta_j X^{n-j} Y^{j-1}$.

Die charakteristische Potenzreihe $\chi(\underline{t},R) = \sum_{\underline{d}} \dim_k(R_{\underline{d}}) \cdot \underline{t}^{\underline{d}}$ des graduierten Invariantenringes R ist nach Folgerung 2 charakteristikunabhängig, wir werden zu ihrer Berechnung char k = 0 voraussetzen. In Charakteristik 0 ist die Translationsinvarianz (2b) einer Form F äquivalent zur infinitesimalen Invarianz, d.h. zur Differentialgleichung

$$(2b)' \qquad \mathcal{D} F = \sum_{i=1}^{n} \beta_i \cdot \partial F/\partial \alpha_i = 0$$

Ebenso genügt eine Invariante F, z.B. wegen (2c), der (mit (1) und (2a) zu (2b)' äquivalenten) Differentialgleichung

$$(2b)'' \qquad \Delta F = \sum_{i=1}^{n} \alpha_i \cdot \partial F/\partial \beta_i = 0$$

Um die Operatoren $\mathfrak{D}$ und Δ besser studieren zu können, bezeichnen wir die vorkommenden Graduierungen und schreiben für natürliche Zahlen a, b ≥ 0

$$T(a,b) = \{ F \in k[\underline{\alpha},\underline{\beta}] \; ; \; F \text{ homogen in } \underline{\alpha} \text{ bzw. } \underline{\beta} \text{ vom Grad a bzw. b} \}$$

sowie für $\underline{d} = (d_1,\dots,d_n)$

$$T_{\underline{d}}(a,b) = \{ F \in T(a,b) \; ; \; F \text{ erfüllt } (1) \}$$

wobei natürlich $a + b = 2w$ mit w aus (3) gelten muß, damit $T_{\underline{d}}^{(a,b)} \neq 0$ ist. Offenbar ist

$$(4) \qquad R_{\underline{d}} = T_{\underline{d}}(w,w) \cap \ker \mathfrak{D}$$

Bezeichnen wir überdies für $w \geq 0$ den Modul der insgesamt homogenen Invarianten vom Totalgrad 2w mit

$$R_{2w} = \sum R_{\underline{d}} \quad ,$$

wobei in der Summation $\underline{d}$ alle Lösungen von (3) durchläuft, so gilt auch

$$(4)' \qquad R_{2w} = T(w,w) \cap \ker \mathfrak{D} \quad .$$

Nun sind $\mathfrak{D} : T(a,b) \to T(a-1,b+1)$ und $\Delta : T(a-1,b+1) \to T(a,b)$ lineare Operatioren, die den Totalgrad erhalten, und für die überdies

$$(5) \qquad \mathfrak{D}\Delta - \Delta\mathfrak{D} = \sum_{i=1}^{n} (\beta_i \, \partial/\partial\beta_i - \alpha_i \, \partial/\partial\alpha_i)$$

gilt, woraus

$$(6) \qquad (\mathfrak{D}\Delta - \Delta\mathfrak{D})F = (b-a)F \qquad \text{für } F \in T(a,b)$$

folgt. Mit Induktion folgen hieraus (vgl. [9] oder [15] , S. 68) die auf T(a,b) gültigen Formeln

$$(6a) \qquad \mathfrak{D}\Delta^{i+1} - \Delta^{i+1}\mathfrak{D} = (i+1)(b-a-i)\Delta^{i}$$

$$(6b) \qquad \mathfrak{D}^{i+1}\Delta - \Delta\mathfrak{D}^{i+1} = (i+1)(b-a+i)\mathfrak{D}^{i}$$

woraus überdies ebenfalls mit Induktion

$$(6c) \qquad \mathfrak{D}^{i+1} \Delta^{i+1} F = (i+1)! \prod_{j=0}^{i} (b-a-j).F \qquad \text{für } F \in T(a,b) \cap \ker \mathfrak{L}$$

folgt. Hieraus folgt

Satz 6: Für $a \leq b$ und char $k = 0$ ist

$$\mathfrak{D} : T(a,b) \to T(a-1, b+1)$$

surjektiv.

Beweis:

Setze $T_i = T(a-i,b+i)$ für $i \geq 0$ und $T_i^o = T_i \cap \ker \mathfrak{L}$. Dann operiert $\mathfrak{L}^i \Delta^i$ als (wegen $a \leq b$ nichtverschwindender) Skalar auf T_i^o nach (6c), also ist $\Delta^i : T_i^o \to T_o$ eine Injektion auf einen Teilraum mit genauem Annullator $\mathfrak{D}^{i+1}$, die $\Delta^i T_i^o$ sind also linear disjunkte Teilräume in T_o, woraus

$$\dim T_o \geq \sum_{i \geq o} \dim T_i^o$$

folgt. Andererseits ist, da $T_i^o = \ker (T_i \overset{\mathfrak{D}}{\to} T_{i+1})$,

$$\dim T_i^o \geq \dim T_i - \dim T_{i+1}$$

Setzt man dies in die obige Gleichung ein, so ergibt sich, daß überall Gleichheitszeichen stehen müssen, das Gleichheitszeichen für $i = 0$ in der zweiten Gleichung liefert die Behauptung.

Satz 6 liefert zusammen mit (4) bzw. (4)' die Grundlage der Cayley-Sylvesterschen Methode der Invariantenzählung (vgl. [3], [19]):

Folgerung 1: Es gilt

$$\dim_k R_{\underline{d}} = \dim_k T_{\underline{d}}(w,w) - \dim_k T_{\underline{d}}(w-1, w+1)$$

$$\dim_k R_{2w} = \dim_k T(w,w) - \dim_k T(w-1, w+1)$$

<u>Bemerkung</u>: Diese Folgerung gilt unabhängig von char k, aber Satz 6 und ebenso die Gleichungen (4) bzw. (4)' werden in jeder Primzahlcharakteristik falsch.

Um Folgerung 1 auszuwerten, haben wir zu bedenken, daß $\dim_k T(a,b)$ die Anzahl der Lösungen der Gleichungen

$$\sum_{i=1}^{n} e_i = a \quad , \quad \sum_{i=1}^{n} f_i = b$$

mit $e_i, f_i \geq 0$ ist und $\dim_k T_{\underline{d}}(a,b)$ diejenigen Lösungen zählt, für die $e_i + f_i = d_i$ $(i=1,\ldots,n)$ ist. Daraus folgt

$$\dim_k T(a,b) = \binom{a+n-1}{n-1}\binom{b+n-1}{n-1}$$

und mit Folgerung 1 also

$$\begin{aligned} \dim_k R_{2w} &= \binom{w+n-1}{n-1}^2 - \binom{w+n-2}{n-1}\cdot\binom{w+n}{n-1} \\ &= \frac{(w+1)(w+n-1)}{(n-1)!(n-2)!} \cdot \prod_{i=2}^{n-2} (w+i)^2 \end{aligned} \tag{7}$$

wodurch nochmals R als Ring der Dimension 2n-3 erkannt ist.

Vergleichen wir den höchsten Koeffizienten dieses charakteristischen Polynoms mit dem des zum Polynomring $k[c_t]$ gehörigen (s.Folgerung 4 zu Satz 5) charakteristischen Polynoms, so erhalten wir (vgl.[10] oder [18] , Lemma 1) analog zu Hilbert

<u>Folgerung 2</u>: Der Invariantenring $R = k[p_{ij}]$ ist ganz über dem Polynomring $k[c_t]$ vom Grad

$$\frac{(2n-4)!}{(n-1)!(n-2)!}$$

Die Dimension von $R_{\underline{d}}$ ist nicht ganz so leicht wie (7) berechenbar. Nach dem oben Gesagten ist $\dim_k T_{\underline{d}}(a,b)$ der Koeffizient von z^a im Polynom

$$\prod_{i=1}^{n} (1+z+\ldots+z^{d_i}) = (1-z)^{-n} \prod_{i=1}^{n} (1-z^{1+d_i}) \quad ,$$

also ist $\dim_k R_{\underline{d}}$ nach Folgerung 1 der Koeffizient von z^w in

$$(8) \qquad \Psi_{\underline{d}}(z) = (1-z)^{1-n} \cdot \prod_{i=1}^{n} (1-z^{1+d_i}) \quad ,$$

die charakteristische Potenzreihe $\chi(\underline{t},R) = \sum_{\underline{d}} \dim_k(R_{\underline{d}})\underline{t}^{\underline{d}}$,

wobei $\underline{t}^{\underline{d}} = \prod_{i=1}^{n} t_i^{d_i}$, ist also verborgen in der Potenzreihe

$$\Psi(z, \underline{t}) = \sum_{\underline{d}} \Psi_{\underline{d}}(z)\underline{t}^{\underline{d}}$$

$$= \frac{1-z}{\prod_{i=1}^{n} [(1-t_i)(1-t_i z)]}$$

In dem speziellen Fall, daß alle $d_i = d$ für $i = 1,\ldots,n$ sind, wir schreiben dafür $\underline{d} = (d)$, wollen wir eine zu (7) analoge Formel für $\dim_k R_{(d)}$ aufstellen. Jetzt geht (3) über in

$$(3)' \qquad n\,d = 2w$$

und es ist

$$\Psi_{(d)}(z) = (1-z)^{1-n}(1-z^{1+d})^n$$

$$= (1-z)^{1-n} \sum_{i=o}^{n} (-1)^i \binom{n}{i} z^{i+id} \quad .$$

Der Koeffizient von z^j in $(1-z)^{1-n}$ ist $\binom{n+j-2}{n-2}$, also hat $\Psi_{(d)}(z)$ bei z^w den Koeffizienten

$$(9) \qquad \dim_k R_{(d)} = \sum_{i\geq o} (-1)^i \binom{n}{i} \binom{w+n-2-id-i}{n-2} \quad ,$$

die Summe hat wegen (3)' etwa $^n/_2$ Glieder. Bei gerader Variablenzahl $n = 2m$ wird insbesondere (3)' zu

$$(3)'_o \qquad w = d\,m$$

und wir haben statt (9)

$$(9)_o \qquad \dim_k R_{(d)} = \sum_{i=o}^{m-1} (-1)^i \binom{2m}{i} \binom{(m-i)d+2m-2-i}{2m-2} \quad .$$

Wegen

$$\sum_{i=o}^{m-1} (-1)^i \binom{2m}{i} (m-i)^{2m-2} = 0$$

für $m > 1$ ist das Polynom auf der rechten Seite von (9) nur vom Grad $n-3$ in d, genauer ist

(9a) $$\dim_k R_{(d)} = \frac{c_o}{(n-3)!} d^{n-3} + \ldots + 1$$

mit

(9b) $$c_o = \frac{1}{2} \sum_{i=o}^{m-1} (-1)^{i+1} \binom{2m}{i} (m-i)^{2m-3}$$

also z.B.

$n = 4$: $\dim_k R_{(d)} = d+1$

$n = 6$: $\dim_k R_{(d)} = 1/2\ (d^3 + 3d^2 + 4d + 2)$

Der bisher betrachtete $(2n-3)$-dimensionale Invariantenring R ist noch zu groß. Lassen wir auf dem Polynomring $k[\underline{\alpha},\underline{\beta}]$ den n-dimensionalen Torus $GL_1(k)^n$ operieren, indem $(\lambda_1,\ldots,\lambda_n) \in GL_1(k)^n$ vermöge $\alpha_i \to \lambda_i\alpha_i$, $\beta_i \to \lambda_i\beta_i$ operiert, so ist diese Operation mit der der $GL_2(k)$ vertauschbar, und $k[\underline{\alpha},\underline{\beta}]$ zerfällt gemäß den Charakteren von $GL_1(k)^n$ in die direkte Summe der Teilräume

$$T_{\underline{d}} = \{ F \in k[\underline{\alpha},\underline{\beta}] \ ; \quad F \text{ erfüllt } (1) \} .$$

Wir betrachten nun den Teilring

$$T = \sum_d T_{(d)}$$

des Polynomrings $k[\underline{\alpha},\underline{\beta}]$, erzeugt von den Polynomen, die den gleichen Grad d in allen (α_i,β_i) haben. T behandelt die Punkte $(P_1,\ldots, P_n)$ eines n-Tupels "gleichwertig" und ist der eigentliche Formenring auf dem n-fachen Produkt $(\mathbb{P}_1)^n = \mathbb{P}_1 \times \ldots \times \mathbb{P}_1$, genauer ist $(\mathbb{P}_1)^n = \mathrm{Proj}(T)$. Ersetzt man alle β_i durch α_o, so wird T der Teilring des Polynomringes $k[\alpha_o,\ldots,\alpha_n]$, der von den Formen $\alpha_1\ldots\alpha_i\,\alpha_o^{n-i}$ $(i=0,\ldots,n)$ und den daraus durch Permutation der Indizes $1,\ldots,n$ hervorgehenden Formen erzeugt wird. T ist nicht faktoriell, aber als Fixring eines (n-1)-dimensionalen Torus ganz abgeschlossen, wenn k ganz abgeschlossener Integritätsbereich ist. Der Fixring von $SL_2(k)$ bei der Operation auf T ist dann

$$\overline{R} = \sum_d R_{(d)} \quad ,$$

der Ring der eigentlichen Invarianten. Wegen (3)', also $nd = 2w$, kann die Graduierung genau so gut nach w erfolgen, in der klassischen Invariantentheorie wird d als Grad und w als Gewicht einer Invariante bezeichnet. Als Folgerung aus Satz 5 und dessen Folgerungen erhalten wir nun

Satz 7: a) Der Ring $\overline{R}$ der eigentlichen Invarianten auf $\mathbb{P}_1^n$ bez. $GL_2(k)$ ist eine endlich erzeugte graduierte k-Algebra der Dimension n-2.

b) Das charakteristische Polynom von $\overline{R}$ ist durch (9) gegeben.

c) Es ist $\overline{R}(k) = \overline{R}(\mathbb{Z}) \otimes k$, jede eigentliche Invariante kommt von eigentlichen Invarianten über $\mathbb{Z}$. $\overline{R}$ ist ganz abgeschlossen in T, für ganz abgeschlossenes k also selbst ganz abgeschlossen.

d) Alle eigentlichen Invarianten (positiven Grades) verschwinden genau dann auf einem n-Tupel $(P_1,\ldots,P_n) \in \mathbb{P}_1^n$, wenn ein P_i entartet oder mehr als n/2 der n Punkte P_i zusammenfallen.

Beweis:

Die endliche Erzeugbarkeit von $\overline{R}$ folgt daraus, daß $\overline{R}$ eine Teilgraduierung der endlich erzeugten k-Algebra R darstellt. Ein explizites Erzeugendensystem, bestehend aus gewissen Produkten $X_{\underline{m}} \in R_{(d)}$, also mit $d(\underline{m}) = (d)$, kann man finden, indem man die aus der letzteren Relation folgenden linearen Gleichungen

$$d_1 = d_2 = \ldots = d_n \tag{10}$$

mit

$$d_i = \sum_{j=1}^{i=1} m(j,i) + \sum_{j=i+1}^{n} (m(i,j)$$

über $\mathbb{N}_o$, d.h. im Bereich $m(i,j) \geq 0$, löst. Durchläuft $\underline{m}$ ein Fundamentalsystem von Lösungen (aus denen sich alle übrigen durch Summation ergeben), so durchläuft $X_{\underline{m}}$ ein Erzeugendensystem von $\overline{R}$. Die Behauptung über dim $\overline{R}$ folgt aus dim $SL_2(k) = 3$ oder durch Gradbetrachtung des charakteristischen Polynoms (9a). Der ganze Abschluß von $\overline{R}$ in T folgt daraus, daß SL_2 keine Untergruppen von endlichem Index hat, also auch keine endlichen Bahnen haben kann. Es bleibt noch Behauptung d) zu zeigen. Ist $X_{\underline{m}} \in R_{(d)}$, so kann nicht $m(i,j) = 0$ für $i,j = 1,\ldots, \frac{n}{2} + 1$ (oder $\frac{n+1}{2}$) sein, weil der Grad von $X_{\underline{m}}$ in dem restlichen α_1 nicht d übersteigen darf, der Gesamtgrad in $\underline{\alpha}$ aber $w = \frac{n}{2}d$ sein muß. Also verschwindet $X_{\underline{m}}$, wenn mehr als $\frac{n}{2}$ Punkte P_i zusammenfallen, bei Entartung natürlich auch. Zum Beweis der Umkehrung nehmen wir ein n-Tupel $(P_1,\ldots P_n)$ nicht entarteter Punkte, von denen höchstens $m = \frac{n}{2}$ bzw. $\frac{n-1}{2}$ zusammenfallen. Dann können sie (vgl.[15], Seite 55), so geordnet werden, daß $P_i \neq P_{i+m}$ für $i \leq m$, für ungerades n zusätzlich $P_1 \neq P_n \neq P_{m+1}$.
Damit verschwindet die Invariante $F = \prod_{i=1}^{m} P_{i\ i+m} \in R_{(1)}$ bzw.

$$F = P_{1\ m+1}\, P_{m+1\ n}\, P_{1n} \cdot \prod_{i=2}^{m} P_{i\ i+m} \in R_{(2)}$$ nicht auf $(P_1,\ldots,P_n)$.

<u>Folgerung:</u> a) Für gerades n ist $\overline{R}$ ganz über dem Ring $k[R_{(1)}]$. Wählt man aus $R_{(1)}$ n-2 algebraisch unabhängige Invarianten geeignet aus, so wird $\overline{R}$ über dem von ihnen erzeugten Polynomring ganz vom Grad c_o mit c_o aus (9b).

b) Für ungerades n ist $R_{(1)} = 0$ und $\overline{R}$ ganz über dem Ring $k[R_{(2)}]$.

Bemerkung: Aus der Folgerung ergibt sich ebenfalls ein Beweis dafür, daß $\overline{R}$ endlich erzeugt ist, wenn k Körper oder = $\mathbb{Z}$ ist, das ist einer der Hilbertschen Beweise für die endliche Erzeugbarkeit des klassischen Invariantenringes, vgl. [10] §§ 3-5.

In Charakteristik 0 läßt sich aus den Formeln (6a), (6b) der Zusammenhang zwischen $\overline{R}$ und T vertiefen, vgl. Hilbert [9], es gilt nämlich

Satz 8: Ist char k = 0, so gilt auf $T_{(d)}$

$$\sum_{i\geq 0} (-1)^i \frac{\mathfrak{D}^i \Delta^i}{i!(i+1)!} = \sum_{i\geq 0} (-1)^i \frac{\Delta^i \mathfrak{D}^i}{i!(i+1)!}$$

(die Summanden verschwinden auf $T_{(d)}$ ab $i = d + 1$) und dieser Differentialoperator liefert eine Projektion von $T_{(d)}$ auf $R_{(d)}$, allgemeiner: eine SL_2 - invariante, also $\overline{R}$ - lineare Projektion von T auf den eigentlichen Invariantenring $\overline{R}$.

Mit Hurwitz (vgl.[15], S. 127 oder [20]) erhält man für $k = \mathbb{C}$ einen solchen Projektionsoperator $\natural$ auch eleganter durch Integration über die kompakte Form SU_2 der SL_2:

$$F^{\natural} = \int_{SU_2} F^{\sigma}\, d\sigma$$

Ein solcher Operator liefert zum einen wieder die endliche Erzeugbarkeit von $\overline{R}$, indem man das von $\sum_{d>0} R_{(d)}$ erzeugte Ideal in T betrachtet - eine Idealbasis ist bereits Erzeugendensystem der k-Algebra $\overline{R}$. Er liefert überdies das Werkzeug, um die "Abbildung" $\mathrm{Proj}(T) \to \mathrm{Proj}(\overline{R})$ auf die Eigenschaften (vgl.[12]) zu untersuchen, die man braucht, um $\mathrm{Proj}(\overline{R})$ bzw. einen Teilraum als Bahnenraum der PGL_2 auf $\mathbb{P}_1^n$, eventuell nach Fortlassen eines Teilraumes mit zu vielen Koinzidenzen, anzusprechen.

In Charakteristik p existiert ein solcher Operator nach Nagata [13] nicht, SL_2 operiert auf T nicht voll reduzibel, der Invariantenring $\overline{R}$ ist kein direkter Summand des $\overline{R}$-Moduls T. Dennoch gilt

Satz 9:

Ist $\mathfrak{a}$ Ideal von $\overline{R}$, so ist

$$T\mathfrak{a} \cap \overline{R} = \mathfrak{a}$$

Beweis:

Wir können k als endlich erzeugten Ring annehmen und daher als Faktorring eines Polynomringes $k_o = \mathbb{Z}[\underline{x}]$. Sei $\mathfrak{a}_o$ das volle Urbild von $\mathfrak{a}$ über k_o. Ist die Behauptung für $\mathfrak{a}_o$ gezeigt, so folgt sie durch Reduktion für $\mathfrak{a}$. Wir können uns daher auf $k \subset \mathbb{C}$ beschränken, wo aus $b \in T\mathfrak{a} \cap \overline{R}$, also

$b = \Sigma t_i a_i \in \overline{R}$ mit $t_i \in T$, $a_i \in \mathfrak{a}$,

$$b = b^{\natural} = \Sigma\, t_i^{\natural}\, a_i \qquad \in \qquad \mathfrak{a} \otimes \mathbb{C} ,$$

und daher $b \in \mathfrak{a}$ folgt.

Folgerung:

Ist Y der Teilraum des $\mathbb{P}_1^n$, in dessen n-Tupeln mehr als n/2 Punkte zusammenfallen, so ist die von $\overline{R} \subset T$ induzierte Abbildung

$$\pi : (P_1^n - Y) \longrightarrow \mathrm{Proj}(\overline{R})$$

surjektiv.

Die PGL_2-Bahnen auf $\mathbb{P}_1^n - Y$ sind 3-dimensional mit folgender Ausnahme: Ist n = 2m gerade, so ist die Bahn eines n-Tupels, das aus zweimal m gleichen Punkten besteht, offenbar nur zweidimensional.

Lemma: Sei n = 2m gerade. Eine Invariante $F \in \overline{R}$, die auf dem n-Tupel (P,...,P,Q,...,Q), jeder Punkt sei m-fach, verschwindet, verschwindet auf allen n-Tupeln, bei denen die m ersten oder die m letzten Punkte zusammenfallen.

Beweis:

Nach Folgerung a) von Satz 7 genügt es, $F \in R_{(1)}$ zu wählen. Für jede Bijektion

$\gamma : \{1,\dots,m\} \to \{m+1,\dots,n\}$ setze

$$X_\gamma = \prod_{i=1}^{m} p_{i\,\gamma(i)}$$

Dann sind die X_γ die einzigen p_{ij} - Produkte in $R_{(1)}$, die nicht auf $(P,\dots,P,Q,\dots,Q)$ verschwinden, wohl aber tun das die Differenzen $X_\gamma - X_{\gamma'}$. Diese Invarianten verschwinden aber auch für die im Lemma angegebenen Punkte.

Alle GL_2-Bahnen der in diesen Lemma auftretenden Punkte werden also durch π in einen einzigen Punkt abgebildet. Die übrigen GL_2-Bahnen werden durch π getrennt, wie man etwa wie im Beweis von Satz 9 durch Zurückführen in Charakteristik 0 und Benutzung des Operators $\natural$ sieht. Damit ergibt sich

Satz 10:

Sei Y_1 der Teilraum des $\mathbb{P}_1^n$, in dessen n-Tupeln mindestens n/2 Punkte zusammenfallen.

a) Ist n ungerade, so ist $Y_1 = Y$, also

$$\pi : \mathbb{P}_1^n - Y_1 \to \mathrm{Proj}(\overline{R})$$

surjektiv, die Fasern sind genau die GL_2-Bahnen auf $\mathbb{P}_1^n - Y_1$.

b) Ist $n = 2m$ gerade, so ist $Y_1 \neq Y$, es gibt eine $\binom{n}{m}$ -elementige Punktmenge M so, daß

$$\pi : \mathbb{P}_1^n - Y_1 \to \mathrm{Proj}(\overline{R}) - M$$

surjektiv ist und die Fasern wieder genau die GL_2-Bahnen auf $\mathbb{P}_1^n - Y_1$ sind.

c) Jeder längs Bahnen konstante Morphismus auf $\mathbb{P}_1^n - Y_1$ faktorisiert über π.

B. Die Invarianten einer binären Form

Auf $\mathbb{P}_1^n$ bzw. auf T operiert die symmetrische Gruppe S_n, indem sie die Ordnung des n-Tupels $(P_1,\dots,P_n)$ bzw. die Variablen $(\alpha_1,\dots,\alpha_n)$ permutiert. Die S_n-Bahnen auf $\mathbb{P}_1^n$ bilden das symmetrisierte n-fache Produkt $\widetilde{\mathbb{P}_1^n}$ der projektiven Geraden $\mathbb{P}_1$, es ist $\widetilde{\mathbb{P}_1^n} = \mathrm{Proj}(S)$, wo S der Invariantenring der Gruppe S_n auf T ist. Betrachten wir die binäre Form

$$\begin{aligned} f(x,y) &= \prod_{i=1}^{n} (\beta_i x - \alpha_i y) \\ &= \alpha_o^n \prod_{i=1}^{n} \left(x - \frac{\alpha_i}{\alpha_o} y\right) \\ &= \sum_{i=o}^{n} u_i x^{n-i} y^i \end{aligned}$$

n-ten Grades mit den Nullstellen $P_i = (\alpha_i : \beta_i)$, so wird $u_o = \alpha_o^n$ und $\frac{u_i}{u_o}$ bis aufs Vorzeichen die i-te elementarsymmetrische Funktion der $\frac{\alpha_1}{\alpha_o}, \dots, \frac{\alpha_n}{\alpha_o}$. Betrachtet man die Erzeugung von T, so sieht man, daß

$$S = k[u_o,\dots,u_n]$$

der Polynomring in den u_i, also der Koordinatenring der binären Form f(x,y) wird. Dabei erhält das Element u_i von T her den $\underline{\alpha}$-Grad i.

Die Operation von S_n auf T ist mit der Operation der GL_2 vertauschbar, GL_2 operiert also auch auf dem Polynomring S bzw. auf dem Raum der binären Formen f(x,y) vom Grad n. Dies ist der Ausgangspunkt der klassischen Invariantentheorie ([4], [15]), die u.a. den Teilring I der Invarianten in S, d.h. den Fixring der SL_2-Operation auf S, untersuchte.

Wir haben bereits den Fixring der SL_2-Operation auf T als Ring $\overline{R}$ der eigentlichen Invarianten auf $\mathbb{P}_1^n$ bestimmt, und erhalten nun den Ring I der klassischen Invarianten auf $\widetilde{\mathbb{P}_1^n} = \mathbb{P}_n$ als Fixring der S_n-Operation auf $\overline{R}$.

<u>Satz 11</u>: Die symmetrische Gruppe S_n operiert auf dem Ring $\overline{R} = \sum_d R_{(d)}$ homogen. Der Fixring dieser Operation

$$I = \sum_d I_{(d)}$$

ist der Ring der klassischen Invarianten einer binären Form bzw. der Ring der Formen, die die geometrische Lage von n (ungeordneten) Punkten in $\mathbb{P}_1^n$ beschreiben.

Da S_n eine endliche Gruppe ist, folgt hieraus mit Satz 7 sofort der zuerst von Gordan [8] in Charakteristik 0 bewiesene Endlichkeitssatz der Invariantentheorie binärer Formen.

<u>Folgerung 1</u>:

a) I ist eine endlich erzeugt k-Algebra der Dimension n-2

b) $\overline{R}$ ist ganz über I vom Grad n!

c) Eine binäre Form f vom Grad n wird genau dann von allen Invarianten (positiven Grades) annulliert, wenn sie eine Nullstelle der Vielfachheit $> \frac{n}{2}$ besitzt.

Da I auch Fixring von SL_2 im Polynomring S ist, folgt wie früher

<u>Folgerung 2</u>: I ist faktoriell und ganz abgeschlossen in S.

Aus Satz 10 ergibt sich, da Quotientenbildung nach einer endlichen Gruppe keine Schwierigkeiten bereitet:

<u>Satz 12</u>: Sei k ein algebraisch abgeschlossener Körper. Sei Y_1 der Teilraum der binären Formen n-ten Grades über k, die eine Nullstelle der Vielfachheit $\geq \frac{n}{2}$ besitzen. Dann ist der Invariantenmorphismus

$$\pi : (\mathbb{P}_n - Y_1) \to \mathrm{Proj}(I)$$

für ungerades n surjektiv und läßt bei geradem n nur einen Punkt aus. Die Fasern von π sind genau die Klassen projektiv äquivalenter binärer Formen (mit Nullstellenvielfachheit $< \frac{n}{2}$).

Folgerung 1: Sei $D = \prod_{i<j} p_{ij}^2 \in I_{(2n-2)}$ die Diskriminante der binären Form f. Dann ist $D \neq 0$ gleichbedeutend mit der Separabilität von f, also mit der Existenz von n verschiedenen Nullstellen.

Sei

$$J = I[D^{-1}]_0$$

der Ring aller Invariantenquotienten mit D-Potenz im Nenner, deren Grad 0 ist. Dann ist J eine endlich erzeugte k-Algebra und wir haben einen Morphismus

$$\pi : (\mathbb{P}_n - Z) \to \operatorname{Spec}(J) \quad ,$$

wo $Z = \{f;\ D(f) = 0\}$ die Hyperfläche der inseparablen Formen bedeutet, und die Fasern wieder PGL_2-Bahnen sind.

Verbinden wir diese Folgerung mit Satz 1, so erhalten wir

Folgerung 2: Sei k ein algebraisch abgeschlossener Körper, char $k \neq 2$, sei $n = 2g + 2$. Dann liefert $\operatorname{Spec}(J)(k) = \operatorname{Hom}_{k\text{-Alg}}(J,k)$ die Menge der Isomorphieklassen hyperelliptischer Funktionenkörper vom Geschlecht g.

Damit haben wir so etwas wie ein affines Modulschema von endlichem Typ (mit faktoriellem Integritätsbereich J) für hyperelliptische Kurven vom Geschlecht g gewonnen. (Die Affinität ist bei Fischer [7] noch nicht ganz klar). Es bleibt die Frage, wie J vom Grundkörper k, insbesondere von der Charakteristik abhängt.

Satz 13:

Sei $k_o = \mathbb{Z}[1/p \; ; \; p \leq n]$. Dann gilt für jede k_o - Algebra k

$$I(k) = I(k_o) \otimes_{k_o} k$$

und dann natürlich auch

$$J(k) = J(k_o) \otimes_{k_o} k$$

Beweis:

In k_o ist die Gruppenordnung $n!$ der S_n invertierbar, damit wird die Darstellung der S_n auf $R_{(d)}$ vollreduzibel, der Projektionsoperator

$$F^{\natural} = \frac{1}{n!} \cdot \sum_{\sigma \in S_n} F^{\sigma}$$

projiziert $R_{(d)}$ auf $I_{(d)}$, woraus die Behauptung mit Satz 7c folgt.

Folgerung 1: Die symmetrisierten Produkte $X_{\underline{m}}^{\natural}$ mit $d(\underline{m}) = (d)$ erzeugen $I_{(d)}$ über k_o.

Folgerung 2: Der Cayley-Sylvestersche Abzählkalkül (vgl.[15]) bzw. die Weylschen Charakterformeln (vgl.[20]) zur Berechnung der charakteristischen Funktion $\dim_k I_{(d)}$ gelten für jede k_o-Algebra, insbesondere also in Charakteristik > n.

Bemerkung: Die S_n-Operation auf $R_{(d)}$ ist i.a. nicht voll reduzibel über $\mathbb{Z}$. Schon bei n = 4 ist $I = k[I_{(2)}, I_{(3)}]$, sobald char k > 3, aber $I = k[I_{(2)}, I_{(6)}]$ für char k = 3, ähnliches findet sich für n = 6 bei Igusa [11].

Literatur:

[1] E. Artin: Algebraic Numbers and Algebraic Functions (Princeton 1950/51), Gordon and Breach 1967

[2] O. Bolza: On binary sextics with linear transformations into themselves Amer.J.Math. 10(1888), 47-70

[3] A. Cayley: A Second Memoir upon Quantics Philos. Transact. 146(1856), 101-126

[4] A. Clebsch Theorie der binären algebraischen Formen Leipzig 1872

[5] H.S.M. Coxeter & W.O.J. Moser: Generators and Relations for Discrete Groups Springer 1972

[6] M. Deuring: Die Typen der Multiplikatorenringe elliptischer Funktionenkörper Abh.Math.Sem.Hamb. 14(1941), 197-272

[7] I. Fischer: The Moduli of Hyperelliptic Curves Transact. of the Am.Math.Soc. 82(1956), 64-84

[8] P. Gordan: Beweis, dass jede Covariante und Invariante einer binären Form eine ganze Function mit numerischen Coeffizienten einer endlichen Anzahl solcher Formen ist. J.f.d.r.u.a. Math. 69(1868), 323-354

[9] D. Hilbert: Über eine Darstellungsweise der invarianten Gebilde im binären Formengebiete Math.Ann. 30(1887), 15-29

[10] D. Hilbert: Über die vollen Invariantensysteme Math.Ann. 42(1893), 313-373

[11] J. Igusa: Arithmetic variety of moduli for genus two Ann. of Math. 72(1960), 612-649

[12] D. Mumford: Geometric Invariant Theory Springer 1965

[13] M. Nagata: Complete reducibility of rational representations of a matric group
J. Math. Kyoto Univ. 1(1961), 89-99

[14] M. Nagata: Invariants of a group in an affine ring
J. Math. Kyoto Univ. 3(1964), 369-377

[15] I. Schur: Vorlesungen über Invariantentheorie
Springer 1968

[16] C.S. Seshadri: Mumford's conjecture for GL(2) and applications
Algebraic Geometry Conference (Bombay 1968)

[17] C.S. Seshadri: Quotient spaces modulo reductive algebraic groups
Ann. of Math. 95(1972), 511-556

[18] T. Shioda: On the graded ring of invariants of binary octics
Amer.J.Math. 89(1967), 1022-1046

[19] J.J. Sylvester: Tables of the Generating Functions and Groundforms for the Binary Quantics of the First Ten Orders
Amer.J.Math. 2(1879), 223-251

[20] H. Weyl: The Classical Groups
Princeton 1939

Deformation kompakter komplexer Räume

Hans Grauert

1. Es seien X eine $(n+d)$-dimensionale komplexe Mannigfaltigkeit, $G \subset \mathbb{C}^d$ ein Gebiet mit $0 \in G$ und $\pi : X \longrightarrow G$ eine eigentliche, reguläre, holomorphe Abbildung. Die Funktionalmatrix von π hat also in jedem Punkt von X den Rang d. Die Urbilder der Punkte $\mathfrak{t} = (t_1, \ldots, t_d) \in G$ sind n-dimensionale kompakte komplexe Mannigfaltigkeiten $X_{\mathfrak{t}}$. Der Raum X ist in diese $X_{\mathfrak{t}}$ gefasert. Bekanntlich sind alle $X_{\mathfrak{t}}$ als differenzierbare (und sogar als reell-analytische Mannigfaltigkeiten) zu einander isomorph, ihre komplexe Struktur hängt jedoch i.a. von $\mathfrak{t}$ ab. Man nennt deshalb das Tripel (X, π, G) eine holomorphe Deformation von $X_{\mathfrak{t}}$ für jedes $\mathfrak{t} \in G$. Die Faser X_0 sei mit Y bezeichnet. (X, π, G) ist also insbesondere eine holomorphe Deformation von Y.

Es sei $G' \subset \mathbb{C}^{d'}$ ein weiterer Bereich, $\varphi : G' \longrightarrow G$ eine holomorphe Abbildung. Das gefaserte Produkt $X \times_G G' = X'$ ist eine komplexe Mannigfaltigkeit. Zu X' gehören holomorphe Projektionen $\pi' : X' \longrightarrow G'$ und $\hat{\varphi} : X' \longrightarrow X$. Dabei ist π' eigentlich und regulär und es gilt $\pi \circ \hat{\varphi} = \varphi \circ \pi'$, d.h. $\hat{\varphi}$ ist fasertreu: das Tripel (X', π', G') ist wieder eine holomorphe Deformation; ist $\varphi(\mathfrak{t}') = \mathfrak{t}$, so bildet $\hat{\varphi}$ die Faser $X'_{\mathfrak{t}'}$ isomorph auf $X_{\mathfrak{t}}$ ab. Über entsprechenden Punkten hat man also isomorphe Fasern. Gilt $0' \in G'$ und $\varphi(0') = 0$, so folgt $X'_{0'} \simeq Y$ (kanonisch isomorph). (X', π', G') ist dann wieder eine holomorphe Deformation von Y.

Definition 1. (X, π, G) ist vollständig in $\mathfrak{t} \in G$, wenn jede andere holomorphe Deformation (X', π', G') mit $X_{\mathfrak{t}} = X'_{0'}$, $0' \in G'$ über einer Umgebung von $0'$ isomorph zu einer holomorphen Liftung (mittels eines φ mit $\varphi(0') = \mathfrak{t}$) von (X, π, G) ist.

Definition 2. (X,π,G) heisst versell (oder auch: semi-universell) in 0, wenn durch (X',π',G') das totale Differential $(d\varphi)(0')$ stets bestimmt ist.

Verselle Deformationen von Y sind, wenn sie existieren, in der Nähe von 0 bis auf eine nahe Isomorphie eindeutig festgelegt. 1961 zeigte Kuranishi den folgenden Satz:

Satz 1. Es sei Y eine beliebige kompakte komplexe Mannigfaltigkeit. Dann gibt es eine holomorphe, in $0 \in B$ verselle Deformation (X,π,B) von Y, die für jedes $\mathfrak{s} \in B$ vollständig ist.

Im Satz von Kuranishi ist B nicht ein Gebiet im $\mathbb{C}^d$, sondern ein (zusammenhängender) komplexer Unterraum eines Gebietes $G \subset \mathbb{C}^d$ mit $0 \in B$. Die Einbettungsdimension von B in 0 ist dabei gleich d und d gleich $\dim_{\mathbb{C}} H^1(Y,\Theta)$, wobei Θ die Tangentialgarbe (Keime holomorpher Vektorfelder) auf Y bezeichnet. Natürlich ist auch X jetzt keine Mannigfaltigkeit mehr, sondern ein komplexer Raum. Die Abbildung π ist jedoch noch in einem verallgemeinerten Sinne regulär: Die Fasern $X_{\mathfrak{s}}$, $\mathfrak{s} \in B$ sind kompakte komplexe Mannigfaltigkeiten. Die Versalität und die Vollständigkeit gelten verschärft für holomorphe Deformationen (X',π',B'), wobei B' wieder ein komplexer Raum ist.

2. In meiner Arbeit, "Der Satz von Kuranishi für kompakte komplexe Räume" (Invent. Math. 1974, 36 p.), ist der Satz 1 auf kompakte komplexe Räume Y übertragen worden. (Vgl. auch die noch nicht veröffentlichten Untersuchungen von Forster-Knorr, Douady-Hubbard, Palamodow und Commichau.) Ist Y ein kompakter komplexer Raum, so ändert sich bei sinnvollen holomorphen Deformationen sogar die lokale (evtl. sogar die topologische Struktur) von Y. Ein einfaches Beispiel erhält man auf folgendem Wege:

Es sei X die disjunkte Vereinigung der $X_t = \{z_3^2 - z_1 \cdot z_2 = t\} \subset P^3$ mit $t \in G = \mathbb{C}$. Die Menge X wird auf natürliche Weise zu einem topologischen und komplexen Raum. Die Abbildung $\pi : X \longrightarrow \mathbb{C}$, die X_t nach t abbildet, ist eigentlich und holomorph. Die Faser X_o hat in $0 \in X_o$ eine isolierte Singularität, alle anderen Fasern sind kompakte komplexe Mannigfaltigkeiten. Man sieht: π ist nicht regulär.

Seit langem ist bekannt, dass die direkte Verallgemeinerung von "regulär" auf den Fall, wo die Fasern komplexe Räume sind, durch den Begriff "platt" gegeben wird. Mit dieser Ersetzung gilt dann der Satz 1 für die Deformation von kompakten komplexen Räumen unverändert.

3. Kuranishi hat seinen Satz 1961 mit Hilfe von fast-komplexen Strukturen und elliptischen Differentialoperatoren bewiesen. Diese Operatoren sind auf komplexen Räumen unbekannt. Der Beweis des verallgemeinerten Satzes benutzt deshalb Methoden völlig anderer Art. Der komplexe Raum Y wird in kleine Teile U_ι, $\iota = 1,\ldots,\iota_*$ zerhackt. Sodann werden die U_ι deformiert und die Resultate der Deformation wieder neu miteinander verheftet, so dass die neue Zusammenfügung ungefähr mit der alten übereinstimmt. Wie dieses Zerhacken und Verkleben vor sich geht, soll im folgenden etwas erläutert werden. Man konstruiert eine Aufbereitung von Y. Das ist ein Hepttupel $(U_\iota, \Phi_\iota, Q_\iota, \mathfrak{f}_\iota, \mathfrak{g}_\iota, H_{\iota_1\iota_2}, A_{\iota_1\iota_2})$ der folgenden Art:

1) Die U_ι, $\iota = 1,\ldots,\iota_*$ sind eine offene Steinsche Überdeckung von Y.

2) Die $Q_\iota \subset \mathbb{C}^N$ sind offene Quader. Die Zahl N hängt dabei nicht von ι ab.

3) $\Phi_\iota : U_\iota \longrightarrow Q_\iota$ ist stets eine biholomorphe Abbildung von U_ι auf einen komplexen Unterraum $Y_\iota \subset Q_\iota$.

4) $\mathfrak{f}_\iota = \begin{vmatrix} f_1^\iota \\ \vdots \\ f_m^\iota \end{vmatrix}$ ist immer ein m-Tupel beschränkter holomorpher Funktionen in Q_ι. Die f_i^ι erzeugen die Idealgarbe von Y_ι. Die Zahl m hängt wieder nicht von ι ab.

5) $\mathfrak{g}_\iota = \begin{vmatrix} \mathfrak{g}_1^\iota \\ \vdots \\ \mathfrak{g}_l^\iota \end{vmatrix}$ ist für alle ι eine beschränkte holomorphe

(l,m)-Matrizenfunktion in Q_ι. Die Zeilen $\mathfrak{g}_j^\iota$ erzeugen die Relationengarbe $\mathcal{R}(\mathfrak{f}_\iota)$ und l hängt wieder nicht von ι ab.

6) Ist $U_{\iota_1\iota_2} = U_{\iota_1} \cap U_{\iota_2} \neq \emptyset$, so sind Holomorphiebereiche $T_{\iota_1\iota_2} \subset Q_{\iota_2}$, $T_{\iota_2\iota_1} \subset Q_{\iota_1}$ mit $Y_{\iota_1\iota_2} = T_{\iota_1\iota_2} \cap Y_{\iota_2} = \Phi_{\iota_2}(U_{\iota_1\iota_2})$, $Y_{\iota_2\iota_1} = T_{\iota_2\iota_1} \cap Q_{\iota_1} = \Phi_{\iota_1}(U_{\iota_1\iota_2})$ und biholomorphe Abbildungen $H_{\iota_1\iota_2} : T_{\iota_1\iota_2} \xrightarrow{\sim} T_{\iota_2\iota_1}$, $H_{\iota_2\iota_1} = H^{-1}_{\iota_1\iota_2} : T_{\iota_2\iota_1} \longrightarrow T_{\iota_1\iota_2}$ gegeben, die die Punkte von $T_{\iota_1\iota_2}$, $T_{\iota_2\iota_1}$ und $Y_{\iota_1\iota_2}$, $Y_{\iota_2\iota_1}$ miteinander identifizieren und aus der disjunkten Vereinigung von Q_{ι_1} und Q_{ι_2} einen Hausdorffraum machen. Ferner gilt $H_{\iota_1\iota_2}\big|Y_{\iota_1\iota_2} = \Phi_{\iota_1} \circ \Phi^{-1}_{\iota_2}$ und $H_{\iota_2\iota_1}\big|Y_{\iota_2\iota_1} = \Phi_{\iota_2} \circ \Phi^{-1}_{\iota_1}$.

7) Die $A_{\iota_1\iota_2}$ sind beschränkte holomorphe (m,m) - Matrixfunktionen über $T_{\iota_1\iota_2}$ mit $f_{\iota_1} \circ H_{\iota_1\iota_2} = A_{\iota_1\iota_2} \circ f_{\iota_2}$.

Aus Zweckmässigkeitsgründen wird sodann noch verlangt, dass die Q_ι in bezug auf $\mathfrak{g}_\iota$, $\mathfrak{f}_\iota$ privilegiert im Sinne von Douady sind. Durch Verheftung der Y_ι erhält man aus den Tripeln $(\mathfrak{f}_\iota, \mathfrak{g}_\iota, H_{\iota_1\iota_2})$ den Raum Y zurück. Die Deformation von Y wird schliesslich durch kleine Abänderung der $\mathfrak{f}_\iota$, $\mathfrak{g}_\iota$, $H_{\iota_1\iota_2}$ bewerkstelligt. Nur im geeigneten Falle entsteht durch die Verheftung wieder ein kompakter komplexer Raum.

4. Der Begriff der infinitesimalen Deformation von Y kann definiert werden. Die infinitesimalen Deformationen bilden einen komplexen Vektorraum V, der in Isomorphieklassen zerfällt. Diese Isomorphieklassen sind ein endlichdimensionaler Vektorraum $\underline{V}$. Es gibt einen endlichdimensionalen Untervektorraum $V_o \subset V$, der durch die Quotientenabbildung $\eta : V \longrightarrow \underline{V}$ isomorph auf $\underline{V}$ geworfen wird. Die Elemente aus V

und damit aus V_o können auch durch Tripel der Art $(\tilde{f}_\iota, \tilde{g}_\iota, \tilde{H}_{\iota_1\iota_2})$ beschrieben werden.

Die Menge $\mathfrak{A} = \mathfrak{A}_\varepsilon = \{(\tilde{f}_\iota, \tilde{g}_\iota, \tilde{H}_{\iota_1\iota_2}) : \|\tilde{f}_\iota - f_\iota\| < \varepsilon, \|\tilde{g}_\iota - g_\iota\| < \varepsilon, \|\tilde{H}_{\iota_1\iota_2} - H_{\iota_1\iota_2}\| < \varepsilon\}$, $\varepsilon > 0$, ist offene Menge eines Banachraumes (für geeignete Betragnormen); die Menge der Tripel, die durch Verheftung zu kompakten komplexen Räumen führen, bilden einen banach-analytischen Unterraum $\mathfrak{b} \subset \mathfrak{A}$. Man sieht, dass Douady's Theorie der banach-analytischen Räume für den Beweis unseres Satzes Bedeutung erlangt!

Es werden zwei holomorphe Operatoren definiert. Γ bringt jedes Element aus $\mathfrak{b}$ möglichst nahe an eine infinitesimale Deformation aus V_o. Dabei wird die Aufbereitung verkleinert (d.h. U_ι, Q_ι etc.). Λ "glättet" dann wieder die verkleinerte Aufbereitung auf die alte Grösse. Die Isomorphieklasse der durch Verheftung entstehenden komplexen Räume ändert sich bei Anwendung von Γ und Λ nicht. Das gleiche gilt für einen holomorphen Operator τ, der im wesentlichen gleich $\Lambda \circ \Gamma$ ist. Dieser Operator ist vollstetig (kompakt). Nach einem Satz von Douady ist seine Fixpunktmenge $B \subset \mathfrak{b}$ ein endlichdimensionaler Raum. Jedem $\mathfrak{s} \in B$ entspricht ein kompakter komplexer Raum $X_\mathfrak{s}$. Man erhält einen komplexen Raum $X = \dot{\bigcup} X_\mathfrak{s}$ und eine eigentliche holomorphe Abbildung $\pi : X \longrightarrow B$ mit $X_\mathfrak{s} \longrightarrow \mathfrak{s}$. Es folgt, dass π auch platt ist. Das liegt an der Existenz der $\tilde{g}_\iota$. Schliesslich beweist man, dass (X, π, B) eine holomorphe Deformation von Y ist, die alle geforderten Eigenschaften hat.

Kurven auf den Hilbertschen Modulflächen und Klassenzahlrelationen

F. Hirzebruch

In meinem Mannheimer Vortrag habe ich die Kurven auf den Hilbertschen Modulflächen betrachtet, die durch schief-hermitesche Matrizen definiert werden (siehe § 3 (11)) und die zum Teil bereits für Klassifikationsfragen ([4], [5]) verwendet wurden. Homologiebeziehungen zwischen den Kurven führen zu Klassenzahlrelationen, da die Schnittzahl zweier Kurven durch eine Summe von Klassenzahlen gegeben werden kann. Die berühmte Hurwitzsche Klassenzahlrelation [6] kann durch eine solche Schnittzahlbetrachtung bewiesen werden, das ist sicher wohlbekannt. Dieser Beweis zeigt, wie man entsprechende Beweise für Kurven auf den Hilbertschen Modulflächen führen und zu neuen Klassenzahlrelationen kommen kann.

Während meines Aufenthalts am Collège de France und am Institut des Hautes Études Scientifiques im März und April 1974 konnte ich diese Kurven auf den Hilbertschen Modulflächen viel systematischer untersuchen. Von grosser Hilfe waren Diskussionen mit den Mathematikern am IHES, insbesondere mit G. Harder und D. Zagier. Manche der Ergebnisse und der Vermutungen wurden gemeinsam mit D. Zagier erarbeitet. Sie schliessen sich an den Mannheimer Vortrag an und wurden deshalb hier mit aufgenommen. Dabei musste ich mich auf Beweisandeutungen beschränken oder auch die Beweise weglassen: "Informal report" im Sinne der lecture notes.

Hinweis auf weitere Arbeiten:

Die Kurven F_N und T_N (vgl. § 3) hatte ich für Zwecke der Klassifikation der Hilbertschen Modulflächen auch bereits während meines Aufenthalts in Berkeley (Sommer 1973) studiert. Man kann zum Beispiel feststellen, welche Kurven F_N in den singularitätenfreien Modellen der Hilbertschen Modulflächen $\overline{\mathfrak{H}^2/SL_2(\mathfrak{o})}$ und $(\overline{\mathfrak{H}^2/SL_2(\mathfrak{o})})/\tau$ (vgl. § 3) exzeptionelle Kurven werden. Mit Hilfe dieser Kurven konnte das Klassifikationsproblem für Primzahldiskriminanten, das noch unerledigt geblieben war (vgl. [4] § 0.2), abgeschlossen werden. Da ich bisher noch keine Gelegenheit hatte, das Ergebnis anzukündigen, soll das hier geschehen, um zugleich die Nützlichkeit der Kurven F_N zu illustrieren.

Es sei p eine Primzahl $\equiv 1 \bmod 4$ und $\mathfrak{o}$ der Ring der ganzen Zahlen in $\mathbb{Q}(\sqrt{p})$. Die singularitätenfreien Modelle von $\overline{\mathfrak{H}^2/SL_2(\mathfrak{o})/\tau}$, wo $\tau : (z_1,z_2) \longrightarrow (z_2,z_1)$ die Involution ist, sind algebraische Flächen folgenden Typs.

Rational für $p < 193$ und $p = 197,229,269,293,317$.

Aufgeblasene K3-Flächen für $p = 193,233,257,277,349,389,397,461,509$.

Aufgeblasene (echt, "honestly") elliptische Flächen für $p = 241,281,353,373,421,557$.

Allgemeiner Typ für alle anderen $p \equiv 1 \bmod 4$.

Hierüber ist eine Arbeit geplant. Das Klassifikationsproblem kann auch für beliebige Körper-Diskriminanten $D > 0$ und die Hilbertschen Modulflächen, die zu $\mathbb{Q}(\sqrt{D})$ gehören, studiert werden. Darüber sind gemeinsame Arbeiten mit D. Zagier in Vorbereitung.

§ 1. Quadratische Formen und Klassenzahlen.

Es sei M ein freier orientierter $\mathbb{Z}$-Modul vom Range 2. Bezüglich einer (mit der Orientierung verträglichen) $\mathbb{Z}$-Basis von M lassen sich die Elemente von M durch Paare (x,y) ganzer Zahlen geben, und eine quadratische Form $S : M \longrightarrow \mathbb{Z}$ lässt sich schreiben als

$$S(x,y) = ax^2 + bxy + cy^2,$$

wo a,b,c ganze Zahlen sind.

Die Diskriminante $\Delta = b^2-4ac$ von S hängt nicht von der Wahl der Basis ab. Als Diskriminanten quadratischer Formen treten die durch 4 teilbaren ganzen Zahlen sowie die Zahlen $\equiv 1 \bmod 4$ auf.

Die Form S ist primitiv, dann und nur dann, wenn $(a,b,c) = 1$.

Die Form S ist positiv-definit, dann und nur dann, wenn $\Delta < 0$ und $a > 0$.

Für gegebenes $\Delta < 0$ bezeichnen wir mit $h(\Delta)$ die Anzahl der Isomorphieklassen primitiver positiv-definiter quadratischer Formen der Diskriminante Δ. (Falls Δ keine Diskriminante ist, wird $h(\Delta)$ gleich 0 gesetzt.) Dabei heissen die Moduln M und M' mit den Formen S und S' isomorph, wenn es einen orientierungstreuen Isomorphismus $M \longrightarrow M'$ gibt, der S und S' ineinander überführt.

Für $\Delta < 0$ setzen wir (vgl. Hurwitz [6])

$$(1) \qquad H(-\Delta) = \sum_{\substack{f \in \mathbb{N} \\ f^2 | \Delta}} \hat{h}(\Delta/f^2) ,$$

wobei $\hat{h}(-3) = \frac{1}{3}$, $\hat{h}(-4) = \frac{1}{2}$ und sonst $\hat{h}(\Delta) = h(\Delta)$.

Für eine Form $S(x,y) = ax^2 + bxy + cy^2$ mit $\Delta = b^2-4ac < 0$ bestimmen wir die komplexe Zahl z in der oberen Halbebene $\mathfrak{H}$ durch $az^2 + bz + c = 0$, d.h.

$$z = \frac{-b + i\sqrt{|\Delta|}}{2a} \in \mathfrak{H} .$$

Die Gruppe $SL_2(\mathbb{Z})$ ist die übliche Modulgruppe. Sie operiert auf $\mathfrak{H}$. Die Isomorphieklassen positiv-definiter quadratischer Formen gegebener negativer Diskriminante stehen in eineindeutiger Korrespondenz zu den Orbiten derartiger Punkte z unter $SL_2(\mathbb{Z})$. Deshalb kann man annehmen, dass z in dem Fundamentalbereich

$$\left\{ z = x + iy \;\middle|\; -\tfrac{1}{2} < x \leq \tfrac{1}{2},\ |z| \geq 1,\ |z| > 1 \text{ für } x < 0 \right\}$$

liegt. Daher gilt für jede natürliche Zahl N die folgende Gleichung

$$H(N) = \#\left\{ (a,b,c) \in \mathbb{Z}^3 \;\middle|\; \begin{array}{l} -a \leq b < a,\ c \geq a \\ c > a \text{ falls } b > 0 \\ N = 4ac-b^2 \end{array} \right\}$$

In dieser Anzahlformel ist ein Tripel $(a,-a,a)$ mit der Vielfachheit $\frac{1}{3}$, ein Tripel $(a,0,a)$ mit der Vielfachheit $\frac{1}{2}$ zu zählen. Diese Tripel entsprechen dem Punkt $z = \frac{1}{2}(1 + i\sqrt{3})$ bzw. $z = i$ im Fundamentalbereich. In diesen Punkten ist die Isotropiegruppe der Aktion von $SL_2(\mathbb{Z})/\{1,-1\}$ auf $\mathfrak{H}$ nicht-trivial. Sie hat die Ordnung 3 bzw. 2. Die obige Anzahlformel für $H(N)$ gilt auch, wenn $-N$ keine Diskriminante ist (d.h. $N \equiv 2 \bmod 4$ oder $N \equiv 1 \bmod 4$). Dann ist $H(N) = 0$.

Wenn $p = 1$ oder eine Primzahl $\equiv 1 \bmod 4$ ist, dann führen wir folgende zahlentheoretische Funktion ein.

$$H_p(N) = \sum_{\substack{s \in \mathbb{Z} \\ 4N-s^2 > 0 \\ 4N-s^2 \equiv 0 \bmod p}} H\left(\frac{4N-s^2}{p}\right) \quad . \tag{2}$$

Die Hurwitzsche Klassenzahlrelation besagt [6]:

<u>Wenn</u> N <u>keine Quadratzahl ist, dann gilt</u>

$$H_1(N) = 2 \sum_{\substack{d \mid N \\ d > \sqrt{N}}} d \quad . \tag{3}$$

Insbesondere ist für eine Primzahl q

$$H_1(q) = 2q \quad . \tag{4}$$

<u>Beispiel:</u>

$$\begin{aligned} H_1(5) &= H(20) + 2H(19) + 2H(16) + 2H(11) + 2H(4) \\ &= h(-20) + 2h(-19) + 2h(-16) + 2\hat{h}(-4) + 2h(-11) \\ &\quad + 2\hat{h}(-4) \\ &= 2 + 2 + 2 + 1 + 2 + 1 \;=\; 10 \quad . \end{aligned}$$

§ 2. Ein Beweis der Hurwitzschen Klassenzahlrelation.

Wir betrachten in der (singularitätenfreien) komplexen Fläche

$$X = \mathfrak{H}/SL_2(\mathbb{Z}) \times \mathfrak{H}/SL_2(\mathbb{Z})$$

die Kurve T_N $(N>0)$, welche durch alle Gleichungen $z_2 = \frac{az_1+b}{cz_1+d}$ mit $a,b,c,d \in \mathbb{Z}$ und $ad-bc = N$ gegeben wird. (Hier ist $(z_1,z_2) \in \mathfrak{H}\times\mathfrak{H}$.) Die Kurve T_1 ist die Diagonale von X, während die (im allgemeinen reduzible) Kurve T_N eine Hecke-Korrespondenz ist. Wenn N keine Quadratzahl ist, dann besteht $T_N \cap T_1 \subset X$ aus endlich vielen Punkten. Was ist die Schnittzahl $T_N \cdot T_1$ der beiden Kurven?

Satz. Die Schnittzahl $T_N \cdot T_1$ in X ist gleich $H_1(N)$.

Beweisandeutung: Ein Punkt $(z,z) \in \mathfrak{H}\times\mathfrak{H}$ kann nur dann einen Schnittpunkt liefern, wenn er einer ganzzahligen quadratischen Gleichung

$$\alpha z^2 + \beta z + \gamma = 0, \quad (\alpha,\beta,\gamma) = 1, \quad \beta^2 - 4\alpha\gamma = \Delta < 0 \tag{5}$$

genügt. Bis auf $SL_2(\mathbb{Z})$ - Äquivalenz gibt es genau $h(\Delta)$ solche Punkte. Falls wir noch $\alpha > 0$ verlangen, dann ist die Gleichung (5) eindeutig bestimmt. Wir betrachten den $\mathbb{Z}$-Modul $\mathfrak{M}$ aller ganzzahligen Matrizen $\begin{pmatrix} a & b \\ c & d \end{pmatrix}$, sodass die Gleichung $z_2 = \frac{az_1+b}{cz_1+d}$ von (z,z) erfüllt wird.

$$\mathfrak{M} = \left\{ \begin{pmatrix} a & b \\ c & d \end{pmatrix} \;\middle|\; cz^2 - az + dz - b = 0 \right\}.$$

Da (5) die primitive Gleichung für z ist, gilt

$$c = x\alpha, \quad d - a = x\beta, \quad -b = x\gamma \quad \text{mit } x \in \mathbb{Z} \quad .$$

Für jede Wahl von x sind c und b sowie $d-a$ bestimmt, während a noch frei ist; wir setzen $a = y$ und erhalten

$$ad - bc = y^2 + \beta xy + \alpha\gamma x^2 .$$

Also ist $\begin{pmatrix} a & b \\ c & d \end{pmatrix} \longrightarrow$ ad-bc eine quadratische Form auf $\mathfrak{M}$, nämlich

(6) $$(x,y) \longrightarrow \alpha\gamma x^2 + \beta xy + y^2 \quad .$$

Die Werte der Form (6) geben an, welche Kurven T_N durch (z,z) gehen.

Die Form (6) hat die Diskriminante Δ, stellt die 1 dar und ist deshalb äquivalent zu der durch die Norm gegebenen Form auf der Ordnung M der Diskriminante Δ im Körper $\mathbb{Q}(\sqrt{\Delta})$. Wir können deshalb $\mathfrak{M}$ mit M identifizieren. (Bekanntlich ist M im Sinne der komplexen Multiplikation der Endomorphismenring des Torus $\mathbb{C}^2/\mathbb{Z}\cdot 1+\mathbb{Z}\cdot z$.) Die auf $\mathfrak{H}\times\mathfrak{H}$ effektiv operierende Gruppe mit unserer komplexen Fläche X als Orbitraum ist

$$G = SL_2(\mathbb{Z})/\{1,-1\} \times SL_2(\mathbb{Z})/\{1,-1\} \quad .$$

Nehmen wir zunächst an, dass G im Punkte (z,z) triviale Isotropiegruppe hat. Dies gilt dann und nur dann, wenn z nicht $SL_2(\mathbb{Z})$ - äquivalent zu i oder zu $\frac{1}{2}(1 + i\sqrt{3})$ ist, d.h. $\Delta \neq -4$ und $\Delta \neq -3$. Jedes Element ξ aus der Ordnung M definiert einen Zweig von T_N mit $N =$ Norm(ξ), der die Diagonale T_1 in (z,z) transversal schneidet. Die Zweige von T_N, die durch den durch (z,z) repräsentierten Punkt von T_1 gehen, entsprechen eineindeutig den Elementen ξ aus M mit Norm$(\xi) = N$ und $\mathrm{Im}(\xi) > 0$, wo Im den Imaginärteil der komplexen Zahl $\xi \in M \subset \mathbb{Q}(\sqrt{\Delta})$ bezeichnet. (Man beachte, dass ξ und $-\xi$ den gleichen Zweig liefern und deshalb $\mathrm{Im}(\xi) > 0$ angenommen werden darf.) Die ganze algebraische Zahl ξ mit $\mathrm{Im}(\xi) > 0$ wird eindeutig festgelegt durch ihre Norm N und ihre Spur s. Sie liegt in der Ordnung M dann und nur dann, wenn $s^2 - 4N = f^2\Delta$ mit einer natürlichen Zahl f gilt.

Falls die Diskriminante Δ der Gleichung (5) gleich -3 oder -4 ist, dann ist die Isotropiegruppe von G zu berücksichtigen. Die Zahlen ξ_1 und ξ_2 aus M bestimmen dann und nur dann den gleichen Zweig, wenn sie durch Multiplikation mit einer Einheit von M auseinander hervorgehen. Jetzt ist es leicht, den Beweis für die Gleichung $T_N \cdot T_1 = H_1(N)$ zu beenden.

Wir kompaktifizieren in bekannter Weise $\mathfrak{H}/SL_2(\mathbb{Z})$ durch Hinzufügung eines Punktes ∞. Wir setzen $\overline{\mathfrak{H}/SL_2(\mathbb{Z})} = \mathfrak{H}/SL_2(\mathbb{Z}) \cup \{\infty\}$. Dies ist eine komplexe projektive Gerade. Die Fläche X wird kompaktifiziert zu

$$\overline{X} = \overline{\mathfrak{H}/SL_2(\mathbb{Z})} \times \overline{\mathfrak{H}/SL_2(\mathbb{Z})} \quad .$$

Es sei $S_1 = \{\infty\} \times \overline{\mathfrak{H}/SL_2(\mathbb{Z})}$ und $S_2 = \overline{\mathfrak{H}/SL_2(\mathbb{Z})} \times \{\infty\}$.

Die Kurven T_N lassen sich abschliessen zu Kurven in $\overline{X}$, <u>die wir ebenfalls mit T_N bezeichnen.</u> Die Kurve T_N schneidet $S_1 \cup S_2$ nur in (∞,∞), und zwar werden ihre Zweige in (∞,∞) durch

$$(7) \qquad z_2 = \frac{az_1 + b}{d}$$

mit $a > 0$ und $ad = N$ und $0 \leq b < (a,d)$ gegeben. In (∞,∞) haben wir die lokalen holomorphen Koordinaten

$$u = e^{2\pi i z_1} \quad , \quad v = e^{2\pi i z_2} \quad .$$

Dann werden für feste Zahlen a,d die durch (7) gegebenen Zweige (ihre Anzahl ist gleich dem grössten gemeinsamen Teiler von a und d) durch die Gleichung

$$(8) \qquad u^a = v^d$$

zusammengefasst. Da S_1 und S_2 durch $u = 0$ bzw. $v = 0$ gegeben werden, gilt für den Schnitt in $\overline{X}$

$$T_N \cdot S_1 = T_N \cdot S_2 = \sum_{d|N} d \quad .$$

Es folgt der Satz

<u>Satz.</u> <u>In</u> $\overline{X}$ <u>gilt die Homologiebeziehung</u>

$$(9) \qquad T_N \sim \Big(\sum_{d|N} d\Big)(S_1 + S_2) \quad .$$

Die Schnittzahl von T_N (jetzt sei N wieder keine Quadratzahl) mit T_1 im Punkte (∞,∞) ist wegen (8) gleich

$$\sum_{d|N} \min(d,\tfrac{N}{d}) = 2 \sum_{\substack{d|N \\ d<\sqrt{N}}} d \ .$$

Da die Schnittzahl von T_N und T_1 in X gleich $H_1(N)$ ist, gilt für die Schnittzahl von T_N und T_1 in $\overline{X}$ die Beziehung

$$(10) \qquad (T_N \cdot T_1)_{\overline{X}} = H_1(N) + 2 \sum_{\substack{d|N \\ d<\sqrt{N}}} d \ .$$

Andererseits ist wegen (9)

$$(T_N \cdot T_1)_{\overline{X}} = 2 \sum_{d|N} d \ .$$

Damit ist die in § 1 angegebene Hurwitzsche Klassenzahlrelation bewiesen.

§ 3. Kurven in den Hilbertschen Modulflächen.

Es sei p eine Primzahl $\equiv 1 \bmod 4$, die festgewählt wird. Wir betrachten den Körper $\mathbb{Q}(\sqrt{p})$ und in ihm den Ring $\mathfrak{o}$ der ganzen algebraischen Zahlen

$$\mathfrak{o} = \mathbb{Z}\cdot 1 + \mathbb{Z}\cdot \frac{1+\sqrt{p}}{2} \ .$$

Die Gruppe $SL_2(\mathfrak{o})$ operiert in bekannter Weise auf $\mathfrak{H}\times\mathfrak{H} = \mathfrak{H}^2$ (vgl. z.B. [4]). Die (nichtkompakte) Hilbertsche Modulfläche

$$X = \mathfrak{H}^2/SL_2(\mathfrak{o})$$

hat endlich viele Quotientensingularitäten, die von den Punkten auf $\mathfrak{H}^2$ herrühren, in denen $SL_2(\mathfrak{o})/\{1,-1\}$ eine nicht-triviale Isotropiegruppe hat. Die Fläche X ist eine rationale Homologiemannigfaltigkeit.

Mit $x \longrightarrow x'$ werde der nicht-triviale Automorphismus des Körpers $\mathbb{Q}(\sqrt{p})$

bezeichnet. Für $a,b \in \mathbb{Z}$ und $\lambda \in \mathfrak{o}$ betrachten wir die "schief-hermitesche" Matrix

$$A = \begin{pmatrix} a\sqrt{p} & \lambda \\ -\lambda' & b\sqrt{p} \end{pmatrix}$$

und die zugehörige Kurve

$$a\sqrt{p}\, z_1 z_2 - \lambda' z_1 + \lambda\, z_2 + b\sqrt{p} = 0 \tag{11}$$

in $\mathfrak{H}^2$, welche für $\det A = abp + \lambda\lambda' > 0$ nicht leer ist. Die Gleichung (11) ist nämlich äquivalent zu

$$z_2 = \frac{\lambda' z_1 - b\sqrt{p}}{a\sqrt{p}\, z_1 + \lambda} . \tag{12}$$

Die Kurve ist also für $\det A > 0$ der Graph einer gebrochen linearen Abbildung $\mathfrak{H} \to \mathfrak{H}$. Eine positive natürliche Zahl N lässt sich genau dann als $\det A = abp + \lambda\lambda' = N$ schreiben, wenn N quadratischer Rest mod p ist. Wir bezeichnen in diesem Fall mit T_N die Menge aller Punkte in $X = \mathfrak{H}^2/SL_2(\mathfrak{o})$, deren Repräsentanten (z_1, z_2) in $\mathfrak{H}^2$ wenigstens einer Gleichung (11) mit $\det A = N$ genügen. Man kann zeigen, dass T_N wirklich eine komplexe Kurve in der komplexen Fläche X ist. Die Kurve T_N ist im allgemeinen nicht irreduzibel. Betrachten wir in X nur diejenigen Punkte, die einer Gleichung (11) mit $\det A = N$ genügen, wobei A primitiv ist (d.h. es gibt keine natürliche Zahl $f > 1$, sodass $\frac{a}{f}, \frac{b}{f} \in \mathbb{Z}$ und $\frac{\lambda}{f} \in \mathfrak{o}$), dann erhalten wir eine Kurve F_N in X, welche man als irreduzibel nachweisen kann. Es ist

$$T_N = \sum_{\substack{f \geq 1 \\ f^2 | N}} F_{N/f^2} . \tag{13}$$

Die Kurve T_N ist analog zu der in § 2 "für p = 1" betrachteten Kurve T_N.

Wir beschränken uns jetzt auf den Fall $N \not\equiv 0 \bmod p$, d.h. $(\frac{N}{p}) = 1$. Die

Zahl N schreiben wir dann in der Form

(14) $$N = N_1 \cdot N_2 \quad ,$$

wo N_1 nur durch Primzahlen q mit $(\frac{q}{p}) = 1$ und N_2 nur durch Primzahlen q mit $(\frac{q}{p}) = -1$ teilbar ist. Die Zahl N_2 ist gleich einer Quadratzahl multipliziert mit dem Produkt einer geraden Anzahl verschiedener Primzahlen

$$N_2 = m^2 q_1 q_2 \cdots q_{2r} \quad .$$

Die Kurve $F_N \subset X$ ist dann Bild von $\mathfrak{H}/\Gamma$ unter einer Abbildung vom Grade 1, wo Γ eine diskrete Untergruppe von $SL_2(\mathbb{R})/\{1,-1\}$ ist, welche zur Einheitengruppe einer Ordnung in der (indefiniten) Quaternionenalgebra über $\mathbb{Q}$ isomorph ist, welche genau an den Primstellen $q_1, q_2, \ldots, q_{2r}$ verzweigt ist. Hieraus kann man schliessen:

<u>Satz.</u> <u>Die Kurve</u> T_N (<u>mit</u> $N = N_1 N_2$) <u>ist dann und nur dann kompakt</u> (als <u>Teilmenge von</u> $\mathfrak{H}^2/SL_2(\mathcal{O})$), <u>wenn</u> N_2 <u>keine Quadratzahl ist</u>, (<u>d.h.</u> $r > 0$).

In $\mathfrak{H}$ (mit der komplexen Koordinate $z = x + iy$) haben wir das invariante Volumenelement

$$\omega = -\frac{1}{2\pi} \frac{dx \wedge dy}{y^2} \quad ,$$

das im Sinne des Gauß-Bonnetschen Satzes normiert ist, d.h. $\int_{\mathfrak{H}/\Gamma} \omega$ ist gleich der Eulerschen Zahl von $\mathfrak{H}/\Gamma$, falls Γ irgendeine diskrete Untergruppe von $SL_2(\mathbb{R})/\{1,-1\}$ ist, <u>welche frei auf</u> $\mathfrak{H}$ <u>operiert und kompakten Quotienten</u> $\mathfrak{H}/\Gamma$ <u>hat</u>. Da F_N Bild eines Quotienten $\mathfrak{H}/\Gamma$ ist (Γ diskret, Γ operiert aber im allgemeinen nicht frei und $\mathfrak{H}/\Gamma$ ist im allgemeinen nicht kompakt), können wir im Sinne der Form ω von dem Volumen von F_N und auch von den Volumen von T_N sprechen, was wir mit $\mathrm{vol}(T_N)$ bezeichnen.

Mit Hilfe einer von Eichler [2] durchgeführten Volumenbestimmung kann man mit einiger Mühe beweisen (für $N \not\equiv 0 \bmod p$)

$$(15) \qquad \mathrm{vol}(T_N) = -\frac{1}{6} \sum_{d|N} \left(\frac{d}{p}\right) d \ .$$

Führt man in $\mathfrak{H}^2$ die Chernsche Form

$$c_1 = -\frac{1}{2\pi} \left(\frac{dx_1 \wedge dy_1}{y_1^2} + \frac{dx_2 \wedge dy_2}{y_2^2} \right)$$

ein, dann ist 2 $\mathrm{vol}(T_N)$ gleich dem Integral von c_1 über T_N.

Bemerkung: Für "p = 1" entspricht (15) der Formel (9) in § 2, denn $\mathrm{vol}(T_1) = \mathrm{vol}(\mathfrak{H}/SL_2(\mathbb{Z})) = -\frac{1}{6}$ und T_N ist (in § 2) in $\overline{X}$ homolog zu einem Vielfachen von T_1, das in (9) angegeben wird, woraus man schliessen kann, dass für "p = 1"

$$\mathrm{vol}(T_N) = -\frac{1}{6} \sum_{d|N} d \ .$$

Falls N keine Quadratzahl ist, dann schneiden sich T_N und T_1 in $X = \mathfrak{H}^2/SL_2(\mathfrak{o})$ in endlich vielen Punkten. Die Kurve T_1 ist Bild der Diagonalen von $\mathfrak{H}^2$ und $\mathfrak{H}/SL_2(\mathbb{Z}) \longrightarrow T_1$ ist bijektiv. Auf T_1 liegen zwei Quotientensingularitäten von X (der Ordnungen 2 und 3). Sie gehören zu den bekannten endlichen Untergruppen von $SL_2(\mathbb{Z})/\{1,-1\} \subset SL_2(\mathfrak{o})/\{1,-1\}$ und lassen sich durch die Punkte (i,i) bzw. $(\frac{1}{2}(1 + i\sqrt{3}), \frac{1}{2}(1 + i\sqrt{3}))$ von $\mathfrak{H}^2$ repräsentieren. Die Zweige der Kurven T_N und T_1 schneiden sich überall transversal. In den Quotientensingularitäten ist jeder Schnitt (im Sinne der Schnitt-Theorie auf rationalen Homologie-Mannigfaltigkeiten) mit der Vielfachheit $\frac{1}{2}$ bzw. $\frac{1}{3}$ zu zählen. In diesem Sinne ist die Schnittzahl $T_N \cdot T_1$ zu verstehen. Das folgende Resultat wird genauso wie der entsprechende Satz in § 2 bewiesen.

Satz. Die Schnittzahl $T_N \cdot T_1$ in der zur Primzahl $p \equiv 1 \bmod 4$ gehörigen Hilbertschen Modulfläche X ist gleich $H_p(N)$.

Die Definition von $H_p(N)$ wurde in (2) angegeben. Im Gegensatz zu p = 1 ist $H_p(N)$ im allgemeinen nicht ganzzahlig, aber $6H_p(N)$ ist eine ganze Zahl.

Die Hilbertsche Modulfläche $X = \mathfrak{H}^2/SL_2(\mathfrak{o})$ lässt eine natürliche Involution $\tau : X \longrightarrow X$ zu, welche durch $(z_1,z_2) \longrightarrow (z_2,z_1)$ induziert wird. Offensichtlich ist $\tau(T_N) = T_N$. Wenn T_N kompakt ist, dann repräsentiert T_N eine Homologieklasse

$$[T_N] \in H_2(X;\mathbb{Q}) \quad ,$$

welche unter τ invariant ist. Es bezeichne $\mathfrak{T}$ den Unterraum von $H_2(X;\mathbb{Q})$, der von den Homologieklassen $[T_N]$ der kompakten Kurven T_N erzeugt wird. Es gilt

$$(16) \qquad \mathfrak{T} \subset H_2(X;\mathbb{Q})^{\tau} \cong H_2(X/\tau;\mathbb{Q}) \ .$$

Die zweite Bettische Zahl von X/τ, d.h. der Rang von $H_2(X/\tau;\mathbb{Q})$, kann mit den Methoden von [4] berechnet werden. Wir geben hier zunächst nur an, dass

$$(17) \qquad \dim_{\mathbb{Q}} H_2(X/\tau;\mathbb{Q}) = \left[\frac{p-5}{24}\right] + 1 \qquad \text{für } p < 193 \ .$$

Die einzigen Primzahlen mit $\dim_{\mathbb{Q}} H_2(X/\tau;\mathbb{Q}) = 1$ sind 5,13,17. In diesen drei Fällen ist also auch $\dim_{\mathbb{Q}} \mathfrak{T} = 1$, und deshalb muss (für kompaktes T_N) die Schnittzahl $T_N \cdot T_1$ ein konstantes Vielfaches des Volumens von T_N sein. Durch Berechnung eines Beispiels lässt sich dieses Vielfache leicht bestimmen, und man erhält:

<u>Satz.</u> <u>Es sei</u> $p = 5$, 13 <u>oder</u> 17 <u>und</u> N <u>eine natürliche Zahl mit</u> $(\frac{N}{p}) = 1$, <u>für die in der Zerlegung</u> (14) <u>die Zahl</u> N_2 <u>kein Quadrat ist, dann gilt:</u>

$$6\,H_5(N) = 5 \sum_{d|N} \left(\frac{d}{5}\right) d$$

$$6\,H_{13}(N) = \sum_{d|N} \left(\frac{d}{13}\right) d$$

$$6\,H_{17}(N) = \frac{1}{2} \sum_{d|N} \left(\frac{d}{17}\right) d$$

<u>Beispiel:</u> $p = 17, \quad N = 42, \quad N_1 = 2, \quad N_2 = 21$

$$H_{17}(42) = 2\, H\left(\frac{4\cdot 42 - 10^2}{17}\right) + 2\, H\left(\frac{4\cdot 42 - 7^2}{17}\right)$$

$$= 2\, H(4) + 2\, H(7) = 3$$

$$\sum_{d|42} \left(\frac{d}{17}\right) d = (1+2)(1-3)(1-7) = 36 \quad .$$

Die Klassenzahlrelationen in dem vorstehenden Satz können auf nichtkompaktes T_N erweitert werden, wenn man den Durchgang der kompaktifizierten Kurve durch die Auflösung der Spitze von $\overline{\mathfrak{H}^2/SL_2(\mathfrak{o})}$ berücksichtigt. Für $p = 5$ wird die Spitze nur in eine einzige Kurve aufgeblasen [4] . In diesem Fall lässt sich das Ergebnis besonders einfach formulieren.

<u>Satz.</u> <u>Es sei</u> $N > 0$ <u>und keine Quadratzahl. Dann gilt für</u> $\left(\frac{N}{p}\right) = 1$

$$6\, H_5(N) = 5 \sum_{d|N} \left(\frac{d}{5}\right) d - 6 \sum_{\substack{x \geq 0 \\ y \geq 0 \\ x^2+3xy+y^2=N}} (x+y) \quad . \tag{18}$$

<u>Hierbei durchlaufen</u> x,y <u>alle Paare ganzer Zahlen mit den angegebenen Bedingungen.</u> (Wenn T_N kompakt ist, dann lässt sich die Gleichung $x^2 + 3xy + y^2 = N$ nicht lösen, es handelt sich dann um die Formel des vorstehenden Satzes.)

Wenn die natürlichen Zahlen N und M teilerfremd sind, dann lässt sich die angegebene Formel für die Schnittzahl von T_N und T_1 verallgemeinern zu einer Formel für die Schnittzahl $T_N \cdot T_M$. Wir setzen dabei ferner voraus, dass N und M nicht beide Quadratzahlen sind (dann ist $T_N \cap T_M$ eine endliche Menge) und dass N und M zu p teilerfremd sind. Es gilt dann

$$T_N \cdot T_M = T_{NM} \cdot T_1 = H_p(NM) \quad . \tag{19}$$

Die Voraussetzung, dass N und M zu p teilerfremd sind, ist wahrscheinlich überflüssig. Mit Hilfe von (19) lässt sich die Dimension von $\mathfrak{F}$ in einigen Fällen bestimmen.

<u>Beispiel:</u> Für p = 89 schneiden wir die kompakten Kurven T_{21}, T_{39}, T_{57}, T_{69}, T_{91} mit den nicht-kompakten Kurven T_1, T_2, T_5, T_{11}, T_{17} und fassen die Schnittzahlen in der folgenden Matrix zusammen.

6 x <u>Schnittzahl</u>:

	T_{21}	T_{39}	T_{57}	T_{69}	T_{91}
T_1	0	0	0	4	0
T_2	0	0	6	6	12
T_5	6	0	6	6	12
T_{11}	0	12	24	24	48
T_{17}	18	12	12	24	24

Da die Determinante der (4×4)-Matrix in der linken oberen Ecke nicht verschwindet und $\dim_{\mathbb{Q}} \mathfrak{F} \leq \left[\frac{p-5}{24}\right] + 1 = 4$ ist (vgl. (16), (17)), gilt $\dim_{\mathbb{Q}} \mathfrak{F} = 4$. Die Homologieklassen $[T_{21}]$, $[T_{39}]$, $[T_{57}]$, $[T_{69}]$ bilden eine Basis des Vektorraums $\mathfrak{F}$. Aus der Schnittmatrix liest man ferner die Homologiebeziehung

$$[T_{91}] = 2\,[T_{57}]$$

ab. Man kontrolliere, dass $\mathrm{vol}(T_{91}) = 2\,\mathrm{vol}(T_{57})$ und z.B.

$$T_{91} \cdot T_{71} = 2\,T_{57} \cdot T_{71}, \text{ d.h. } H_{89}(6461) = 2\,H_{89}(4047).$$

§ 4. Bemerkungen über weitere Resultate und Vermutungen.

Wir betrachten wie in § 3 die Hilbertsche Modulfläche $X = \mathfrak{H}^2/SL_2(\mathfrak{o})$ für die Primzahl $p \equiv 1 \mod 4$. Sie kann kompaktifiziert werden durch Hinzufügung endlich vieler Spitzen zu einer komplexen Fläche $\overline{X}$ mit endlich vielen Singularitäten (vgl. [4]). Auch auf $\overline{X}$ operiert die Involution τ. Der Quotient $\overline{X}/\tau$ ist eine Fläche mit endlich vielen Singularitäten, die alle in minimaler Weise aufgelöst werden sollen. Wir erhalten dann eine reguläre singularitätenfreie algebraische Fläche, die hier mit V bezeichnet werden soll. In [4] wurde ein etwas anderes singularitätenfreies Modell für $\overline{X}/\tau$ benutzt und $Y^o(p)/\tau$ genannt.

Es sei n die Anzahl der irreduziblen Kurven auf V, in die die Singularitäten von $\overline{X}/\tau$ aufgeblasen wurden. Für die 2-dimensionale Cohomologie von V betrachten wir die Hodge-Zerlegung

$$H^2(V;\mathbb{C}) = H^{2,o}(V) \oplus H^{1,1}(V) \oplus H^{o,2}(V) \quad .$$

Mit Hilfe der Methoden von [4] kann man ausrechnen, dass

$$\text{(20)} \qquad \dim_{\mathbb{C}} H^{1,1}(V) = n + \chi(p),$$

wo $\chi(p)$ das arithmetische Geschlecht von $\overline{X}$ ist (vgl. [4] § 5.6 (20)).

Es sei $\mathcal{Y}$ der $\mathbb{C}$-Vektorraum der holomorphen Spitzenformen $a(z_1,z_2)dz_1 \wedge dz_2$ für die Gruppe $SL_2(\mathfrak{o})$. Es sei ε die Grundeinheit für den Körper $\mathbb{Q}(\sqrt{p})$ mit $\varepsilon > 0$, $\varepsilon' < 0$. Dann ist

$$\text{(21)} \qquad a(\varepsilon z_1,\varepsilon'\bar{z}_2)dz_1 \wedge d\bar{z}_2 + a(\varepsilon z_2,\varepsilon'\bar{z}_1)dz_2 \wedge d\bar{z}_1$$

eine Form vom Typ (1,1) in $\mathfrak{H} \times \mathfrak{H}$, invariant unter der um die Involution τ erweiterten Gruppe $SL_2(\mathfrak{o})$. Mit $\mathcal{Y}^{1,1}$ werde der komplexe Vektorraum bezeichnet, der von den Formen (21) sowie von der Chernschen Form c_1 (siehe § 3) aufgespannt wird. Nach einer Mitteilung von Harder kann $\mathcal{Y}^{1,1}$ in natürlicher Weise mit dem Unterraum von $H^{1,1}(V)$ identifiziert werden, der aus allen Cohomologieklassen besteht, die auf sämtlichen Kurven, die durch die Auflösung entstanden sind, verschwinden. Es ist

$$\dim_{\mathbb{C}} \mathcal{Y}^{1,1} = \dim_{\mathbb{C}} \mathcal{Y} + 1 = \chi(p) \quad .$$

Dies erklärt (20). Der Poincarésche Isomorphismus bildet $\mathfrak{F} \otimes \mathbb{C}$ (vgl. (16)) auf einen Unterraum von $\mathcal{Y}^{1,1}$ ab, der ebenfalls mit $\mathfrak{F} \otimes \mathbb{C}$ bezeichnet werde.

Auf $\mathfrak{H}^2$ haben wir die komplexe Konjugation

$$\varkappa : (z_1, z_2) \longrightarrow (-\bar{z}_1, -\bar{z}_2) \ .$$

Der Raum $\mathcal{Y}^{1,1}$ zerfällt bezüglich $\varkappa$ in die Unterräume zu den Eigenwerten +1, -1.

$$\text{(22)} \qquad \mathcal{Y}^{1,1} = \mathcal{Y}^{1,1}_{+} \oplus \mathcal{Y}^{1,1}_{-} \ .$$

Der Raum $\mathfrak{F} \otimes \mathbb{C}$ ist offensichtlich in $\mathcal{Y}^{1,1}_{-}$ enthalten. Auch die Form c_1 gehört zu $\mathcal{Y}^{1,1}_{-}$. Vermöge (21) ist $\mathcal{Y}$ als Unterraum von $\mathcal{Y}^{1,1}$ anzusehen. Die Involution $\varkappa$ operiert auf $\mathcal{Y}$ (als Unterraum von $\mathcal{Y}^{1,1}$) in gleicher Weise wie $\tau : (z_1, z_2) \longrightarrow (z_2, z_1)$ auf $\mathcal{Y}$ operiert. Deshalb ist

$$\dim_{\mathbb{C}} \mathcal{Y}^{1,1}_{+} = \dim_{\mathbb{C}} \mathcal{Y}_{+} = \chi_{\tau}(p) - 1.$$

Hier bezeichnet $\chi_{\tau}(p)$ das arithmetische Geschlecht von V. Nach [4] (§ 5.6 (21)) ist

$$\chi_{\tau}(p) = \frac{1}{2} \left(\chi(p) - \left[\frac{p-29}{24}\right] \right) ,$$

woraus folgt

$$\text{(23)} \qquad \dim_{\mathbb{C}} \mathcal{Y}^{1,1}_{-} = \dim_{\mathbb{C}} \mathcal{Y}^{1,1}_{+} + \left[\frac{p-5}{24}\right] + 1 \ .$$

Unterhaltungen mit G. Harder und D. Zagier am IHES glaube ich entnehmen zu können, dass es für die Hilbertsche Modulfläche für jedes Ideal $\mathfrak{n} \subset \mathfrak{o}$ eine Hecke-Korrespondenz $A_{\mathfrak{n}}$ gibt und man in $\mathcal{Y}^{1,1}$ einen Unterraum U definieren kann, der aus allen Elementen besteht, die unter $A_{\mathfrak{n}} - A_{\mathfrak{n}'}$ für alle $\mathfrak{n}$ verschwinden. Es ist $U \subset \mathcal{Y}_{-}^{1,1}$ und $\mathcal{Y}_{-}^{1,1}/U = \mathcal{Y}_{+}^{1,1}$. Dies erklärt (23) und beweist zugleich als Folge von (23), dass

$$\dim_{\mathbb{C}} U = \left[\frac{p-5}{24}\right] + 1 .$$

Man kann ferner zeigen, dass $\mathfrak{F} \otimes \mathbb{C} \subset U$ und kommt so zu dem Satz.

<u>Satz.</u> <u>Die Dimension des $\mathbb{Q}$-Vektorraumes</u> $\mathfrak{F}$, <u>der von den Homologieklassen der kompakten Kurven</u> T_N <u>erzeugt wird</u>, <u>ist kleiner oder gleich</u> $\left[\frac{p-5}{24}\right] + 1$.

<u>Vermutung:</u> $\dim_{\mathbb{Q}} \mathfrak{F} = \left[\frac{p-5}{24}\right] + 1$.

Diese Vermutung konnte für viele p mit Hilfe von Schnittzahlberechnungen (vgl. § 3) bestätigt werden. So liessen sich für p = 193 in der Tat 8 linear unabhängige kompakte Kurven finden, nämlich T_{55}, T_{65}, T_{85}, T_{95}, T_{143}, T_{185}, T_{187}, T_{209}.

Für jedes Element $K \in \mathfrak{F} \otimes \mathbb{C}$ und jede Kurve T_N (kompakt oder nichtkompakt) ist die Schnittzahl $T_N \cdot K \in \mathbb{C}$ in $X = \mathfrak{H}^2/SL_2(\mathfrak{o})$ im homologischen Sinne erklärt. (Das Symbol "K" soll hier an <u>kompakter</u> Zyklus erinnern.) Auch das Volumen von K ist durch lineare Erweiterung wohldefiniert.

Wie üblich ist $\Gamma_0(p)$ die Untergruppe derjenigen Elemente $\begin{pmatrix} a & b \\ c & d \end{pmatrix}$ von $SL_2(\mathbb{Z})$, für die $c \equiv 0 \bmod p$.

<u>Vermutung.</u> <u>Es sei</u> $K \in \mathfrak{F} \otimes \mathbb{C}$. <u>Dann ist</u>

$$f(z) = \frac{1}{2}\,\mathrm{vol}(K) + \sum_{\substack{N=1 \\ (\frac{N}{p}) \neq -1}}^{\infty} (T_N \cdot K)\, e^{2\pi i N z} \tag{24}$$

<u>die Fourier-Entwicklung einer Modulform für die Gruppe</u> $\Gamma_o(p)$ <u>vom Gewicht</u> 2 <u>und "Nebentypus", d.h.</u>

$$f\left(\frac{az+b}{cz+d}\right) = \left(\frac{a}{p}\right)\cdot(cz+d)^2\cdot f(z)$$

<u>für</u> $\begin{pmatrix} a & b \\ c & d \end{pmatrix} \in \Gamma_o(p)$.

Nach Hecke hat der Raum der Modulformen für $\Gamma_o(p)$ vom Gewicht 2 und Nebentypus die Dimension $2\left(\left[\frac{p-5}{24}\right]+1\right)$. Der Unterraum der Formen, deren Fourierreihen $\sum_{n=o}^{\infty} a_n e^{2\pi inz}$ verschwindende Koeffizienten a_n haben für alle n mit $\left(\frac{n}{p}\right) = -1$, hat die halbe Dimension. Es wird auch vermutet, dass die Zuordnung $K \longrightarrow f$ (siehe (24)) ein Isomorphismus von $\mathfrak{F} \otimes \mathbb{C}$ auf diesen Unterraum ist.

Ein Zusammenhang zwischen Modulformen für $\Gamma_o(p)$ und Hilbertschen Modulformen ist wohlbekannt (vgl. [1]). Die Ergebnisse von D. Zagier [7] werden es vielleicht ermöglichen, die vorstehende Vermutung zu beweisen.

<u>Bemerkung.</u> Die Chernsche Differentialform c_1 ist Poincaré-Dual eines Elementes $K \in \mathfrak{F} \otimes \mathbb{C}$. Wendet man hierauf (24) an, dann erhält man (bis auf einen konstanten Faktor) die Summe der von Hecke ([3] S. 818) angegebenen Eisenstein-Reihen E_1 und E_2.

Literatur

[1] K. Doi and H. Naganuma, On the functional equation of certain Dirichlet series, Inventiones Math. 9, 1-14 (1969).

[2] M. Eichler, Über die Einheiten der Divisionsalgebren, Math. Ann. 114, 635-654 (1937).

[3] E. Hecke, Mathematische Werke, Göttingen 1970.

[4] F. Hirzebruch, Hilbert modular surfaces, L'Enseignement mathématique 19, 183-281 (1973).

[5] F. Hirzebruch and A. Van de Ven, Hilbert modular surfaces and the classification of algebraic surfaces, Inventiones Math. 23, 1-29 (1974).

[6] A. Hurwitz, Mathematische Werke, Bd. II, Basel und Stuttgart 1963. Siehe: Über Relationen zwischen Klassenzahlen binärer quadratischer Formen von negativer Determinante (Math. Ann. 25 (1885)).

[7] D. Zagier, erscheint in C.R. Acad. Sci. Paris.

PICARD SCHEMES OF FORMAL SCHEMES; APPLICATION TO RINGS WITH DISCRETE DIVISOR CLASS GROUP

Joseph Lipman[(1)]

Introduction.

We are going to apply scheme-theoretic methods - originating in the classification theory for codimension one subvarieties of a given variety - to questions which have grown out of the problem of unique factorization in power series rings.

Say, with Danilov [D2], that a normal noetherian ring A has discrete divisor class group (abbreviated DCG) if the canonical map of divisor class groups $\bar{i}:C(A) \to C(A[[T]])$ is bijective[(2)]. In §1, a proof (due partially to J.-F. Boutot) of the following theorem is outlined:

THEOREM 1. *Let A be a complete normal noetherian local ring with algebraically closed residue field. If the divisor class group C(A) is finitely generated (as an abelian group), then A has DCG.*

---

(1) Supported by National Science Foundation grant GP-29216 at Purdue University.

(2) For the standard definition of $\bar{i}$, cf. [AC, ch. 7, §1.10]. (Note that the formal power series ring A[[T]] is noetherian [AC, ch. 3, §2.10, Cor. 6], integrally closed [AC, ch. 5, §1.4], and flat over A [AC, ch. 3, §3.4, Cor. 3].)

The terminology DCG is explained by the fact that in certain cases (cf. [B];[SGA 2, pp. 189-191]) with A complete and local, C(A) can be made into a locally algebraic group over the residue field of A, and this locally algebraic group is discrete (i.e. zero-dimensional) if and only if $\bar{i}$ is bijective.

A survey of results about rings with DCG is given in [F, ch. V].

Recall that A is <u>factorial</u> if and only if $C(A) = (0)$ [AC, ch. 7, §3]. Also, A local ⇒ A[[T]] local, with the same residue field as A; and A complete ⇒ A[[T]] complete [AC, ch. 3, §2.6]. Hence (by induction):

COROLLARY 1. *If A (as in Theorem 1) is factorial, then so is any formal power series ring* $A[[T_1, T_2, \ldots, T_n]]$.

When the singularities of A are resolvable, more can be said:

THEOREM 1'. *Let A be as in Theorem 1, with C(A) finitely generated, and suppose that there exists a proper birational map* $X \to \mathrm{Spec}(A)$ *with X a regular scheme (i.e. all the local rings of points on X are regular). Let B be a noetherian local ring and let* $f: A \to B$ *be a local homomorphism making B into a formally smooth A-algebra (for the usual maximal ideal topologies on A and B).*[3] *Then B is normal, and the canonical map* $C(A) \to C(B)$ *is bijective.*

Some brief historical remarks are in order here. Corollary 1 was conjectured by Samuel [S2, p. 171];[4] however Samuel did not

---

[3] "Formal smoothness" means that the completion $\hat{B}$ is A-isomorphic to a formal power series ring $\bar{A}[[T_1, T_2, \ldots, T_n]]$, where $\bar{A}$ is a complete local noetherian flat A-algebra with maximal ideal generated by that of A (cf. [EGA 0_{IV}, §§19.3, 19.6, 19.7]). In particular, B is flat over A.

[4] For some earlier work on unique factorization in power series rings cf. [S1] and [K].

assume that the residue field of A was algebraically closed, and without this assumption, the conjecture was found by Salmon to be false [SMN]. Later, a whole series of counterexamples was constructed by Danilov [D1] and Grothendieck [unpublished].[5] Danilov's work led him to the following modification of Samuel's conjecture [D1, p. 131]:

<u>If A is a local ring which is "geometrically factorial"</u> (i.e. the strict henselization of A is factorial) <u>then also A[[T]] is geometrically factorial.</u>

In this general form, the conjecture remains open, though some progress has been made by Boutot [unpublished].

The study of Samuel's conjecture evolved into the study of rings with DCG. A complete normal noetherian local ring A has been shown to have DCG in the following cases[6]:

(i) (Scheja [SH]). A is factorial and depth $A \geq 3$.

(ii) (Storch [ST2]) A contains a field, and the residue field of A is algebraically closed and uncountable, with cardinality greater than that of C(A).

[Actually, for such A, Storch essentially proves Theorem 1' *without* needing any desingularization $X \to \mathrm{Spec}(A)$. Storch's proof uses a theorem of Ramanujam-Samuel (cf. proof of Theorem 1' in §1) and an elementary counting argument.]

---

[5] In these counterexamples the locally algebraic group of footnote [2] above has dimension > 0, but has <u>just one point</u> - namely zero - <u>rational over the residue field of A.</u>

[6] For some investigations in the context of analytic geometry, cf. [ST1] and [P].

(iii) (Danilov [D3]) If

either (a) A contains a field of characteristic zero

or (b) A contains a field, the residue field of A is separably closed, and there exists a projective map $g: X \to \mathrm{Spec}(A)$ with X a regular scheme, such that g induces an isomorphism

$$X - g^{-1}(\{\underline{m}\}) \xrightarrow{\approx} \mathrm{Spec}(A) - \{\underline{m}\}$$

($\underline{m}$ = maximal ideal of A)

then C(A) finitely generated ⇒ A has DCG.

[Danilov uses a number of results from algebraic geometry, among them the theory of the Picard scheme of schemes proper over a field, and the resolution of singularities (by Hironaka in case (a), and by assumption in case (b)).]

Significant simplifications have been brought about by Boutot. His lemma (§1) enabled him to eliminate all assumptions about resolution of singularities in the above-quoted result of Danilov, and also to modify the proof of Theorem 1' to obtain the proof of Theorem 1 which appears in §1 below.

Our proof of Theorem 1' is basically a combination of ideas of Danilov and Storch, except that in order to treat the case when A does not contain a field, we need a theory of Picard schemes for schemes proper over a complete local ring of mixed characteristic. This theory - which is the main underlying novelty in the paper - is given in §§2-3.

§1. Proofs of Theorems 1 and 1'.

The two theorems have much in common, and we will prove them together. Let A, B be as in Theorem 1'; for Theorem 1 we will simply take $B = A[[T]]$. Since A is local and B is faithfully flat over A, the canonical map $C(A) \to C(B)$ is <u>injective</u> [F, Prop. 6.10]; so we need only show that $C(A) \to C(B)$ is surjective.

Both B and its completion $\hat{B}$ are normal: when $B = A[[T]]$ this is clear; and under the assumption of Theorem 1', since B and $\hat{B}$ are formally smooth over A, it follows from the existence of the "desingularization" $X \to \mathrm{Spec}(A)$ [L1, Lemma 16.1]. As above, since $\hat{B}$ is faithfully flat over B, $C(B) \to C(\hat{B})$ is injective, and consequently we may assume that $B = \hat{B}$ ($= \bar{A}[[T_1, T_2, \ldots, T_n]]$, cf. footnote (3) in the Introduction). [Note here that if $R \subseteq S \subseteq T$ are normal noetherian rings with S flat over R and T flat over S (and hence over R), then the composition of the canonical maps

$$C(R) \to C(S) \to C(T)$$

<u>is</u> the canonical map $C(S) \to C(T)$.]

Let M be the maximal ideal of A. Then $M\bar{A}$ is the maximal ideal of $\bar{A}$, and by the theorem of Ramanujam-Samuel [F, Prop. 19.14],

$$C(B) \to C(B_{MB})$$

is bijective. Furthermore [EGA 01, p. 170, Cor. (6.8.3)], there exists a complete local noetherian flat B_{MB}-algebra B^* such that B^*/MB^* is an algebraically closed field. B^* is formally smooth over A (footnote (3) above) so under the hypotheses of Theorem 1', B^* is normal; furthermore B^* is faithfully flat over B_{MB}, so that, as before

$$C(B_{MB}) \to C(B^*)$$

is injective. Thus for Theorem 1' it suffices to show that $C(A) \to C(B^*)$ is surjective.

To continue the proof of Theorem 1', let U_A be the domain of definition of the rational map inverse to $X \to \mathrm{Spec}(A)$. Then U_A is isomorphic to an open subscheme of X, so we have a surjective map $\mathrm{Pic}(X) \to \mathrm{Pic}(U_A)$ [EGA IV, (21.6.11)]; furthermore the codimension of $\mathrm{Spec}(A) - U_A$ in $\mathrm{Spec}(A)$ is ≥ 2, so there is a natural isomorphism $\mathrm{Pic}(U_A) \xrightarrow{\sim} C(A)$ [ibid, (21.6.12)]. Similar considerations hold with B^* in place of A, and $X^* = X \otimes_A B^*$ in place of X. (The projection $X^* \to \mathrm{Spec}(B)$ is proper and birational, and X^* is a regular scheme [L1, Lemma 16.1].) There results a commutative diagram

$$\begin{array}{ccccc} \mathrm{Pic}(X) & \longrightarrow & \mathrm{Pic}(U_A) & \xrightarrow{\sim} & C(A) \\ \downarrow & & \downarrow & & \downarrow \\ \mathrm{Pic}(X^*) & \longrightarrow & \mathrm{Pic}(U_{B^*}) & \xrightarrow{\sim} & C(B^*) \end{array}$$

Since $\mathrm{Pic}(X^*) \to \mathrm{Pic}(U_{B^*})$ is surjective, it will be more than enough to show that $\mathrm{Pic}(X) \to \mathrm{Pic}(X^*)$ is bijective.

The corresponding step in the proof of Theorem 1 is more involved, and goes as follows. Let $B = A[[T]]$, let B^* be as above, and let I be a divisorial ideal in B. We will show below that there exists an open subset U_A of $\mathrm{Spec}(A)$ whose complement has codimension ≥ 2, and such that, with

$$U_B = (U_A)\otimes_A B \quad (\subseteq \mathrm{Spec}(B)), \qquad U^* = (U_A)\otimes_A B^* \ (\subseteq \mathrm{Spec}(B^*))$$

we have that

(i) <u>IB_q is a principal ideal in B_q for all prime ideals $q \in U_B$, and</u>

(ii) <u>the canonical map $\nu: \mathrm{Pic}(U_B) \to \mathrm{Pic}(U^*)$ is injective.</u>

Now there is a natural commutative diagram

$$\begin{array}{ccc} \mathrm{Pic}(U_A) & \xrightarrow{\lambda} & \mathrm{Pic}(U_B) \\ {\scriptstyle \mu_A}\downarrow & & \downarrow{\scriptstyle \mu_B} \\ C(A) & \longrightarrow & C(B) \end{array}$$

cf. [EGA IV, (21.6.10)]. Since B is flat over A, it is immediate (from the corresponding property for U_A) that the complement of U_B in $\mathrm{Spec}(B)$ has codimension ≥ 2; hence <u>(i) signifies that the element of $C(B)$ determined by I is of the form $\mu_B(\xi)$ for some $\xi \in \mathrm{Pic}(U_B)$.</u> So if we could show that ξ lies in the image of λ, then we would have the desired surjectivity of $C(A) \to C(B)$.

At this point we need:

LEMMA (J.-F. Boutot)[1]. <u>There exists a projective birational map $\phi: X \to \operatorname{Spec}(A)$ such that ϕ induces an isomorphism $\phi^{-1}(U_A) \tilde{\to} U_A$, and such that ξ lies in the image of the canonical map $\operatorname{Pic}(X \otimes_A B) \to \operatorname{Pic}(U_B)$.</u>

(Here X may be taken to be <u>normal</u>, but not necessarily regular.) Setting $X^* = X \otimes_A B^*$, we have a natural commutative diagram

$$\begin{array}{ccccc} \operatorname{Pic}(X) & \longrightarrow & \operatorname{Pic}(X \otimes_A B) & \longrightarrow & \operatorname{Pic}(X^*) \\ \downarrow & & \downarrow & & \downarrow \\ \operatorname{Pic}(U_A) & \xrightarrow{\lambda} & \operatorname{Pic}(U_B) & \xrightarrow{\nu} & \operatorname{Pic}(U^*) \end{array}$$

with ν injective (cf. (ii) above). A simple diagram chase shows then that for ξ to lie in the image of λ, it more than suffices that <u>$\operatorname{Pic}(X) \to \operatorname{Pic}(X^*)$ be bijective</u>.

Let us finish off this part of the argument by constructing U_A satisfying (i) and (ii). [It will then remain - for proving both Theorems 1 and 1' - to examine the map $\operatorname{Pic}(X) \to \operatorname{Pic}(X^*)$.]

Let

$$U_A = \{p \in \operatorname{Spec}(A) \mid A_p \text{ is a regular local ring}\}.$$

By a theorem of Nagata [EGA IV (6.12.7)], U_A is open in $\operatorname{Spec}(A)$; and certainly, A being normal, the codimension of $\operatorname{Spec}(A) - U_A$ in $\operatorname{Spec}(A)$ is ≥ 2. Since the fibres of $\operatorname{Spec}(B) \to \operatorname{Spec}(A)$ are regular [EGA IV, (7.5.1)], therefore B_q is regular for all $q \in U_B$ [EGA 0_{IV}, (17.3.3)], and (i) follows.

---

[1] The proof, which will appear in Boutot's thèse, was presented at a seminar at Harvard University in January, 1972.

As for (ii), setting $U' = U_A \otimes_A B_{MB}$ (M = maximal ideal of A) we have the commutative diagram

$$\begin{array}{ccc} \mathrm{Pic}(U_B) & \longrightarrow & \mathrm{Pic}(U') \\ \downarrow & & \downarrow \\ C(B) & \longrightarrow & C(B_{MB}) \end{array}$$

in which the vertical arrows are <u>isomorphisms</u> [EGA IV, (21.6.12)], and also $C(B) \to C(B_{MB})$ is an isomorphism (cf. above); so we have to show that <u>$\mathrm{Pic}(U') \to \mathrm{Pic}(U^*)$ is injective.</u> Since $\mathrm{Pic}(U')$ is isomorphic to $C(B_{MB})$, this injectivity amounts to the following statement:

(#) <u>Let I be a divisorial ideal of B_{MB}, and let $\mathscr{I}^*$ be the coherent ideal sheaf on $\mathrm{Spec}(B^*)$ determined by the ideal IB^*. If $\mathscr{I}^*|U^* \cong \mathcal{O}_{U^*}$, then I is a principal ideal.</u>

Since B_{MB} is local, and B^* is faithfully flat over B_{MB}, we have

$$I \text{ principal} \Leftrightarrow I \text{ invertible} \Leftrightarrow IB^* \text{ invertible.}$$

Now I is a reflexive B_{MB}-module [CA, p. 519, Ex. (2)], and therefore IB^* is a reflexive B^*-module [<u>ibid</u>, p. 520, Prop. 8]. Since B^* is flat over B_{MB}, it follows (from the corresponding property of U') that for every prime ideal P in B^* such that $P \notin U^*$, the local ring B^*_P has depth ≥ 2. This being so, if $i: U^* \to \mathrm{Spec}(B^*)$ is the inclusion map, then the natural map

$$\mathcal{O}_{\mathrm{Spec}(B^*)} \to i_*(\mathcal{O}_{U^*})$$

is an <u>isomorphism</u> [EGA IV, (5.10.5)]. Since IB^* is reflexive, application of $\mathrm{Hom}_{B^*}(\cdot, B^*)$ to a "finite presentation"

$$(B^*)^n \to (B^*)^m \to \mathrm{Hom}_{B^*}(IB^*, B^*) \to 0,$$

gives an exact sequence

$$0 \to IB^* \to (B^*)^m \to (B^*)^n,$$

whence a commutative diagram, with exact rows,

$$\begin{array}{ccccccc} 0 & \longrightarrow & \mathcal{J}^* & \longrightarrow & \mathcal{O}^m & \longrightarrow & \mathcal{O}^n \\ & & \downarrow & & \downarrow \wr & & \downarrow \wr \\ 0 & \to & i_*(\mathcal{J}^*|U^*) & \to & i_*(\mathcal{O}^m_{U^*}) & \to & i_*(\mathcal{O}^n_{U^*}) \end{array} \qquad [\mathcal{O} = \mathcal{O}_{\mathrm{Spec}(B^*)}]$$

from which we conclude that the canonical map

$$\mathcal{J}^* \to i_*(\mathcal{J}^*|U^*)\,[\cong i_*(\mathcal{O}_{U^*})]$$

is an isomorphism. Thus $\mathcal{J}^*$ is isomorphic to $\mathcal{O}_{\mathrm{Spec}(B^*)}$, and (ii) is proved.

The rest of the discussion applies to both Theorems (1 and 1'). We must now examine the map $\mathrm{Pic}(X) \to \mathrm{Pic}(X^*)$.

The kernel of the surjective map $\mathrm{Pic}(X) \to \mathrm{Pic}(U_A)$ consists of the linear equivalence classes of those divisors on X which are supported on $X - U_A$; hence (X being assumed to be normal) this kernel is isomorphic to a subgroup of the free abelian group generated by those irreducible components of $X - U_A$ having codimension one in X; since $\mathrm{Pic}(U_A) \subseteq C(A)$, and $C(A)$ is finitely generated, therefore Pic(X) is finitely generated.

Let k (resp. k^*) be the residue field of A (resp. B^*). There is an obvious map $k \to k^*$. In §2 we will show that

(1.1) There exists a k-group-scheme P and a commutative diagram

$$\begin{array}{ccc} P(k) & \longrightarrow & P(k^*) \\ \downarrow \wr & & \downarrow \wr \\ \mathrm{Pic}(X) & \longrightarrow & \mathrm{Pic}(X^*) \end{array}.$$

Here $P(k) \to P(k^*)$ is the map from k-valued points of P to k^*-valued points corresponding to the map $k \to k^*$; and the vertical maps are isomorphisms.

Furthermore, in §3 it will be shown that

(1.2) There exists a closed irreducible k-subgroup P^o of P, whose underlying subspace is the connected component of the zero point of P, and such that:

(i) P^o is the inverse limit of its algebraic (= finite type over k) quotients; moreover if $\bar{P}$ is such a quotient, then $P(k) \to \bar{P}(k)$ is surjective.

(ii) $P/P^o = \varprojlim_{n>0} Q_n$, <u>where Q_n is a discrete</u> (= reduced and zero-dimensional) <u>locally algebraic k-group; moreover</u> $P(K) \to (P/P^o)(K)$ <u>is surjective for any algebraically closed field</u> $K \supseteq k$.

To show that $\mathrm{Pic}(X) \to \mathrm{Pic}(X^*)$ is bijective, it will then suffice to show that <u>P^o is infinitesimal</u> [in other words, every algebraic quotient of P^o is zero-dimensional, so that $P^o(k) = P^o(k^*) = 0$, whence $\mathrm{Pic}(X) \to \mathrm{Pic}(X^*)$ can be identified with the map

$$\varprojlim_{n} (Q_n(k) \to Q_n(k^*))$$

which is obviously bijective].

But since $P^o(k) \subseteq P(k)$ is finitely generated, so is $\bar{P}(k)$ for any algebraic quotient $\bar{P}$ of P^o. By the structure theorem for connected reduced commutative algebraic groups over an algebraically closed field, we know that $\bar{P}_{red}$ has a composition series whose factors are multiplicative groups, additive groups, and abelian varieties. It follows easily that if $\bar{P}(k) = \bar{P}_{red}(k)$ is finitely generated, then $\bar{P}(k) = 0$, i.e. $\bar{P}$ is zero-dimensional.

§2. The Picard Scheme of a Formal Scheme.

In this section we establish the existence of a <u>natural group-scheme structure on $\mathrm{Pic}(\mathfrak{X})$</u> for certain formal schemes $\mathfrak{X}$. (If

$p\mathcal{O}_{\mathfrak{X}} = (0)$ (cf. (2.2)) there will be nothing new here. For the case $p\mathcal{O}_{\mathfrak{X}} \neq (0)$, most of the work is carried out in [L2], whose results will be quoted and used.) From this we will obtain (1.1). However, for completeness, we prove more general results than are required in the proof of Theorems 1 and 1'.

DEFINITION (2.1). A formal scheme $(\mathfrak{X}, \mathcal{O}_{\mathfrak{X}})$ is weakly noetherian if $\mathfrak{X}$ has a fundamental system of ideals of definition $\mathcal{J}_0 \supseteq \mathcal{J}_1 \supseteq \mathcal{J}_2 \supseteq \cdots$ such that for each $n \geq 0$ the scheme $(\mathfrak{X}, \mathcal{O}_{\mathfrak{X}}/\mathcal{J}_n)$ is noetherian.

It amounts to the same thing to say: in the category of formal schemes,

$$\mathfrak{X} = \varinjlim_{n \geq 0} X_n$$

where $X_0 \to X_1 \to X_2 \to \cdots$ is a sequence of immersions of noetherian schemes X_n, the underlying topological maps being homeomorphisms (cf. [EGA 01, §10.6, pp. 411-413]).

Any noetherian formal scheme is weakly noetherian [ibid, middle of p. 414].

If $\mathfrak{X}$ is weakly noetherian and $\mathcal{J}$ is any ideal of definition, then $(\mathfrak{X}, \mathcal{O}_{\mathfrak{X}}/\mathcal{J})$ is a noetherian scheme; indeed, $\mathcal{J} \supseteq \mathcal{J}_n$ for some n (since $\mathfrak{X}$ is quasi-compact) so that $(\mathfrak{X}, \mathcal{O}_{\mathfrak{X}}/\mathcal{J})$ is a closed subscheme of the noetherian scheme $(\mathfrak{X}, \mathcal{O}_{\mathfrak{X}}/\mathcal{J}_n)$. In particular, taking $\mathcal{J}$ to be the largest ideal of definition of

$\mathfrak{X}$, we see that we may - and, for convenience, we always will - assume that <u>the scheme</u> $\mathfrak{X}_{red} = (\mathfrak{X}, \mathcal{O}_{\mathfrak{X}}/\mathcal{J}_0)$ <u>is reduced</u>. (Cf. [EGA 01, p. 172 (7.1.6)].)

Next, let k be a perfect field of characteristic $p \geq 0$. For $p > 0$ let $W(k)$ be the ring of (infinite) Witt vectors with coefficients in k; and for $p = 0$ let $W(k)$ be the field k itself. $W(k)$ is complete for the topology defined by the ideal $pW(k)$; the corresponding formal scheme $Spf(W(k))$ will be denoted by $\mathfrak{W}_k$.

(2.2) <u>In what follows we consider a triple</u> $(\mathfrak{X}, k, f)$ <u>with</u>:

(i) $\mathfrak{X}$ a weakly noetherian formal scheme.

(ii) k a perfect field of characteristic $p \geq 0$.

(iii) $f: \mathfrak{X} \to \mathfrak{W}_k$ a morphism of formal schemes such that for every ideal of definition $\mathcal{J}$ of $\mathfrak{X}$, the induced map of schemes

$$f_{\mathcal{J}}: (\mathfrak{X}, \mathcal{O}_{\mathfrak{X}}/\mathcal{J}) \to Spec(W(k))$$

is <u>proper</u>[(1)].

<u>Remarks.</u> Morphisms $f: \mathfrak{X} \to \mathfrak{W}_k$ are in one-one correspondence with <u>continuous homomorphisms</u> $i: W(k) \to H^0(\mathfrak{X}, \mathcal{O}_{\mathfrak{X}})$ [EGA 01, p. 407, (10.4.6)][(2)]. The above map $f_{\mathcal{J}}$ corresponds to the composed

(1) For (iii) to hold it suffices that $f_{\mathcal{J}}$ be proper for <u>one</u> $\mathcal{J}$ (cf. (2.6) below).

(2) The existence of such an i implies that p is topologically nilpotent in $H^0(\mathfrak{X}, \mathcal{O}_{\mathfrak{X}})$ (since the image of a topologically nilpotent element under a continuous homomorphism is again topologically nilpotent). On the other hand, if p is topologically nilpotent in $H^0(\mathfrak{X}, \mathcal{O}_{\mathfrak{X}})$, then clearly <u>every</u> ring homomorphism $W(k) \to H^0(\mathfrak{X}, \mathcal{O}_{\mathfrak{X}})$ is continuous.

homomorphism

$$W(k) \xrightarrow{\ i\ } H^{o}(\mathfrak{X}, \mathcal{O}_{\mathfrak{X}}) \xrightarrow{\text{canonical}} H^{o}(\mathfrak{X}, \mathcal{O}_{\mathfrak{X}}/\mathcal{J}).$$

It is practically immediate that <u>$f_{\mathcal{J}}(\mathfrak{X})$ is supported in the closed point of Spec(W(k)).</u>

<u>Example.</u> Let R be a complete noetherian local ring with maximal ideal M and residue field k (perfect, of characteristic $p \geq 0$); let $g:X \to \mathrm{Spec}(R)$ be a proper map; and let $\mathfrak{X}$ be the formal completion of X along the closed fibre $g^{-1}(\{M\})$. The structure theory of complete local rings gives the existence of a (continuous) homomorphism $W(k) \to R$; composing with the map

$$R \to H^{o}(\mathfrak{X}, \mathcal{O}_{\mathfrak{X}}) \; [= H^{o}(X, \mathcal{O}_{X})]$$

determined by g, we obtain $i:W(k) \to H^{o}(\mathfrak{X}, \mathcal{O}_{\mathfrak{X}})$, whence a triple $(\mathfrak{X}, k, f)$ as above.

(2.3) For any k-algebra A let $W_n(A)$ (resp. $W(A)$) be the ring of Witt vectors of length n (resp. of infinite length) with coefficients in A. ($W_n(A) = W(A) = A$ if $p = 0$.) We consider $W_n(A)$ to be a discrete topological ring, and give $W(A)$ the topology for which $K_1 \supseteq K_2 \supseteq K_3 \supseteq \cdots$ is a fundamental system of neighborhoods of 0, K_n being the kernel of the canonical map $W(A) \to W_n(A)$ $(n \geq 1)$; then, in the category of topological rings,

$$W(A) = \varprojlim_{n\geq 1} W_n(A).$$

It is not hard to see that $K_1^2 = pK_1$, whence

$$K_1^{n+1} = p^n K_1 \subseteq K_n \;;$$

so $W(A)$ is an "admissible" ring, and we may let $\mathfrak{W}_A$ be the affine formal scheme

$$\mathfrak{W}_A = \mathrm{Spf}(W(A)).$$

In particular, for $A = k$, we get the same $\mathfrak{W}_k$ as in (2.1). If B is an A-algebra, then $W(B)$ is in an obvious way a topological $W(A)$-algebra, so that $\mathfrak{W}_A$ varies functorially with A.

With $f: \mathfrak{X} \to \mathfrak{W}_k$ as in (2.2), we set

$$\mathfrak{X}_A = \mathfrak{X} \times_{\mathfrak{W}_k} \mathfrak{W}_A = \mathfrak{X} \hat{\otimes}_{W(k)} W(A)$$

(product in the category of formal schemes). We have then the <u>covariant functor of k-algebras</u>

$$A \to \mathrm{Pic}(\mathfrak{X}_A).$$

What we show below is that <u>the fpqc sheaf P associated to this functor is a k-group scheme,</u> and that furthermore <u>the canonical map $\mathrm{Pic}(\mathfrak{X}_A) \to P(A)$ is bijective if A is an algebraically closed field.</u>

<u>Example</u> (continued from (2.2)). Suppose that $\mathfrak{X}$ is obtained from a proper map $g: X \to \mathrm{Spec}(R)$ as in the example of (2.2). For

any k-algebra A, setting $R_A = R\hat{\otimes}_{W(k)}W(A)$ (completed tensor product, R being topologized as usual by its maximal ideal M), we have

$$\mathfrak{X}_A = \mathfrak{X}\hat{\otimes}_{W(k)}W(A) = \mathfrak{X}\hat{\otimes}_R R_A.$$

Now if A is a perfect field, then R_A has the following properties, which characterize R_A as an R-algebra (up to isomorphism): R_A is a complete local noetherian flat R-algebra such that $R_A/MR_A \cong A$ (cf. [EGA 01, p. 190, (7.7.10)] and [EGA 0_{IV}, (19.7.2)]). Furthermore, $\mathfrak{X}_A$ is then the completion of the scheme $X_A = X\otimes_R R_A$ along the closed fibre of the projection $g_A : X_A \to \mathrm{Spec}(R_A)$. Hence Grothendieck's algebrization theorem [EGA III, (5.1.6)] gives that "completion" is an equivalence from the category of coherent $\mathcal{O}_{X_A}$-modules to the category of coherent $\mathcal{O}_{\mathfrak{X}_A}$-modules. Since an $\mathcal{O}_X$-module is invertible if and only if so is its completion[3], we deduce a <u>natural isomorphism</u>

$$\mathrm{Pic}(X_A) \cong \mathrm{Pic}(\mathfrak{X}_A).$$

Hence, restricting our attention to those A which are <u>algebraically closed fields</u>, we will have an A-functorial isomorphism

$$\mathrm{Pic}(X_A) \cong P(A).$$

---

[3] This follows easily from the fact that the completion $\hat{B}_I$ of a noetherian ring B w.r.t. an ideal I is faithfully flat over the ring of fractions B_{1+I}, so that if J is a B-ideal with $J\hat{B}_I$ a projective $\hat{B}_I$-module, then JB_{1+I} is a projective B_{1+I}-module.

This gives us the diagram (1.1) which is needed in the last step of the proof of Theorems 1 and 1'.

(2.4) We fix a fundamental system $\mathcal{J}_0 \supseteq \mathcal{J}_1 \supseteq \mathcal{J}_2 \supseteq \cdots$ of defining ideals of $\mathfrak{X}$, and for $n \geq 0$ let X_n be the scheme $(\mathfrak{X}, \mathcal{O}_{\mathfrak{X}}/\mathcal{J}_n)$. For any k-algebra A, let $X_{n,A}$ be the scheme

$$X_{n,A} = X_n \otimes_{W(k)} W_n(A).$$

The ringed spaces $X_{0,A}, X_{1,A}, \ldots, X_{n,A}, \ldots$ and $\mathfrak{X}_A$ all have the same underlying topological space, say X, and on this space X we have $\mathcal{O}_{\mathfrak{X}_A} = \varprojlim_n \mathcal{O}_{X_{n,A}}$. Hence there is a natural map

$$(*) \qquad \operatorname{Pic}(\mathfrak{X}_A) \to \varprojlim_n \operatorname{Pic}(X_{n,A}).$$

LEMMA. <u>Let A be a k-algebra, and if $p > 0$ assume that $A^p = A$</u> (i.e. the Frobenius endomorphism $x \to x^p$ of A is surjective). <u>Then the above map (*) is bijective.</u>

<u>Remark.</u> When $p > 0$ and $A^p = A$, or when $p = 0$, then $X_{n,A} = X \otimes_{W(k)} W(A)$.

<u>Proof of Lemma.</u> Say that an open subset U of X is <u>affine</u> if $(U, \mathcal{O}_{\mathfrak{X}_A}|U)$ is an affine formal scheme. The affine open sets form a base for the topology of X.

For each n, let $\mathcal{F}_n$ be the sheaf of multiplicative units in the sheaf of rings $\mathcal{O}_{X_{n,A}}$ (on the topological space X) and let

$$\mathcal{F} = \varprojlim_n \mathcal{F}_n = \text{sheaf of units in } \mathcal{O}_{\mathfrak{X}_A}.$$

For $m \geq n$, the kernel of $\mathcal{O}_{X_{m,A}} \to \mathcal{O}_{X_{n,A}}$ is nilpotent; so a simple argument ([L2, Lemma (7.2)], with the Zariski topology in place of the étale topology) shows that for affine U the canonical maps

$$H^i(U, \mathcal{F}_m) \to H^i(U, \mathcal{F}_n)$$

are <u>bijective</u> if $i > 0$, and <u>surjective</u> if $i = 0$. Applying [EGA 0_{III}, (13.3.1)], we deduce that for all $i > 0$, the maps

$$H^i(X, \mathcal{F}) \to \varprojlim_n H^i(X, \mathcal{F}_n)$$

are <u>surjective</u>. Furthermore, in order that

$$\begin{array}{ccc} H^1(X, \mathcal{F}) & \to & \varprojlim_n H^1(X, \mathcal{F}_n) \\ \| & & \| \\ \operatorname{Pic}(\mathfrak{X}_A) & & \varprojlim_n \operatorname{Pic}(X_{n,A}) \end{array}$$

be <u>bijective</u>, it is sufficient that the inverse system $H^o(X, \mathcal{F}_n)_{n\geq 0}$ satisfies the Mittag-Leffler condition (ML); and for this it is enough that the inverse system $H^o(X, \mathcal{O}_{X_{n,A}})$ should satisfy (ML); that is, for each fixed n, if I_{mn} $(m \geq n)$ is the image of $H^o(X, \mathcal{O}_{X_{m,A}}) \to H^o(X, \mathcal{O}_{X_{n,A}})$, then the sequence

(**) $$I_{n,n} \supseteq I_{n+1,n} \supseteq I_{n+2,n} \supseteq \cdots$$

should stabilize (i.e. $I_{N,n} = I_{N+1,n} = I_{N+2,n} = \dots$ for some N).

For $p > 0$ it is shown in [L2, Corollary (0.2) and Theorem (2.4)] that the fpqc sheaf $\underline{\underline{H}}_n$ associated to the functor

$$A \to H^o(X, \mathcal{O}_{X_{n,A}})$$

(of k-algebras A) is an affine algebraic k-group; furthermore [ibid, Corollary (4.4)] the canonical map

$$H^o(X, \mathcal{O}_{X_{n,A}}) \to \underline{\underline{H}}_n(A)$$

is bijective whenever $A^p = A$; and finally, for $m \geq n$, if $\underline{\underline{I}}_{mn}$ is the image (in the category of algebraic k-groups) of the natural map $\underline{\underline{H}}_m \to \underline{\underline{H}}_n$, and if $A^p = A$, then the canonical map

$$\underline{\underline{H}}_m(A) \to \underline{\underline{I}}_{mn}(A)$$

is surjective, so that $I_{mn} = \underline{\underline{I}}_{mn}(A)$ [cf. ibid, last part of proof of (6.3)]. Similar facts when $p = 0$ are well-known (and more elementary).

Now the sequence

$$\underline{\underline{I}}_{n,n} \supseteq \underline{\underline{I}}_{n+1,n} \supseteq \underline{\underline{I}}_{n+2,n} \supseteq \cdots$$

of closed subgroups of $\underline{\underline{H}}_n$ must stabilize, whence so must the sequence (**). Q.E.D.

(2.5) Before stating the basic existence theorem we need some more notation. For any scheme Y, $Br(Y)$ will be the <u>cohomological Brauer group</u> of Y:

$$Br(Y) = H^2_{\text{étale}}(Y, \text{multiplicative group}).$$

For any ring R we set:

$$Br(R) = Br(Spec(R))$$
$$Pic(R) = Pic(Spec(R))$$
$$R_{red} = R/\text{nilradical of } R.$$

For any defining ideal $\mathcal{J}$ of $\mathfrak{X}$ and any k-algebra A:

$$\mathfrak{X}_{\mathcal{J}} = \text{the scheme } (\mathfrak{X}, \mathcal{O}_{\mathfrak{X}}/\mathcal{J})$$

$$\mathfrak{X}_{\mathcal{J},A} = \mathfrak{X}_{\mathcal{J}} \otimes_{W(k)} W(A).$$

Finally, we set

$$k_0 = H^0(\mathfrak{X}_{red}, \mathcal{O}_{\mathfrak{X}_{red}}).$$

Since $\mathfrak{X}_{red}$ is proper over k (cf (2.2)), therefore k_0 is a finite product of finite field extensions of k.

Now for any $\mathcal{J}$, we have (cf (2.2)) a proper map

$$f_{\mathcal{J}}: \mathfrak{X}_{\mathcal{J}} \to Spec(W(k))$$

whose image is supported in the closed point of $Spec(W(k))$.

Hence, when $p > 0$, [L2, Theorem (7.5)] gives us a k-group-scheme $P_{\mathcal{J}}$ and, for all k-algebras A with $A^p = A$, an exact A-functorial sequence

$$0 \to \mathrm{Pic}(k_0 \otimes_k A_{red}) \to \mathrm{Pic}(\mathfrak{X}_{\mathcal{J},A}) \to P_{\mathcal{J}}(A)$$
$$\to \mathrm{Br}(k_0 \otimes_k A_{red}) \to \mathrm{Br}(\mathfrak{X}_{\mathcal{J},A}) .$$

A similar result is well-known for $p = 0$, or more generally when $p\mathcal{O}_{\mathfrak{X}_{\mathcal{J}}} = (0)$, with no condition on A, since then $\mathfrak{X}_{\mathcal{J}}$ is proper over the field k (cf [GR, Cor. 5.3]).

Also, if $\mathcal{J} \subseteq \mathcal{J}'$, then the canonical map

$$P_{\mathcal{J}} \to P_{\mathcal{J}'}$$

is <u>affine</u> ([SGA 6, Expose XII, Prop. (3.5)] when $p = 0$, and [L2, Prop. (2.5)] when $p > 0$). Thus $P = \varprojlim_{\mathcal{J}} P_{\mathcal{J}}$ <u>exists as a k-group-scheme</u> (cf. [EGA IV, §8.2]).

Now, in view of Lemma (2.4), a simple passage to inverse limits gives the desired result:

THEOREM. <u>There exists a k-group scheme P, and for k-algebras A such that $A^p = A$</u> (the condition $A^p = A$ is vacuous when $p = 0$) <u>an exact sequence, varying functorially with A,</u>

$$0 \to \mathrm{Pic}(k_0 \otimes_k A_{red}) \to \mathrm{Pic}(\mathfrak{X}_A) \to P(A) \to$$
$$\to \bigcap_{\mathcal{J}} \ker[\mathrm{Br}(k_0 \otimes_k A_{red}) \to \mathrm{Br}(\mathfrak{X}_{\mathcal{J},A})].$$

COROLLARY. *If A is an algebraically closed field, then the above map $\mathrm{Pic}(\mathfrak{X}_A) \to P(A)$ is bijective.*

For, then $\mathrm{Pic}(k_0 \otimes_k A_{red}) = \mathrm{Br}(k_0 \otimes_k A_{red}) = (0)$.[4]

Remarks. 1. The k-group-scheme P is *uniquely determined* by the requirements of the Theorem. Indeed, since for every k-algebra A there exists a faithfully flat A-algebra $\bar{A}$ with $\bar{A}^p = \bar{A}$ [L2, Lemma (0.1)], and since every element in $\mathrm{Pic}(k_o \otimes_k A_{red})$ or in $\mathrm{Br}(k_o \otimes_k A_{red})$ is locally trivial for the étale topology on A, it follows easily that *P is the fpqc sheaf associated to the functor* $A \to \mathrm{Pic}(\mathfrak{X}_A)$ *of k-algebras A.*

2. P^o, the connected component of zero in P, is described in (3.2) below. The remarks following (1.2) suggest that the following conjecture - or some variant - should hold:

Conjecture: *P^o is infinitesimal if and only if the natural (split injective) map*

$$\mathrm{Pic}(\mathfrak{X}) \to \mathrm{Pic}(\mathfrak{X} \hat{\otimes}_W W[[T]]) \qquad (W = W(k))$$

is bijective.

---

[4] The Corollary, which is what we need for Theorems 1 and 1', could be proved more directly, using [L2, §1, comments on part II]; then we could do without our Lemma (2.4), and without introducing "Br". In a similar vein it can be deduced from the Theorem - or shown more directly - that if K is a normal algebraic field extension of k such that every connected component of $\mathfrak{X}_{red}$ has a K-rational point, and if A is any perfect field containing K, then $\mathrm{Pic}(\mathfrak{X}_A) \to P(A)$ is bijective.

(2.6) Appendix to §2. The following proposition is meant to give a more complete picture of how our basic data $(\mathfrak{X}, k, f)$ can be defined. It will not be used elsewhere in this paper.

To begin with, observe that if $(\mathfrak{X}, k, f)$ is as in (2.2), then f induces a proper map

$$f_{\mathcal{J}_0}:(\mathfrak{X}, \mathcal{O}_{\mathfrak{X}}/\mathcal{J}_0) = \mathfrak{X}_{red} \to \operatorname{Spec}(k)$$

(cf. (2.2)). Hence $H^o(\mathfrak{X}, \mathcal{O}_{\mathfrak{X}_{red}})$ is a finite k-module (equivalently: a finite $W(k)$-module) and - *a fortiori* - a finite $H^o(\mathfrak{X}, \mathcal{O}_{\mathfrak{X}})$ - module. Conversely:

PROPOSITION. *Let $\mathfrak{X}$ be a weakly noetherian formal scheme, and assume that the $H^o(\mathfrak{X}, \mathcal{O}_{\mathfrak{X}})$-module $H^o(\mathfrak{X}, \mathcal{O}_{\mathfrak{X}_{red}})$ is finitely generated. Let k be a perfect field of characteristic $p \geq 0$, and let*

$$f_0:\mathfrak{X}_{red} \to \operatorname{Spec}(k)$$

be a proper map of schemes. Then f_o extends (uniquely, if $p > 0$) to a map of formal schemes $f:\mathfrak{X} \to \mathfrak{W}_k$. Furthermore, all the maps $f_{\mathcal{J}}$ (cf. (2.2)) are proper.

Proof. (Sketch) f_o corresponds to a homomorphism $i_o:k \to H^o(\mathfrak{X}, \mathcal{O}_{\mathfrak{X}_{red}})$; the problem is to lift i_o to a continuous homomorphism

$$i:W(k) \to H^o(\mathfrak{X}, \mathcal{O}_{\mathfrak{X}}).$$

Let $\mathcal{J}_0 \supseteq \mathcal{J}_1 \supseteq \mathcal{J}_2 \supseteq \ldots$ be a fundamental system of defining ideals of $\mathfrak{X}$ (cf. (2.1)), and let $H_o = H^o(\mathfrak{X}, \mathcal{O}_{\mathfrak{X}})/H^o(\mathfrak{X}, \mathcal{J}_o)$. We will show below that:

(*) <u>the canonical map</u> $H_o \xrightarrow{\pi} H^o(\mathfrak{X}, \mathcal{O}_{\mathfrak{X}_{red}})$ <u>is bijective.</u>

Then the existence of the lifting i follows (since W(k) is formally smooth over its subring $\mathbb{Z}_{p\mathbb{Z}}$) from [EGA 0_{IV}, (19.3.10)] (with $\mathcal{I} = H^o(\mathfrak{X}, \mathcal{J}_o)$). For the uniqueness when $p > 0$, cf. [<u>loc. cit.</u> (20.7.5) or (21.5.3)(ii)]. {Or else note that $H^o(\mathfrak{X}, \mathcal{O}_{\mathfrak{X}_{red}})$, being reduced and finite over k, is perfect, and argue as in [SR, p. 48, Prop. 10], using the following easily proved fact in place of [<u>ibid</u>., p. 44, Lemme 1]:

If $a, b \in H^o(\mathfrak{X}, \mathcal{O}_{\mathfrak{X}})$ satisfy $a \equiv b \pmod{H^o(\mathfrak{X}, \mathcal{J}_n)}$, then for some N depending only on n we have

$$a^{p^N} \equiv b^{p^N} \pmod{H^o(\mathfrak{X}, \mathcal{J}_{n+1})}.\}$$

Now (*) simply says that $H^o(\mathfrak{X}, \mathcal{O}_{\mathfrak{X}}) \to H^o(\mathfrak{X}, \mathcal{O}_{\mathfrak{X}_{red}})$ is surjective, and to prove this we may assume that $\mathfrak{X}$ is <u>connected</u>; then $H^o(\mathfrak{X}, \mathcal{O}_{\mathfrak{X}_{red}})$, being finite over k, is a perfect field, as is its subring H_o (since $H^o(\mathfrak{X}, \mathcal{O}_{\mathfrak{X}_{red}})$ is finite over H_o, by assumption), say $H_o = K$. As above, the identity map $K \to K$ lifts to a homomorphism $W(K) \to H^o(\mathfrak{X}, \mathcal{O}_{\mathfrak{X}})$, and thereby, for every ideal of definition $\mathcal{J}$, the scheme $(\mathfrak{X}, \mathcal{O}_{\mathfrak{X}}/\mathcal{J})$ is a W(K)-scheme. For $\mathcal{J} = \mathcal{J}_o$ the structural map $(\mathfrak{X}, \mathcal{O}_{\mathfrak{X}}/\mathcal{J}_o) \to \mathrm{Spec}(W(K))$ factors as

$$(\mathfrak{X}, \mathcal{O}_{\mathfrak{X}}/\mathcal{J}_o) = \mathfrak{X}_{red} \to \mathrm{Spec}(H^o(\mathfrak{X}, \mathcal{O}_{\mathfrak{X}_{red}})) \xrightarrow{\text{finite}} \mathrm{Spec}(K) \hookrightarrow \mathrm{Spec}(W(K)).$$

Note that $\mathfrak{X}_{red}$, being proper over k, is proper over $H^o(\mathfrak{X}, \mathcal{O}_{\mathfrak{X}_{red}})$, and hence also over K. Arguing as below, we see that $(\mathfrak{X}, \mathcal{O}_{\mathfrak{X}}/\mathcal{J}_n)$ is proper over $W(K)$, whence the kernel of $\pi_n : H^o(\mathfrak{X}, \mathcal{O}_{\mathfrak{X}}/\mathcal{J}_n) \to H^o(\mathfrak{X}, \mathcal{O}_{\mathfrak{X}}/\mathcal{J}_o)$ is a $W(K)$-module of <u>finite length</u>. So by [EGA 0_{III}, (13.2.2)], $\pi = \varprojlim \pi_n$ will be surjective if π_n is surjective for all n. Let us show more generally for any scheme map $\phi : X \to \mathrm{Spec}(W(K))$ that if ϕ induces a <u>proper</u> map

$$Y = X_{red} \to \mathrm{Spec}(K) \subseteq \mathrm{Spec}(W(K))$$

then $H^o(X, \mathcal{O}_X) \to H^o(Y, \mathcal{O}_Y)$ is surjective.

Let $\bar{K}$ be an algebraic closure of K. Then $W(\bar{K})$ is a <u>faithfully flat</u> $W(K)$-algebra. In view of [EGA III, (1.4.15)]. (Künneth formula for flat base change) and the fact that

$$Y \otimes_{W(K)} W(\bar{K}) = Y \otimes_K \bar{K}$$

is reduced (K being perfect), we may replace X by $X \otimes_{W(K)} W(\bar{K})$, i.e. we may assume that K is algebraically closed. But then $H^o(Y, \mathcal{O}_Y)$ is a product of copies of K, one for each connected component of Y, so the assertion is obvious.

It remains to be shown that the maps $f_{\mathcal{J}}$ are all proper. $(\mathfrak{X}, \mathcal{O}_{\mathfrak{X}}/\mathcal{J})$ is noetherian, and $\mathfrak{X}_{red} = (\mathfrak{X}, \mathcal{O}_{\mathfrak{X}}/\mathcal{J})_{red}$. By [EGA II (5.4.6) and EGA 01, p. 279, (5.3.1)(vi)] it suffices to show that $f_{\mathcal{J}}$ is locally of finite type; so what we need is that if A is a noetherian $W(k)$-algebra with a nilpotent ideal N such that A/N is finitely generated over $W(k)$, then also A is finitely generated over $W(k)$. But if $a_1, a_2, \ldots, a_r$ in A are such that their images in A/N are $W(k)$-algebra generators of A/N, and if $b_1, b_2, \ldots, b_s$ are A-module generators of N, then it is easily seen that

$$A = W(k)[a_1, a_2, \ldots, a_r, b_1, b_2, \ldots, b_s].$$

Q.E.D.

§3. Structure of inverse limits of locally algebraic k-groups.

In this section, we establish (1.2)- and a little more- for any group-scheme P of the form $\varprojlim P_n$, where (P_n, f_{mn}) (n, m, non-negative integers, $n \geq m$) is an inverse system of locally algebraic k-groups (k a field), the maps $f_{mn}: P_n \to P_m$ ($n \geq m$) being affine (cf. [EGA IV, §8.2]). (Note that the group-scheme P of §(2.5) is of this form.) This is more or less an exercise, and the results are presumably known, but I could not find them recorded anywhere.

(3.1) By [SGA 3, p. 315], $f_{mn}: P_n \to P_m$ ($n \geq m$) factors uniquely as

$$P_n \xrightarrow{u} P_{mn} \xhookrightarrow{v} P_n$$

where v is a closed immersion and u is affine, faithfully flat, and finitely presented. (P_{mn} is the image, or coimage, of f_{mn}.) For $n_1 \geq n_2$, P_{mn_1} is a closed subgroup of P_{mn_2}, and we can set

$$\bar{P}_m = \bigcap_{n\geq m} P_{mn} = \varprojlim_{n\geq m} P_{mn}.$$

$\bar{P}_m$ is a closed subgroup of P_m, its defining ideal in $\mathcal{O}_{P_m}$ being the union of the defining ideals of the P_{mn}. Clearly f_{mn} induces a map $\bar{f}_{mn}:\bar{P}_n \to \bar{P}_m$, so we have an inverse system $(\bar{P}_n, \bar{f}_{mn})$.

PROPOSITION. (i) _P (together with the natural maps_ $\bar{f}_n:P \to \bar{P}_n$) _is equal to_ $\varprojlim \bar{P}_n$.

(ii) _The maps_ $\bar{f}_{mn}:\bar{P}_n \to \bar{P}_m$ _and_ $\bar{f}_m:P \to \bar{P}_m$ _are affine, faithfully flat, and universally open._

(iii) _If K is any algebraically closed field containing k, then_

$$\bar{f}_m(K):P(K) \to \bar{P}_m(K)$$

is surjective.

(iv) $\ker(\bar{f}_{mn})$ _is a closed subgroup of_ $\ker(f_{mn})$.

Proof. (i) and (iv) are left to the reader. It is clear that all the maps $\bar{f}_{mn}$ and $\bar{f}_m$ are affine. We show below that $\bar{f}_m$ is faithfully flat for all m. Since $\bar{f}_m = \bar{\bar{f}}_{mn} \circ \bar{f}_n$ for $n \geq m$, it will follow that $\bar{f}_{mn}$ is faithfully flat [EGA IV, (2.2.13)]. This implies that $\bar{f}_{mn}$ is universally open [EGA IV,

(2.4.6)] and hence so is $\bar{f}_m$ [EGA IV, (8.3.8)], proving (ii). As for (iii), since $\bar{f}_{st}$ is locally of finite type and surjective, it follows that $\bar{f}_{st}(K)$ is surjective for all $t \geq s$; in particular, $\bar{f}_{n,n+1}(K)$ is surjective for all $n \geq m$, so any element of $\bar{P}_m(K)$ can be lifted back to $P(K) = \varprojlim_{n \geq m} \bar{P}_n(K)$, i.e. $\bar{f}_m(K)$ is surjective.

So let us show that $\bar{f}_m$ is faithfully flat. Let $y \in \bar{P}_m$, and let U be an affine open neighborhood of y in P_m. Since U is noetherian, we see that for some n_o

$$\bar{P}_m \cap U = P_{mn} \cap U \qquad \text{for all } n \geq n_o.$$

But f_{mn} induces a faithfully flat map

$$P_n \times_{P_m} U \to P_{mn} \times_{P_m} U = \bar{P}_m \cap U \qquad (n \geq n_o).$$

Since $P_n \times_{P_m} U$ and $\bar{P}_m \cap U$ are affine, and since for any ring R an inductive limit of faithfully flat R-algebras is still a faithfully flat R-algebra, we conclude that

$$P \times_{\bar{P}_m} (\bar{P}_m \cap U) = P \times_{P_m} U = \varprojlim_{n \geq n_o} (P_n \times_{P_m} U)$$

is faithfully flat over $\bar{P}_m \cap U$. Thus $\bar{f}_m$ is faithfully flat.

(3.2) Because of Proposition (3.1), we can <u>assume from now on that</u> $P_m = \bar{P}_m$ (so that all the maps f_{mn} $(= \bar{f}_{mn})$ are faithfully flat etc. etc.). Furthermore, certain additional conditions which may be imposed on the original f_{mn} (for example the condition that $\ker(f_{mn})$ be unipotent) will not be destroyed by this replacement of P_m by $\bar{P}_m$ (because of (iv) in Prop. (3.1)).

We examine now the connected component of the zero-point of P. Let P_n^o be the open and closed subgroup of P_n supported by the connected component of zero in P_n (cf. [DG, ch. II, §5, no. 1]). Then $f_{mn}:P_n \to P_m$ $(n \geq m)$ induces a map $f_{mn}^o:P_n^o \to P_m^o$, so we have an inverse system (P_n^o, f_{mn}^o). Set $P^o = \varprojlim P_n^o$.

PROPOSITION. (i) <u>The maps f_{mn}^o are affine, faithfully flat and finitely presented; and $\ker(f_{mn}^o)$ is a closed subgroup of $\ker(f_{mn})$.</u>

(ii) <u>P^o is a closed irreducible subgroup of P, and the underlying subspace of P^o is the connected component of zero in P. Furthermore, if $x \in P^o$, then the canonical map of local rings $\mathcal{O}_{P,x} \to \mathcal{O}_{P^o,x}$ is bijective.</u>

<u>Proof.</u> (i) is immediate except perhaps for the surjectivity of f_{mn}^o, which follows from the fact that the (topological) image of f_{mn}^o is open [EGA IV, (2.4.6)] and closed [DG, p. 249, (5.1)].

As for (ii), it is clear that P^o is a closed subgroup of P; and if Q is any connected subspace of P containing zero, then $f_n(Q) \subseteq P_n^o$ for all n ($f_n:P \to P_n$ being the natural map) whence $Q \subseteq P^o$ (since $P^o = \varprojlim P_n^o$ in the category of

<u>topological spaces</u> [EGA IV, 8.2.9]). So for the first assertion of (ii), it remains to be shown that P^o is irreducible (hence connected). For this it suffices to show that P^o is covered by open irreducible subsets, any two of which have a non-empty intersection. P^o_o, being irreducible, has such a covering by irreducible affine subsets, and we can cover P^o by their inverse images. Since all the maps f^o_{on} are affine and each P^o_n is irreducible, we need only check that <u>a direct limit of rings with irreducible spectrum has irreducible spectrum.</u> But this is easily seen, since "A has irreducible spectrum" means that "for a, b ∈ A, ab is nilpotent ⇔ either a or b is nilpotent".

Finally, for $x \in P^o$, we have

$$\mathcal{O}_{P,x} = \varinjlim \mathcal{O}_{P_n, f_n(x)} = \varinjlim \mathcal{O}_{P^o_n, f_n(x)} = \mathcal{O}_{P^o,x} \quad .$$

Q.E.D.

<u>Remark.</u> Though P^o is not algebraic over k in general, it may nevertheless have certain finite-dimensional structural features. For example, when k is perfect, if A_n is the abelian variety which is a quotient of $(P^o_n)_{red}$ by its maximal linear subgroup L_n (structure theorem of Chevalley) then f_{mn} $(n \geq m)$ induces an <u>epimorphism</u> $A_n \to A_m$, with infinitesimal kernel. If furthermore the kernel of f_{mn} is unipotent (as would be the case, e.g. in (2.5) [L2; Cor. (2.11)]), then, writing

$$L_n = M_n \times U_n \qquad (M_n \text{ multiplicative, } U_n \text{ unipotent})$$

we find that f_{mn} induces an <u>isomorphism</u> $M_n \to M_m$.

(3.3) For each n, let $\pi_o(P_n)$ be the étale k-group P_n/P_n^o (cf. [DG, p. 237, Prop. (1.8)]). The natural map $q_n : P_n \to \pi_o(P_n)$ is faithfully flat and finitely presented (loc. cit). f_{mn} induces a map $\pi_o(f_{mn}) : \pi_o(P_n) \to \pi_o(P_m)$, so we have an inverse system $(\pi_o(P_n), \pi_o(f_{mn}))$. We set $\pi_o(P) = \varprojlim \pi_o(P_n)$.

PROPOSITION. (i) The maps $\pi_o(f_{mn})$ are finite, étale, surjective; and $\ker(\pi_o(f_{mn}))$ is a quotient of $\ker(f_{mn})$.

(ii) The canonical map $q : P \to \pi_o(P)$ is faithfully flat and quasi-compact, with kernel P^o (so that the sequence

$$0 \to P^o \to P \to \pi_o(P) \to 0$$

is exact in the category of fpqc sheaves). The (topological) fibres of $P \to \pi_o(P)$ are irreducible, and they are the connected components of P. For any $x \in P$, the canonical map of local rings $\mathcal{O}_{P,x} \to \mathcal{O}_{q^{-1}q(x),x}$ is bijective. If K is an algebraically closed field containing k, then $P(K) \to \pi_o(P)(K)$ is surjective.

Proof. (i) Consider the commutative diagram (with $n \geq m$):

$$\begin{array}{ccccccccc}
0 & \longrightarrow & P_n^o & \longrightarrow & P_n & \xrightarrow{q_n} & \pi_o(P_n) & \longrightarrow & 0 \\
 & & \downarrow{\scriptstyle f_{mn}^o} & & \downarrow{\scriptstyle f_{mn}} & & \downarrow{\scriptstyle \pi_o(f_{mn})} & & \\
0 & \longrightarrow & P_m^o & \longrightarrow & P_m & \xrightarrow{q_m} & \pi_o(P_m) & \longrightarrow & 0
\end{array} .$$

The maps in the rows are the natural ones, and the rows are exact in the category of fppf sheaves (when we identify k-groups with functors of k-algebras...). Since f^o_{mn} is an epimorphism of fppf sheaves (Prop. (3.2)), so therefore is the natural map $\ker(f_{mn}) \to \ker(\pi_o(f_{mn}))$, and we have the second assertion of (i).

f_{mn}, q_m, and q_n are all faithfully flat - hence surjective - and quasi-compact, and then so is $\pi_o(f_{mn})$. Since $\pi_o(P_n)$ and $\pi_o(P_m)$ are étale over k, therefore the map $\pi_o(f_{mn})$ is étale. Thus the kernel of $\pi_o(f_{mn})$-being quasi-compact and étale over k - is finite over k, and it follows that the map $\pi_o(f_{mn})$ is finite.

(ii) For the last assertion, note that we have an inverse system of exact sequences

$$0 \to P^o_n(K) \to P_n(K) \to \pi_o(P_n)(K) \to 0$$

and that $P^o_{n+1}(K) \to P^o_n(K)$ is <u>surjective</u> for all n (Prop. (3.2)); so on passing to the inverse limit we obtain an exact sequence

$$0 \to P^o(K) \to P(K) \to \pi_o(P)(K) \to 0 \quad .$$

The exactness of $0 \to P^o \to P \overset{q}{\to} \pi_o(P)$ is straightforward. To show that q is flat let $x \in P$, $y = q(x)$, and let x_n, y_n be their images in P_n, $\pi_o(P_n)$ respectively. Then $\mathcal{O}_{P_n,x_n}$ is flat over $\mathcal{O}_{\pi_o(P_n),y_n}$, and passing to inductive limits, we see that $\mathcal{O}_{P,x}$ is flat over $\mathcal{O}_{\pi_o(P),y}$. Next let $z \in \pi_o(P)$, let

z_n be the image of z in $\pi_o(P_n)$, and let $Q = q^{-1}(z)$, $Q_n = q_n^{-1}(z)$. Note that Q_n is irreducible, and is a connected component of P_n. The Q_n form an inverse system of schemes, in which the transition maps are affine, and

$$Q = \varprojlim Q_n.$$

We show next that $Q_n \to Q_m$ <u>is surjective;</u> then it follows that Q is non-empty (so that q is surjective - hence faithfully flat) and the proof of Prop. (3.2) (ii) can be imitated to give all the assertions about the fibres of q.

Let $\bar{k}$ be the algebraic closure of k. By a simple translation argument, we deduce from the surjectivity of $P_n^o \to P_m^o$ that every component of $Q_n \otimes_k \bar{k}$ maps surjectively onto a component of $Q_m \otimes_k \bar{k}$; since every component of $Q_m \otimes_k \bar{k}$ projects surjectively onto Q_m, we find that $Q_n \to Q_m$ is indeed surjective.

It remains to be seen that q is quasi-compact. The fibres of the maps $\pi_o(f_n):\pi_o(P) \to \pi_o(P_n)$ $(n \geq 0)$ form a basis of open sets on $\pi_o(P)$ (since $\pi_o(P_n)$ is discrete as a topological space); furthermore these fibres are quasi-compact (since $\pi_o(f_n)$ is an affine map), and their inverse images in P are quasi-compact (the affine map $P \to P_n$ and the finitely presented map $P_n \to \pi_o(P_n)$ are both quasi-compact, so the composed map $P \to \pi_o(P_n)$ is quasi-compact); it follows that q is quasi-compact. Q.E.D.

Remarks.

1. Say that a k-group Q is pro-étale if it is of the form $\varprojlim Q_n$, where (Q_n, g_{mn}) is an inverse system of the type we have been considering, with all the Q_n étale over k. For example $\pi_o(P)$ is pro-étale. It is immediate that if Q is pro-étale and $f:G \to Q$ is a map of k-groups, with G connected, then f is the zero-map. From this we see that, with P as above, every map of P into a pro-étale k-group factors uniquely through $P \to \pi_o(P)$.

2. Let (P_n, f_{mn}) be as above, and assume that the kernel of f_{mn} is unipotent for all m,n. Set $Q_n = \pi_o(P_n)$, $g_{mn} = \pi_o(f_{mn})$; by (i) of Proposition (3.3), the kernel of g_{mn} is étale and also unipotent (i.e. annihilated by p^t for some t, with p = char. of k). Assume also that the abelian group $Q_n(\bar{k})$ ($\bar{k}$ = algebraic closure of k) is finitely generated (for each n). (These assumptions hold in the situation described in (2.5), cf. [L2; Prop. (2.7), Cor. (2.11)].)

Let Q_n^t be the kernel of multiplication by p^t in Q_n. Then $Q_n^o \subseteq Q_n^1 \subseteq Q_n^2 \subseteq \ldots$, and since $Q_n(\bar{k})$ is finitely generated, we have, for large t, $Q_n^t = Q_n^{t+1} = \ldots$; so we can set

$$Q_n^{(p)} = \bigcup_t Q_n^t = Q_n^t \quad \text{for large } t.$$

Clearly $Q_n^{(p)}$ is finite étale over k, and unipotent; and the quotient $R_n = Q_n/Q_n^{(p)}$ is étale over k. Consider the commutative diagram $(n \geq m)$:

$$\begin{array}{ccccccccc} 0 & \to & Q_n^{(p)} & \to & Q_n & \to & R_n & \to & 0 \\ & & \downarrow & & \downarrow & & \downarrow & & \\ 0 & \to & Q_m^{(p)} & \to & Q_m & \to & R_m & \to & 0 \end{array} .$$

Straightforward arguments give that:

(i) Multiplication by p in R_n is a monomorphism.

(ii) $R_n \to R_m$ is an isomorphism.

(iii) $Q_n^{(p)} \to Q_m^{(p)}$ is an epimorphism.

Then, passing to the inverse limit, we obtain:

There exists an exact sequence

$$0 \to Q^{(p)} \to \pi_o(P) \to R \to 0$$

$Q^{(p)}$ = inverse limit of unipotent finite étale k-groups.

R = étale k-group such that the abelian group $R(\bar{k})$ ($\bar{k}$ = algebraic closure of k) is finitely generated and without p-torsion.

Here R is already determined by P_1.

REFERENCES

EGA A. GROTHENDIECK, J. DIEUDONNÉ, Éléments de Géométrie Algébrique:

— 01 Springer-Verlag, Heidelberg, 1971.

— I, II, III(0_{III}), IV(0_{IV}), Publ. Math. I.H.E.S. 4,8,...

SGA A. GROTHENDIECK et. al., Séminaire de Géométrie Algébrique:

— 2 Cohomologie locale des faisceaux cohérents..., North-Holland, Amsterdam, 1968.

— 3 Schémas en groupes I, Lecture Notes in Mathematics no. 151, Springer-Verlag, Heidelberg, 1970.

— 6 Théorie des intersections et théorème de Riemann-Roch, Lecture Notes in Mathematics no. 225, Springer-Verlag, Heidelberg, 1971.

[AC] N. BOURBAKI, Algèbre Commutative, Hermann, Paris. (English Translation, 1972).

[B] J.-F. BOUTOT, Schéma de Picard local, C. R. Acad. Sc. Paris, 277 (Série A) (1973), 691-694.

[D1] V. I. DANILOV, On a conjecture of Samuel, Math. USSR Sb. 10 (1970), 127-137. (Mat. Sb. 81 (123) (1970), 132-144.)

[D2] ____________, Rings with a discrete group of divisor classes, Math. USSR Sb. 12 (1970), 368-386. (Mat. Sb. 83 (125) (1970), 372-389.)

[D3] ____________, On rings with a discrete divisor class group, Math. USSR Sb. 17 (1972), 228-236. (Mat. Sb. 88 (130) (1972), 229-237.)

[DG] M. DEMAZURE, P. GABRIEL, Groupes Algébriques (Tome I), North- Holland, Amsterdam, 1970.

[F] R. M. FOSSUM, The divisor class group of a Krull domain, (Ergebnisse der Math., vol. 74), Springer-Verlag, Heidelberg, 1973.

[GR] A. GROTHENDIECK, Groupe de Brauer III, in Dix exposés sur la cohomologie des schémas, North-Holland, Amsterdam, 1968.

[K] W. KRULL, Beiträge zur Arithmetik kommutativer Integritätsbereiche V, Math. Z. 43 (1938), 768-782.

[L1] J. LIPMAN, Rational Singularities ..., Publ. Math. I.H.E.S. no. 36 (1969), 195-279.

[L2] ________, The Picard group of a scheme over an Artin ring, to appear (preprint available).

[P] D. PRILL, The divisor class groups of some rings of holomorphic functions, Math. Z. 121 (1971), 58-80.

[S1] P. SAMUEL, On unique factorization domains, Illinois J. Math. 5 (1961), 1-17.

[S2] ________, Sur les anneaux factoriels, Bull. Soc. Math. France 89 (1961), 155-173.

[SMN] P. SALMON, Su un problema posto da P. Samuel, Atti. Accad. Naz. Lincei Rend. Cl. Sci. Fis. Mat. Natur. (8) 40 (1966), 801-803.

[SH] G. SCHEJA, Einige Beispiele faktorieller lokaler Ringe, Math. Ann. 172 (1967), 124-134.

[SR] J.-P. SERRE, Corps locaux, Hermann, Paris, 1968.

[ST1] U. STORCH, Über die Divisorenklassengruppen normaler komplexanalytischer Algebren, Math. Ann. 183 (1969), 93-104.

[ST2] _________, Über das Verhalten der Divisorenklassengruppen normaler Algebren bei nicht-ausgearteten Erweiterungen, Habilitationsschrift, Univ. Bochum, (1971).

Modifications of complex varieties and the Chow Lemma

Boris Moishezon

Par. 1. A-spaces.

It is well known now that the theory of compact complex spaces of dimension n with n algebraically independent meromorphic functions is a part of the (general) theory of algebraic spaces.

We will define in this paragraph complex spaces of a more general type for which it is unknown if they also can be included into an algebraic theory.

Definition 1. (Relative algebraic dimension)

Let $f:X \longrightarrow S$ be a proper surjective morphism of irreducible complex varieties. (A complex variety is a reduced complex space.) Let $s_o \in S$ be a point and let the germ of X along the fibre $f^{-1}(s_o)$ be irreducible. Denote by $\mathcal{M}_X$ the sheaf of germs of meromorphic functions on X and by $\mathcal{M}_X|f^{-1}(s_o)$ the topological restriction of $\mathcal{M}_X$ to $f^{-1}(s_o)$. Let $\mathcal{M}_{X,f^{-1}(s_o)} = \Gamma(f^{-1}(s_o),\mathcal{M}_X|f^{-1}(s_o))$ and denote by $\mathcal{M}_{S,s_o}$ the field of germs of meromorphic functions on S at the point s_o. Then the transcendental degree

$$\text{trans.deg.}(\mathcal{M}_{X,f^{-1}(s_o)}/\mathcal{M}_{S,s_o})$$

of the field extension $\mathcal{M}_{X,f^{-1}(s_o)}/\mathcal{M}_{S,s_o}$ is called the <u>algebraic dimension</u> of X over S at s_o and is denoted by $a.\dim(f,s_o)$.

The following generalisation of Siegel's theorem holds.

Theorem. $a.\dim(f,s_o) \leq \dim_{\mathbb{C}}X - \dim_{\mathbb{C}}S$.

Definition 2. Let $f:X \longrightarrow S$ be a proper morphism of complex varieties. We say that X is an <u>A-space over S at a point</u> $s_o \in f(X)$ if there exist neighborhoods U_S and $U_X = f^{-1}(U_S)$ of s_o and $f^{-1}(s_o)$, respectively, such that

1) Every irreducible component $U_{X,i}$ of U_X defines an irreducible germ along $U_{X,i} \cap f^{-1}(s_o)$.
2) Let $f(U_{X,i}) = S_i$ and $f_i = f|U_{X,i} : U_{X,i} \longrightarrow S_i$. Then for all i, $s_o \in S_i$ and $\text{a.dim}(f_i, s_o) = \dim_{\mathbb{C}} U_{X,i} - \dim_{\mathbb{C}} S_i$.

If X is an A-space over S at any point $s_o \in f(X)$ we say that X is an <u>A-space over S.</u>

The following two theorems generalise corresponding statements for n-dimensional compact complex varieties with n algebraically independent meromorphic functions.

<u>Theorem 1.</u> Let $\begin{array}{ccc} X & \xrightarrow{g} & X' \\ & f\searrow \quad \swarrow f' & \\ & S & \end{array}$ be a commutative diagram of morphisms of complex varieties. If X is an A-space over S at a point $s_o \in f(X)$, then g(X) is also an A-space over S at s_o.

<u>Theorem 2.</u> Let $f: X \longrightarrow S$ be a proper morphism and X be an A-space over $s_o \in S$. Let Y be a closed complex subvariety of X such that $s_o \in f(Y)$. Let $S_Y = f(Y)$ and $f_Y = f|Y : Y \longrightarrow S_Y$. Then Y together with f_Y is an A-space over S_Y at s_o.

<u>Corollary.</u> Let $f: X \longrightarrow Y$ be a proper surjective morphism of complex varieties with $\dim_{\mathbb{C}} X = \dim_{\mathbb{C}} Y$ and let W be the subvariety of Y consisting of all points $y \in Y$ with $\dim_{\mathbb{C}} f^{-1}(y) > 0$. Let $V = f^{-1}(W)$. Then V is an A-space over W.

This corollary shows that if $f: X \longrightarrow Y$ is a proper modification, the analysis of the "bad points" of f leads to consider A-spaces.

<u>Definition 3.</u> Let $f: X \longrightarrow S$ be a morphism of complex spaces and let $s_o \in S$. We say that X is an <u>f-algebraic space (f-scheme)</u> at the point $s_o \in f(X)$ if there exists an algebraic space (scheme) $\mathfrak{X}$ and a morphism of finite type $f: \mathfrak{X} \longrightarrow \operatorname{Spec} O_{S,s_o}$ such that for some

sufficiently small neighborhood U_{s_o} of $s_o \in S$, the complex space X' and the morphism of complex spaces $f':X' \longrightarrow U_{s_o}$ which canonically corresponds to $\mathfrak{X}$ and $f:\mathfrak{X} \longrightarrow \operatorname{Spec} O_{S,s_o}$, coincide with $f^{-1}(U_{s_o})$ and $f\big|_{f^{-1}(U_{s_o})}:f^{-1}(U_{s_o}) \longrightarrow U_{s_o}$.

The following question is natural here.
Let X be an A-space over S at s_o and $f:X \longrightarrow S$ the corresponding proper morphism. Is X an f-algebraic space at s_o?
I don't know the answer to this question. A partial result in this direction will be formulated after the following definition.

<u>Definition 4.</u> Let $f:X \longrightarrow S$ be a morphism of complex spaces and $s_o \in f(X)$. We say that X is locally f-algebraisable at the point s_o if for every $x \in f^{-1}(s_o)$ there exists an open neighborhood U_x of x in X with the following property: There exists a morphism of complex spaces $g_x:Z_x \longrightarrow S$, an open subset $V_x \subset Z_x$ and an isomorphism $\alpha_x:V_x \longrightarrow U_x$ such that

a) $s_o \in g_x(Z_x)$,
b) $g_x\big|_{V_x} = (f\big|_{V_x}) \circ \alpha_x$ and
c) Z_x is a g_x-scheme at the points s_o.

<u>Theorem 3.</u> Let $f:X \longrightarrow S$ be a proper surjective morphism of complex manifolds with $\dim_{\mathbb{C}} X = \dim_{\mathbb{C}} S$ and let $E_f(X)$ be the complex subvariety of X consisting of all points of X at which f is not a local isomorphism. Let $s_o \in S$ and suppose that every irreducible component E_j of $E_f(X)$ is a $f\big|_{E_j}$-scheme over $f(E_j)$ at s_o. Then X is locally f-algebraisable at s_o.

The proof of this theorem uses the following lemma which is similar to M. Artin's approximation theorem but much more elementary.

<u>Lemma.</u> Let $C = \mathbb{C}\{z_1,\dots,z_n\}$ be the $\mathbb{C}$-algebra of all convergent

power series of the variables $z_1,\dots,z_n$ and ζ_1,ζ_2 two endomorphisms of the algebra C. Let $\zeta_1(z_i) = f_i(z_1,\dots,z_n)$, $i = 1,2,\dots,n$, and J_1 be the ideal in C which is generated by the Jacobian determinant $\left\|\frac{\partial f_i}{\partial z_j}\right\|$. Suppose that $J_1 \neq C$ and that ζ_1 and ζ_2 coincide modulo J_1^4 (that is, for any $a \in C$ we have $\zeta_1(a) \equiv \zeta_2(a)$ (modulo J_1^4)). Then there exists an automorphism $\theta : C \longrightarrow C$ of the algebra C such that $\zeta_2 = \theta \circ \zeta_1$.

Par. 2. Chow lemma for complex spaces.

We call the following statement the global Chow lemma for complex spaces.

For every proper bimeromorphic morphism of complex varieties $f: X \longrightarrow Y$ there exists a morphism $\phi: Y' \longrightarrow Y$ which is locally a monoidal transformation with nowhere dense center and a morphism $h: Y' \longrightarrow X$ such that $f \circ h = \phi$; in particular, any bimeromorphic morphism can be dominated by a projective morphism which is almost everywhere an isomorphism.

In [1] H. Hironaka announced this result for the case of n-dimensional complex varieties with n algebraically independent meromorphic functions and for the local case, that is, when for every point $y_0 \in Y$ there exists an open neighborhood $U \ni y_0$, a monoidal transformation $\phi_U : U' \longrightarrow U$ with nowhere dense center in U and a morphism $h: U' \longrightarrow f^{-1}(U)$ such that $\phi_U = (f|_{f^{-1}(U)}) \circ h$.

The proof of the local result was published in [2]. Then I, using Hironaka's method, published in [3] a proof of the Chow lemma for a class of complex varieties which is a little larger than the class of complex varieties of dimension n with n independent meromorphic functions.

But about three years ago, I discovered that both of the proofs in [2] and [3] are not complete. In [4] H. Hironaka notes the same. And so this non completeness stimulated new activities concerning the Chow lemma and created proofs of the global Chow lemma even for arbitrary complex varieties.

In [4] H. Hironaka obtained by new methods a more general result than the Chow lemma, proving that, for a proper morphism $f: X \longrightarrow Y$ of analytic spaces, there exists a monoidal transformation Y' of Y such that the induced morphism $f': X' \longrightarrow Y'$ is flat. (For the exact formulation see [4] .)

My way was to complete the proof of the Chow lemma using the previous method of Hironaka. I did so in 1972 and could prove the global Chow lemma in the general form formulated above.

This is an outline of the proof.

First, theorem 3, formulated above, permits us to complete the proofs in [2] and [3] and to obtain the local Chow lemma. Then the global Chow lemma is deduced from the local one. The key step in the deduction is as follows.

Let

$$\begin{array}{ccc} X' & \xrightarrow{f'} & Y' \\ \psi \downarrow & & \downarrow \varphi \\ X & \xrightarrow{f} & Y \end{array}$$

be a commutative diagram of proper morphisms of complex varieties where f is bimeromorphic, φ is a monoidal transformation and f' and ψ are induced by f and φ . Let $S_{f'}(Y')$ be the minimal closed complex subvariety of Y' such that

$$f' \Big|_{X'-f'^{-1}(S_{f'}(Y'))} : X'-f'^{-1}(S_{f'}(Y')) \longrightarrow Y'-S_{f'}(Y')$$

is an isomorphism and let B be an irreducible component of $\varphi(S_{f'}(Y'))$. We must show that there exists a monoidal transformation $\varphi: Y'' \longrightarrow Y'$ with center in $S_{f'}(Y')$ such that if

$$\begin{array}{ccc} X'' & \xrightarrow{f''} & Y'' \\ \varphi' \downarrow & & \downarrow \varphi' \\ X' & \xrightarrow{f'} & Y' \end{array}$$

is the corresponding commutative diagram, then $\varphi\varphi'(S_{f''}(Y'')) \not\supset B$.

Let $y_o \in B$. From the local Chow lemma it follows that there exists a relatively compact open neighborhood U of y_o on Y and a coherent sheaf of ideals $\mathcal{J}'$ on $U' = \varphi^{-1}(U)$ such that

1) $\operatorname{supp}[O_{U'}/\mathcal{J}'] = S_{f'}(Y') \cap U'$ and

2) if $\tilde{\varphi} : \tilde{U} \longrightarrow U'$ is the monoidal transformation corresponding to $\mathcal{J}'$ and

$$\begin{array}{ccc} \tilde{U}_X & \xrightarrow{\tilde{f}} & \tilde{U} \\ \tilde{\varphi} \downarrow & & \downarrow \tilde{\varphi} \\ U'_X = f'^{-1}(U') & \xrightarrow[f'|U'_X]{} & U' \end{array}$$

is the corresponding commutative diagram, then $\tilde{f}$ is an isomorphism.

Let $\mathcal{J}$ be the sheaf of ideals on Y corresponding to φ and $J = \varphi^{-1}(\mathcal{J})$. Then J is invertible. Let $[J]$ be the corresponding line bundle on Y' and let $\varphi_U = \varphi|_{U'} : U' \longrightarrow U$, $J_{U'} = J|_{U'}$. Let $N > 0$ be an integer such that the canonical homomorphisms

$$\varphi_U^* \varphi_{U*}(O_{U'}[J_{U'}]^{\otimes N}) \longrightarrow O_{U'}[J_{U'}]^{\otimes N} \quad \text{and}$$

$$\varphi_U^* \varphi_{U*}(\mathcal{J}'[J_{U'}]^{\otimes N}) \longrightarrow \mathcal{J}'[J_{U'}]^{\otimes N}$$

are epimorphisms. (N exists because U is relatively compact.)

Let $\mathcal{L}$ be the set of coherent subsheaves L of $O_{Y'}[J]^{\otimes N}$ which have the following two properties:

a) $L|_{U'} \supseteq \mathcal{J}'[J_{U'}]^{\otimes N}$ and

b) $\varphi_U^* \varphi_{U*}(L|_{U'}) \longrightarrow L|_{U'}$ is an epimorphism.

Let b be a smooth point of $B \cap U$ which is not in the closure of $\varphi(S_{f'}(Y')) - B$ in Y.

Let $\mathcal{O}$ be the localisation of $O_{Y',b}$ at the prime ideal p_b which corresponds to B and M_L the localisation of the $O_{Y',b}$-module $R_b = \left[\varphi_{U*}(L/_{U'})/\varphi_{U*}(\mathcal{J}'^2[J_{U'}]^{\otimes N})\right]_b$ at p_b.

As the varieties B and $\varphi(S_{f'}(Y'))$ coincide in a neighborhood of b on Y, there exists an integer $m>0$ such that $p_b^m \cdot R_b = 0$. This yields that the $\mathcal{O}$-modul M_L is of finite length d(L). Let L' be an element of $\mathcal{L}$ such that length d(L') is minimal. (Note that $\mathcal{L}$ is not empty as $O_{Y'}[J]^{\otimes N} \in \mathcal{L}$.) Then L' defines a certain coherent sheaf of ideals $\mathcal{J}''$ on Y'. If $\phi: Y'' \longrightarrow Y'$ is the monoidal transformation determined by $\mathcal{J}''$ we can prove that $\phi: Y'' \longrightarrow Y'$ has the desired properties.

References.

[1] H. Hironaka, Resolution of singularities of an algebraic variety over a field of characteristic zero. Ann. of Math. 79 (1964), 109-326.

[2] H. Hironaka, A fundamental lemma on point modifications. Proc. Conf. Complex Analysis, Minneapolis.

[3] B.G. Moishezon, Resolution theorems for compact complex spaces with sufficiently large field of meromorphic functions. Izvestia AN SSSR, sez. math., 31(1967), 1385-1414.

[4] H. Hironaka, Flattening theorem in complex analytic geometry. Preprint.

SOME RESULTS ON CUBIC THREEFOLDS

J.P. MURRE

INTRODUCTION.

The first part of this paper is the written version of my lecture at the Mannheim conference; it consists of sections I-IV and is of an expository nature, details can be found in my papers [12] and [13]. The second part, i.e. sections V-VII, consists of material which is not contained in [12] and [13], proofs are given in detail here. Section V contains a characterization of the Prym variety, associated with a cubic threefold, by means of algebraic families of 1-dimensional cycles. Section VI deals with the relation between the Prym of the cubic and the Albanese variety of the Fano surface. Section VII contains a result which Clemens and Griffiths call the "theorem of Abel" on the cubic ([6], 0.8); this theorem is proved in [6] by transcendental and topological methods. In this paper we work over an algebraically closed field of characteristic not 2, and all methods are of a geometric and/or algebraic nature.

The inspiration for this paper came from the fundamental works of Clemens and Griffiths on the cubic threefold [6] and of Mumford on the theory of Prym varieties [11]. For a discussion of these works we mention also the interesting paper [14].

In the following X denotes a non-singular 3-dimensional variety of degree 3 in 4-dimensional projective space $\mathbb{P}_4$ (abbreviated: X is a n.s. cubic threefold).

I. SOME RESULTS RELATED TO THE LINES ON X.

We start by recalling the following basic and classical fact concerning the cubic threefold, due to Fano [7]:

<u>Theorem 1</u> (Fano). There is a 2-dimensional algebraic family of lines on X, parametrized by an irreducible non-singular surface S, contained in the Grassmannian $Gr(4,1)$. Moreover, there exists a non-empty open set W on X such that for $P \in W$ there are exactly 6 lines on X going through P ([12] , prop. 1.25).

The above surface S is called the <u>Fano surface</u> of X.

Fix <u>once for all</u> a sufficiently general (see [12], 1.25) line ℓ_o on X, i.e. $\ell_o \in S$; all our constructions are with respect to this line ℓ_o.

Consider the tangentbundle $T(X)$ of X, let $Proj(T(X))$ be the associated projective bundle (of 1-dimensional linear subspaces). The restriction to ℓ_o of this bundle $Proj(T(X))$ is denoted by X^*, i.e.

$$X^* = Proj(T(X))_{/\ell_o}$$

Let $j: X^* \to \ell_o$ be the structure morphism. There exists a rational transformation

$$\Psi: X^* \to X$$

defined as follows: let $x^* \in X^*$, then x^* is a line in the tangent-space T_{X,x_o}, where $x_o = j(x^*)$, and as such we have an intersection

$$X.x^* = 2x_o + x$$

with $x \in X$, now define Ψ by $\Psi(x^*) = x$.

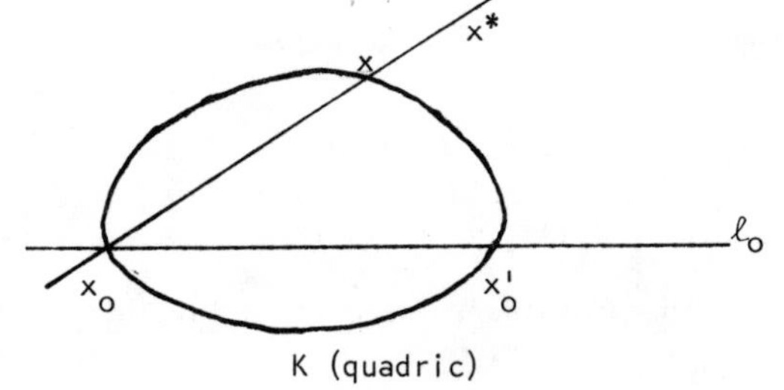

Fig. 1

We have the following properties:

i) $\Psi: X^* \to X$ is a rational transformation, generically 2-1, because $x \in X$ corresponds to tangentlines through x_o and x_o' (see figure 1).

ii) X is unirational because X^* is clearly a rational variety.

iii) The <u>fundamental locus</u> of Ψ consists of two curves Y' and Y'' on X^*, namely

$$Y' = \{x^* \in X^*;\ x^* \leftrightarrow \ell \in S,\ \ell \cap \ell_o \neq \emptyset,\ \ell \neq \ell_o\}$$

$$Y'' = \{x^* \in X^*;\ x^* \leftrightarrow \ell_o \in S\}.$$

Clearly we have rational maps $Y' \to \ell_o$ and $Y'' \to \ell_o$ which are generically 5 - 1, respectively birational. In particular Y'' is a rational curve.

<u>Lemma 1</u> ([12], 2.5). Y' and Y'' are non-singular curves on X^* and $Y' \cap Y'' = \emptyset$.

Let X' be the variety obtained by blowing X^* up along Y' and Y''. We have the following commutative diagram:

(1)

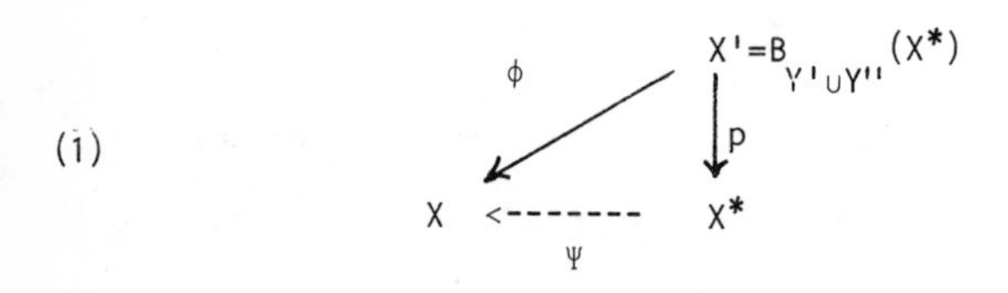

<u>Lemma 2.</u> i) X' is non-singular (by lemma 1).

ii) ϕ is a morphism, generically 2-1 ([12]. 4.2).

Consider on the Fano surface S the curve

$$\Delta' = \{\ \ell \in S;\ \ell \cap \ell_o \neq \emptyset\},$$

then $\Delta' \simeq Y'$ biregularly ([12], 9.2); in the following we usually identify Δ' and Y' with each other. There is an <u>involution</u>:

(2) $$\sigma\colon \Delta' \to \Delta' ,$$

namely for $\ell \in \Delta'$ consider the 2-dimensional space L spanned by ℓ_o and ℓ; then we have

$$L.X = \ell_o + \ell + \ell'$$

with $\ell' \in \Delta'$ and put $\sigma(\ell) = \ell'$. Put

(3) $$q\colon \Delta' \to \Delta'/\sigma = \Delta ,$$

then this curve Δ can be obtained as follows. Take a 2-dimensional linear space $N \subset \mathbb{P}_4$ such that $N \cap \ell_o = \emptyset$; N parametrizes the 2-dimensional linear spaces L going through ℓ_o and for $T \in N$ we denote the linear space spanned by ℓ_o and T by L_T. Clearly

(4) $$L_T.X = \ell_o + K_T \qquad (K_T \text{ quadric curve}) .$$

Then up to a biregular transformation we have

$$\Delta = \{ T \in N;\ K_T \text{ degenerates}\}.$$

Lemma 3 ([12], 1.25):

i) Δ is a non-singular irreducible curve and $q\colon \Delta' \to \Delta$ is an étale, 2-1 covering.

ii) In N we have $\deg(\Delta) = 5$; hence the genus $g(\Delta) = 6$ and (using the Hurwitz formula) $g(\Delta') = 11$.

II. PRYM VARIETIES (theory of Mumford).

Let C and C' be irreducible, non-singular curves and $q\colon C' \to C$ an étale, 2 - 1 covering. Let $\sigma\colon C' \to C'$ be the involution determined by this covering. For the corresponding Jacobian varieties we have homomorphisms

$$q^* : J(C) \to J(C')$$

and $$\sigma^* : J(C') \to J(C') ,$$

where σ^* is an involution. For such a situation Mumford [11] has proved the following facts:

i) $Im(q^*) = \{\eta \in J(C');\ \sigma^*(\eta) = \eta\}^o$ (component of the identity) and $q^*: J(C) \to Im(q^*)$ is an isogeny of degree 2.

ii) Put

(5) $$P(C'/C) = \{ \eta \in J(C') ; \sigma^*(\eta) = -\eta\}^o .$$

This abelian subvariety $i : P(C'/C) \to J(C')$ is called the <u>Prym variety of C' over C</u>. There is a Poincaré decomposition

(6) $$J(C') = Im(q^*) + P(C'/C)$$
$$Im(q^*) \cap P(C'/C) \text{ finite.}$$

This intersection is contained in the points of order 2 i.e. in $J(C')_2$.

iii) Using the natural polarization Θ' on $J(C')$ there is a canonical polarization on $P(C'/C)$:

(7) $$i^* (\Theta') \equiv 2. \Xi \qquad (\equiv \text{ means numerical equivalence})$$

with Ξ a <u>principal</u> divisor on $P(C'/C)$.

Returning to the case of a n.s. cubic threefold, the above theory can be applied to the curves $q: \Delta' \to \Delta$ described above in section I. We obtain a <u>principally polarized abelian variety</u> $(P(\Delta'/\Delta), \Xi)$ which we call the <u>Prym variety associated with the cubic threefold</u>. Note that - à priori - this Prym variety depends on the choice of the line ℓ_o, but later we see that it is independent of ℓ_o. From lemma 3 and (6) we see $\dim P(\Delta'/\Delta) = 5$.

III. A RESULT ON THE CHOW RING OF X ([12]).

$A(X)$ denotes the Chow ring of rational equivalence classes of cycles on X (see [5] or [4]), $A^i(X)$ means those classes which are of codimension i on X and $A_{alg}(X) \subset A(X)$ means those (rational equivalence) classes which are algebraically equivalent to zero. Here we are primarly concerned with $A^2_{alg}(X)$; i.e. with the classes of 1-dimensional cycles algebraically equivalent to zero.

Since X' is obtained from X^* by blowing up along Y' and Y'', it is well-known that

$$A(X') = A(X^*) \oplus A(Y') \oplus A(Y''),$$

with a change in degrees for $A(Y')$ and $A(Y'')$. There is a similar formula for $A_{alg}(-)$. In fact we get

$$A^2_{alg}(X') = A^2_{alg}(X^*) \oplus A^1_{alg}(Y') \oplus A^1_{alg}(Y'').$$

By [8] we have $A^2_{alg}(X^*)= o$; also $A^1_{alg}(Y'')= o$ since Y'' is rational. Therefore

$$A^2_{alg}(X') = A^1_{alg}(Y') = J(Y') = J(\Delta'), \tag{8}$$

where $J(\Delta')$ is the Jacobian variety of Δ'.

Using the 2-1 morphism $\phi: X' \to X$ (lemma 2) we get

$$A^2_{alg}(X) \xrightarrow{\phi^*} A^2_{alg}(X') \xrightarrow[\phi_*]{} A^2_{alg}(X) \tag{9}$$

with

$$\phi_* \cdot \phi^* = 2 \qquad \text{(multiplication by 2)}.$$

The question is now: what is the $\mathrm{Im}(\phi^*)$ in (9)? For this, consider the involution

$$\tau : X' \longrightarrow X' \tag{10}$$

obtained from $\phi: X' \to X$. Note that τ is a birational transformation only (and not a morphism). Nevertheless we get an involution

$$\tau^* : A^2_{alg}(X') \to A^2_{alg}(X')$$

of additive groups ([5], th 3 p 468, see [12], 10.2). The following lemma is crucial, it relates the involutions τ^* on $A(X')$ and σ^* on $J(\Delta')$.

Lemma 4 ([12], 10.6). For $\xi \in J(\Delta') = A^2_{alg}(X')$ we have

$$\tau^*(\xi) = -\sigma^*(\xi).$$

Since clearly $\mathrm{Im}(\phi^*)$ is invariant under τ^*, we get

Corollary. $\mathrm{Im}(\phi^*) \subset P(\Delta'/\Delta)$.

In fact one has the following more precise result:

Theorem 2 ([12], 10.10): Let X be a n.s. cubic threefold, Δ' and Δ the curves introduced in section I and $P(\Delta'/\Delta)$ the Prym variety associated with X. Then

$$A^2_{alg}(X) = P(\Delta'/\Delta) \oplus T,$$

where T is a 2-torsion group.

Problem: Does this group T actually appear? Note that, if so, this would give a new proof of the non-rationality of X by using a same type of argument as in the paper [2] of Artin and Mumford.

IV. THE POLARIZATION OF THE PRYM ([13]).

By Mumford's theory every Prym has a natural principal polarization Ξ (section II iii). On the other hand in the classical case $k = \mathbb{C}$, the intermediate Jacobian variety of the cubic threefold has a principal polarization coming from Poincaré duality on $H^3(X,\mathbb{Q})$. Therefore the question arises whether Ξ on $P(\Delta'/\Delta)$ is related to the Poincaré duality on the 3-dimensional ℓ-adic cohomology group of X. Let ℓ be a prime number, with $\ell \neq p = \text{char.}(k)$ and let

$$H^i(X) = H^i(X,\mathbb{Q}_\ell) = \varprojlim_n H^i(X, \mathbb{Z}/(\ell^n)) \otimes_{\mathbb{Z}_\ell} \mathbb{Q}_\ell .$$

For the cohomology one has relations similar as for the Chow ring:

$$H^3(X') = \underset{\substack{\| \\ 0}}{H^3(X^*)} \oplus H^1(Y') \oplus \underset{\substack{\| \\ 0}}{H^1(Y'')} .$$

Hence

$$(11) \qquad H^3(X') = H^1(Y') = H^1(\Delta') .$$

Using the involution σ on Δ', coming from $q: \Delta' \to \Delta$ (lemma 3), we get an involution σ^* on $H^1(\Delta')$ and hence a decomposition, similar as in (6), into an invariant and an anti-invariant part

$$(12) \qquad H^1(\Delta') = H^1_+(\Delta') \oplus H^1_-(\Delta').$$

Using $\phi: X' \to X$ from diagram (1) we have

$$(13) \qquad H^3(X) \xrightarrow{\phi^*} H^3(X') \xrightarrow[\phi_*]{} H^3(X) ,$$

with $\phi_* . \phi^* = 2$ (multiplication by 2).

If A is an abelian variety, then there is the Tate group

$$T_\ell(A) = \varprojlim_n A_{\ell^n},$$

where A_{ℓ^n} are the points of order ℓ^n on A. If D is a divisor(class) on A then there is a bilinear form (see [10], p. 186)

$$e^D: T_\ell(A) \times T_\ell(A) \longrightarrow \mathbb{Z}_\ell .$$

Let C be a curve and $J(C)$ its Jacobian variety, then canonically ([1], cor. 4.7):

$$J(C)_{\ell^n} \simeq H^1(C, \mu_{\ell^n}), \tag{14}$$

hence "canonically" (after a coherent choice of roots of unity):

$$T_\ell(J(C)) \simeq H^1(C, \mathbb{Z}_\ell) \subset H^1(C, \mathbb{Q}_\ell) . \tag{15}$$

Using this isomorphism one gets ([15], p. 198)

$$e^\Theta(\xi,\eta) = \xi \cup \eta \qquad (\xi,\eta \in T_\ell(J(C))), \tag{16}$$

where Θ is the canonical divisor on $J(C)$ and $\xi \cup \eta$ is the cup product in $H^1(C,\mathbb{Q}_\ell)$.

Applying these things in our situation we get from (11) and (14)

$$H^3(X') \simeq H^1(\Delta') \simeq T_\ell(J(\Delta') \otimes_{\mathbb{Z}_\ell} \mathbb{Q}_\ell$$

and the decomposition (12) corresponds to the decomposition

$$T_\ell(J(\Delta')) \otimes \mathbb{Q}_\ell = \{T_\ell(J(\Delta)) \otimes \mathbb{Q}_\ell\} \oplus \{T_\ell(P(\Delta'/\Delta)) \otimes \mathbb{Q}_\ell\}, \tag{17}$$

coming from II i) and ii) (see (6)). From the involution $\tau: X' \to X'$ (see (10)) we get an involution

$$\tau^*: H^3(X') \to H^3(X')$$

and using (11), (15), lemma 4 of section III and [13] lemma 1, we get

<u>Lemma 5</u>. For $\xi \in H^3(X') = H^1(\Delta')$ we have $\tau^*(\xi) = -\sigma^*(\xi)$.

From the decomposition (12) = (17) and counting dimensions, we finally get for the $\mathrm{Im}(\phi^*)$ in (13):

Corollary: Using the morphism $\phi: X' \to X$ we have

$$T_\ell(P(\Delta'/\Delta)) \otimes_{\mathbb{Z}_\ell} \mathbb{Q}_\ell \xrightarrow{\sim} H^3(X).$$

Theorem 3 Let $(P(\Delta'/\Delta), \Xi)$ be the polarized Prym variety associated with the n.s. cubic threefold X. Using the isomorphism from the above corollary, we have

$$e^{\Xi}(\xi,\eta) = - \xi \cup \eta ,$$

where $\xi,\eta \in T_\ell(P(\Delta'/\Delta))$ and where the right hand side is the cup product in $H^3(X)$

Remark. The - sign comes from [13], lemma 4.

Indication of the proof. The proof is, essentially, contained in [13]; cf. with the proof of lemma 9 in that paper. Firstly, since $i^*(\Theta') \equiv 2\,\Xi$ (see (7)), we have

$$2e^{\Xi}(\xi,\eta) = e^{\Theta'}(\xi,\eta). \tag{18}$$

On the other hand, since $\phi: X' \to X$ is 2-1, we have

$$2\,\xi \cup \eta = \phi^*(\xi) \cup \phi^*(\eta), \tag{19}$$

where the cup products are on $H^3(X)$ and $H^3(X')$ respectively. Finally, using [13], lemma 4 (ii) and the equality (16) applied to $J(\Delta')$, we see that the right hand sides of (18) and (19) are equal except for a sign. This gives the proof.

Remark. Theorem 3 may be expressed in a loosely, but somewhat more expressive way, by saying: the natural polarization on $P(\Delta'/\Delta)$ comes from the Poincaré duality on $H^3(X)$.

From the usual behaviour of the Chow ring and of cohomology under monoïdal transformations, one gets:

<u>Corollary</u> ([13], theor. 3.11): The rationality assumption for a n.s. cubic threefold implies that the associated Prym variety $(P(\Delta'/\Delta),\Xi)$ is isomorphic, as principally polarized abelian variety, to a product of canonically polarized Jacobian varieties of curves.

This, together with Mumford's detailed study of the singularities of Ξ ([11], § 7, in particular the last paragraph preceding the appendix; cf. also [14], § 4), gives:

<u>Theorem 4</u> The n.s. cubic threefold is not rational.

V. CHARACTERIZATION OF THE PRYM BY MEANS OF ALGEBRAIC FAMILIES OF 1-DIMENSIONAL CYCLES.

Let T be a non-singular variety. A mapping $\Psi: T \to A^2_{alg}(X)$ is called <u>algebraic</u> if it is induced by a cycleclass $\zeta \in A^2\ (T\times X)$, i.e. if

$$\Psi(t) = \zeta(t) - \zeta(t_o),$$

with a fixed point $t_o \in T$.

Using the decomposition of theorem 2 we have a homomorphism of groups

$$\lambda_o : A^2_{alg}(X) \longrightarrow P(\Delta'/\Delta), \tag{20}$$

and for any algebraic map Ψ the composite map $\lambda_o.\Psi: T \longrightarrow P(\Delta'/\Delta)$ is a morphism of varieties ([12], 10.5).

<u>Theorem 5 (universal property of $P(\Delta'/\Delta)$)</u>: For every couple (A,λ), where A is an abelian variety and $\lambda: A^2_{alg}(X) \to A$ is a homomorphism of groups, which has the property that for every algebraic map $\Psi: T \to A^2_{alg}(X)$ the composite map $\lambda.\Psi$ is a morphism, we

have a unique homomorphism of abelian varieties $\lambda': P(\Delta'/\Delta) \to A$ such that the following diagram is commutative:

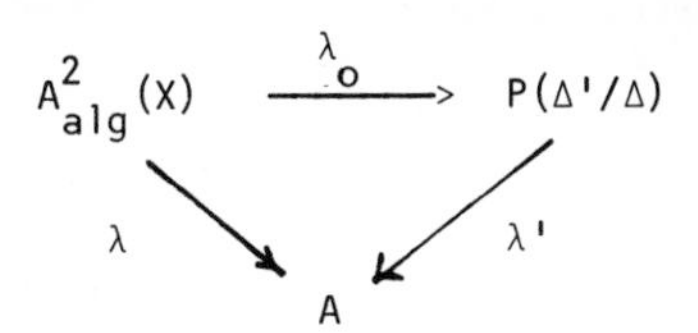

Proof: [13] appendix.

Next we study the polarization Ξ on $P(\Delta'/\Delta)$ with respect to algebraic families of cycles. Introduce the following notations. Abbreviate $P= P(\Delta'/\Delta)$; for $t \in P$, let Ξ_t be the translate of Ξ by t and let Ξ^* be the divisor on $P\times P$ defined by the formula:

(21) $$p_2 \{(t \times P) . \Xi^*\} = \Xi_t ,$$

where p_2 is the projection on the second factor.

Let $\zeta \in A^2(T \times X)$ be a cycleclass defining an algebraic map $\Psi: T \to A^2_{alg}(X)$ and let $f: T \to P$ be the morphism $\lambda_o . \Psi$ with λ_o from (20). Consider furthermore the divisor class ${}^t\zeta.\zeta$ on $T \times T$, defined via $T \times X \times T$ by the formula:

(22) $${}^t\zeta.\zeta = (p_{13})_* \{p^*_{23} ({}^t\zeta) . p^*_{12} (\zeta)\} .$$

<u>Theorem 6</u>. With the above notations, assume that ζ is such that $f: T \to P= P(\Delta'/\Delta)$ is onto of finite degree d. Then

$$\frac{1}{d^2} (f \times f)_* ({}^t\zeta.\zeta) \equiv \Xi^* \qquad \text{in } \frac{A^1(P \times P)}{p_1^*A^1(P) + p_2^*A^1(P)} \otimes_{\mathbb{Z}} \mathbb{Q},$$

where $\equiv$ means algebraic (or numerical) equivalence.

Remarks:

1. Theorem 6 answers, in the affirmative, a question raised to me by Mumford.
2. It follows from theorem 5 and 6 that $(P(\Delta'/\Delta), \Xi)$ is independent of the choice of the line ℓ_o on X.

For the proof of theorem 6 we need some preliminary results. Let V be a non-singular 3 dimensional variety and Y a non-singular curve on V, let V' be the blow up of V along Y and Y' the inverse image of Y.

$$\begin{array}{ccc} Y' & \xrightarrow{j} & V' = B_Y(V) \\ q \downarrow & & \downarrow p \\ Y & \xrightarrow[i]{} & V \end{array}$$

Lemma 6. With the above notations, assume $A^2_{alg}(V) = 0$ and $H^3(V) = 0$. Let $\zeta \in A^2(T \times V')$ with T non-singular. Fix $t_o \in T$ and consider the morphism $f\colon T \to J(Y)$ defined by

$$f(t) = \zeta(t) - \zeta(t_o) \in A^2_{alg}(V') = A^1_{alg}(Y) = J(Y).$$

Then we have

$$(f \times f)^*(\Theta^*) \equiv {}^t\zeta.\zeta \qquad \text{in} \quad \frac{A^1(T \times T)}{p_1^* A^1(T) + p_2^* A^1(T)} \otimes_{\mathbb{Z}} \mathbb{Q}$$

where $\equiv$ means numerical equivalence and Θ^* is the divisor on $J(Y) \times J(Y)$ defined by the Θ-divisor on $J(Y)$ similar as in (21).

Proof: It suffices to prove equality for the two classes modulo homological equivalence (with respect to ℓ-adique cohomology, see [9],1.2.3.). Write $J = J(Y)$. In

$$H^2(J \times J) = H^2(J) + H^2(J) + H^1(J) \otimes H^1(J)$$

write

$$\Theta^* = \Theta_{20} + \Theta_{02} + \Theta_{11} \cdot$$

On $H^1(J)$ choose a basis e_i $(1 \leq i \leq 2g,\ g=$ genus $Y)$ such that

$$e_i \cup e_j = \begin{cases} 0 & i \not\equiv j \mod g \text{ or } i = j \\ +1 & i - j = -g \\ -1 & i - j = g \end{cases}$$

Since Θ_{11} defines the cup product on $H^1(Y)$ ([15], p. 198) we have

$$\Theta_{11} = \sum_{i=1}^{g} (e_i \otimes e_{i+g} - e_{i+g} \otimes e_i) \tag{23}$$

Next consider the morphism $f: T \to J(Y)$ determined by the algebraic family

$$(1 \times q)_* (1 \times j)^* (\zeta) \cdot$$

Denote by $\phi: Y \to J(Y)$ the canonical morphism, then $(1 \times \phi)^*(\Theta^*)$ is the Poincaré divisor on $J(Y) \times Y$ and hence

$$(f \times \phi)^*(\Theta^*) = (1 \times q)_* (1 \times j)^*(\zeta) \qquad \text{in } \frac{A^1(T \times Y)}{p_1^* A^1(T) + p_2^* A^1(Y)} \tag{24}$$

Let $\{d_\ell\}$ be a basis for $H^1(T)$ and let $f^*: H^1(Y) = H^1(J) \to H^1(T)$ be given by

$$f^*(e_i) = \sum_\ell a_{\ell i}\, d_\ell \qquad (i = 1, \ldots, 2g), \tag{25}$$

then, using (23), we see that the left-hand side of (24) is given in $H^1(T) \otimes H^1(Y)$, by

(26) $$\sum_{\substack{i,\ell \\ i=1,\ldots,g}} (a_{\ell i} d_\ell \otimes e_{i+g} - a_{\ell,i+g}\, d_\ell \otimes e_i)$$

On the other hand, the element $\zeta \in A^2(T \times V')$ gives in $H^4(T \times V')$ an element z with Künneth decomposition

$$z = z_{40} + z_{31} + z_{22} + z_{13} + z_{04}\,.$$

Recall that $H^i(V') = H^i(V) \oplus H^{i-2}(Y)$ and $q_* \, j^* \, j_* \, q^* = -1$ ([13] p. 67, formula (3)). Therefore, putting $z'_{ab} = (1 \times q)_* \, (1 \times j)^* \, (z_{a,b+2})$, we have

$$(1 \times q)_* \, (1 \times j)^*(z) = z'_{20} + z'_{11} + z'_{02}$$

and we **are** particularly interested in $z'_{11} \in H^1(T) \otimes H^1(Y)$. Using $H^3(V) = 0$ we can write

$$z_{13} = \sum_{\substack{i,\ell \\ i=1,\ldots,2g}} c_{\ell i}\, d_\ell \otimes j_* \, q^*(e_i)$$

and hence

(27) $$z'_{11} = -\sum_{\substack{i,\ell \\ i=1,\ldots,2g}} c_{\ell i}\, d_\ell \otimes e_i$$

Comparing this with (26) we get from (24):

(28) $$\left.\begin{aligned} c_{\ell i} &= a_{\ell,i+g} \\ c_{\ell,i+g} &= -a_{\ell,i} \end{aligned}\right\} \quad i=1,\ldots,g$$

After these preparations we compute both sides of the assertions in in the lemma. Note however that we are <u>only interested</u> in terms in $H^1(T) \otimes H^1(Y)$.

From (23) and (25) we get

$$(f \times f)^*(\Theta_{11}) = \sum_{\ell,n} \sum_{i=1}^{g} (a_{\ell i}\, a_{n,i+g} - a_{\ell,i+g}\, a_{n,i})\, d_\ell \otimes d_n. \tag{29}$$

Next look to $(p_{13})_* \{p_{23}^* ({}^t\zeta) . p_{12}^* (\zeta)\}$ and note that only components of type (-,6,-) contribute in $(p_{13})_*$. Therefore the only interesting contribution comes from $p_{23}^*({}^t z_{31}) . p_{12}^* (z_{13})$; therefore we get by (27):

$$\sum_{\ell,n} \sum_{i=1}^{g} (c_{\ell i}\, c_{n,i+g} - c_{\ell,i+g}\, c_n)\, d_\ell \otimes d_n$$

Using (28) this gives

$$(p_{13})_* \{ p_{23}^* ({}^t\zeta) . p_{12}^* (\zeta)\} = \sum_{\ell,n} \sum_{i=1}^{g} (-a_{\ell,i+g}\, a_{ni} + a_{\ell i}\, a_{n,i+g}) d_\ell \otimes d_n \tag{30}$$

Comparing (29) and (30) completes the proof of the lemma.

<u>Lemma 7</u>. Let again X be a n.s. cubic threefold and $\zeta \in A^2(T \times X)$ an algebraic family of 1-cycles on X. Let $(P(\Delta'/\Delta), \Xi)$ be the associated Prym variety and $f: T \to P(\Delta'/\Delta)$ the morphism determined by ζ (see above). Then with the notations of (21) and (22) we have

$$(f \times f)^*(\Xi^*) \equiv {}^t\zeta.\zeta \quad \text{in} \quad \frac{A^1(T \times T)}{p_1^*A^1(T)+p_2^*A^1(T)} \otimes \mathbb{Q}$$

($\equiv$ means numerical equivalence).

<u>Remark</u>: Let us put in $A^1(T \times T)$:

(31) $$\mathcal{D} = {}^t\zeta . \zeta \quad ,$$

then $\mathcal{D}$ is the class of the <u>incidence divisor</u> in the sense of [6], 2.4. Namely write

(32) $$\mathcal{D}^* = p_{23}^* ({}^t\zeta) . p_{12}^*(\zeta) \quad \text{in } A^4(T \times X \times T),$$

then if $Z_1, Z_2 \in \zeta$ are such that

$$D^* = p_{23}^* ({}^tZ_2) . p_{12}^* (Z_1)$$

is defined, we have for $t_1, t_2 \in T$, on X:

(33) $$D^*(t_1,t_2) = Z_1(t_1) . Z_2(t_2)$$

<u>Proof of the lemma</u>: Consider the usual 2-1 morphism

$\phi: X' \longrightarrow X$ from (1). Put

$$\zeta' = (1 \times \phi)^*(\zeta) \quad \in A^2(T \times X')$$

and introduce via $T \times X' \times T$ divisor (classes) $\mathcal{D}'$, $\mathcal{D}'^*$ and D'^* similar as in (31), (32) and (33) respectively.

<u>Claim</u>: $\mathcal{D}' = 2\,\mathcal{D}$.

Namely take in ζ representatives Z_1, Z_2 as above, then with obvious notations

$$D^{*'}(t_1,t_2) = \phi^*(Z_1)(t_1) . \phi^*(Z_2)(t_2) = \phi^*\{Z(t_1).Z(t_2)\} = \phi^*D^*(t_1,t_2)$$

i.e., since ϕ is 2-1, $D' = 2D$.

Since $X' = B_{Y' \cup Y''}(X^*)$ (see (1)) we can apply lemma 6.

ζ' gives a morphism $g: T \to J(\Delta') = J(Y')$ given by

$$g(t)= \phi^* \{\zeta(t) - \zeta(t_o)\}= \zeta'(t) - \zeta'(t_o) ,$$

i.e. looking to

$$T \times T \xrightarrow[f \times f]{} P(\Delta'/\Delta) \times P(\Delta'/\Delta) \xrightarrow[i \times i]{} J(\Delta') \times J(\Delta')$$

we have $g= i.f$, and hence by lemma 6 we have

$$2D= D' \equiv (g \times g)^*(\Theta^*) \equiv (f \times f)^*(i \times i)^*(\Theta^*) \equiv (f \times f)^*(2\Xi^*),$$

hence $D \equiv (f \times f)^*(\Xi^*)$.

Proof of theorem 6: Let the assumptions (and notations) be as in the theorem. Consider $f \times f: T \times T \to P \times P$, this has degree d^2. Hence

$$(f \times f)_* (f \times f)^*(\Xi^*) = d^2\Xi^* .$$

On the other hand, by lemma 7, this is

$$(f \times f)_* (f \times f)^*(\Xi^*) \equiv (f \times f)_*({}^t\zeta.\zeta).$$

VI. RELATION WITH THE ALBANESE VARIETY OF THE FANO SURFACE.

As before, in section I, let S denote the Fano surface of the n.s. cubic threefold X; we have fixed $\ell_o \in S$ and

$$\Delta'= \{\ell \in S ; \ell \cap \ell_o \neq \phi\}$$

is the corresponding curve on S. We have the following commutative diagram

$$\begin{array}{ccccccc} & & \Delta' & \hookrightarrow & S & & \\ & & \downarrow h & & \downarrow g & \searrow f & \\ P(\Delta'/\Delta) & \xrightarrow[i]{} & J(\Delta') & \xrightarrow[\alpha]{} & A\ell b(S) & \xrightarrow[\beta]{} & P(\Delta'/\Delta). \end{array} \tag{34}$$

The cannonical vertical maps are normalized by a point $s_o \in \Delta'$ such that $h(s_o) = g(s_o) = o$. If $s \in S$ we usually write ℓ_s for the corresponding line on X (but sometimes, for $\ell \subset X$, we write also shortly $\ell \in S$!). The morphism f comes from the algebraic family

$$s \longmapsto \text{class } \{\ell_s - \ell_{s_o}\} \in A^2_{alg}(X) \;,$$

From the universal property of the Albanese variety we get the homomorphism of abelian varieties $\beta \colon \mathrm{Alb}(S) \longrightarrow P(\Delta'/\Delta)$ which is the so-called <u>Abel-Jacobi mapping</u> ([6] p. 284).

<u>Theorem 7</u>. β is an isomorphism.

<u>Proof</u>:

<u>Step 1</u>: $\beta.\alpha.i = 2$ (multiplication by 2).

Proof: Consider $\phi \colon X' \longrightarrow X$ from (1), let $\sigma \colon \Delta' \longrightarrow \Delta'$ be the involution (see (2)). Let $\xi \in P(\Delta'/\Delta)$, by II ii this can be written as

$$\xi = \sum_j h\,\{s_j - \sigma^*(s_j)\} \qquad\qquad (s_j \in \Delta') \;.$$

Hence

$$\alpha.i(\xi) = \sum_j g\{s_j - \sigma^*(s_j)\} \;,$$

and hence, by the definition of β via f,

$$\beta\alpha i(\xi) = \text{class } \phi^*\Big(\sum_j (\ell_{s_j} - \ell_{\sigma(s_j)}\Big) \;. \tag{35}$$

Now we have to return to [12] to diagram A on page 188 (where $\phi \colon X' \to X$ has the same meaning as here; note also that the curve $\widetilde{H}$ there is essential the curve Δ' here), to lemma 8.1 on page 192 and to formula (47') on page 1 With the notation of that paper [12] we have

(36) $$\phi^*(\ell_s) = \Gamma(s) + \Omega(\sigma s) \quad . \qquad (s \in \Delta')$$

Moreover in $A^2(X')$ we have by the equation (46) on page 192 of [12] that

(37) $$\Gamma(s) + \Omega(s) = \phi^*(K_{T_o})$$

where K_{T_o} is a fixed quadric obtained via a fixed 2-dimensional linear space L_{T_o} (see equation (4) of section I).

Substituting this in formula (35) we get, using (36) and (37), in $A^2_{alg}(X') = J(\Delta')$:

$$\beta\alpha i(\xi) = \text{Class } \{\sum_j (\Gamma(s_j) + \Omega(\sigma s_j) - \Gamma(\sigma s_j) - \Omega(s_j))\} =$$

$$\text{Class } \{\sum_j (\Gamma(s_j) + \phi^*(K_{T_o}) - \Gamma(\sigma s_j) - \Gamma(\sigma s_j) - \phi^*(K_{T_o}) + \Gamma(s_j))\} =$$

$$2 \text{ Class } \{\sum_j (\Gamma(s_j) - \Gamma(\sigma s_j))\} = 2\xi$$

For the last equality in the above expression see [12] p 196 below.

Hence we have $\beta.\alpha.i = 2$

<u>Step 2</u>: β is an isogeny

Proof:

dim Alb(S) = 5 (see [3]), dim $P(\Delta'/\Delta) = 5$ and $\beta.\alpha.i = 2$ is onto.

Hence β is onto and hence an isogeny.

<u>Step 3</u>: β is an isomorphism.

Proof: By step 1 and 2 it suffices to prove now that for the points of order 2 on $P(\Delta'/\Delta)$ we have

(38) $$P(\Delta'/\Delta)_2 \subset \text{Ker}(\alpha.i)$$

Consider the étale covering $q: \Delta' \to \Delta$ (see (3)). By Mumford's theory we have (see (5) and Ii) that

$$P(\Delta'/\Delta)_2 \subset q^*J(\Delta) \quad .$$

Hence it suffices to prove

$$q^*J(\Delta) \subset \mathrm{Ker}(\alpha) \quad .$$

For this, since β is an isogeny it suffices to prove

$$q^*J(\Delta) \subset \mathrm{Ker}(\beta\,\alpha) \quad .$$

Similarly as in step 1, let $\eta \in q^*J(\Delta)$ then

$$\eta = \sum_j \pm \; (h(s_j) + h(\sigma s_j)) \qquad\qquad (s_j \in \Delta')$$

and where the total degree of the cycle is zero. Similarly as in step 1 we get

$$\beta.\alpha(\eta) = \text{ class } \phi^*\{\sum_j \pm (\ell_{s_j} + \ell_{\sigma(s_j)})$$

and using (36) and (37) we get

$$\text{class } \phi^*(\ell_{s_j} + \ell_{\sigma(s_j)}) = \text{class } \{\Gamma(s_j) + \Omega(\sigma s_j) + \Gamma(\sigma s_j) + \Omega(s_j)\} =$$

2 class (K_{T_o}). Since the total degree is zero we get $\beta.\alpha(\eta) = o$.

VII. THE THEOREM OF ABEL.

As before S is the Fano surface and $\zeta \in A^2(S \times X)$ is the class defined by the family of lines. Then there are homomorphisms of abelian varieties (cf [6] , o.6):

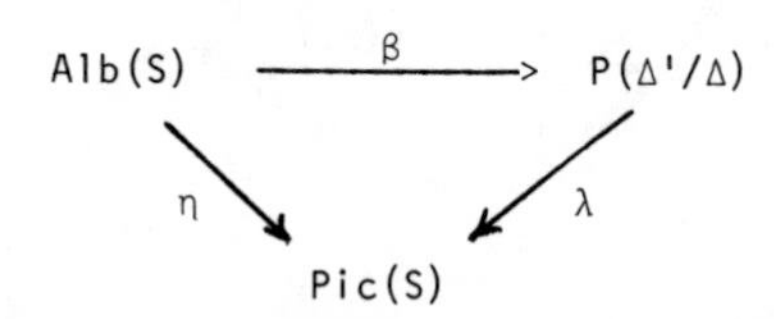

Def. β: The Abel-Jacobi map, see VI.

Def. λ: Let $\pi \in P(\Delta'/\Delta)$, then we may consider $\pi \in A^2_{alg}(X)$; put

$$\lambda(\pi) = {}^t\zeta(\pi) = \text{Class } pr_S \{Z.(S \times \pi)\},$$

for suitable $Z \in \zeta$.

Def. η: $\eta(s) = \mathcal{D}(s) - \mathcal{D}(s_o)$, where $\mathcal{D}$ is from (31), i.e. $\mathcal{D}$ is the class of the incidence divisor on $S \times S$.

Theorem 8. The above diagram is commutative and β, λ and η are isomorphisms.

Proof:

The commutativity is immediate since β, λ and η are defined by the correspondences $\zeta, {}^t\zeta$ and $\mathcal{D}$ respectively and since by definition

$$\mathcal{D} = {}^t\zeta \cdot \zeta \quad .$$

Also we know already that β is an isomorphism (theorem 7). It suffices to prove, therefore, that η is an isomorphism. Consider

$$S \times S \xrightarrow[g \times g]{} Alb(S) \times Alb(S) \xrightarrow[\beta \times \beta]{} P(\Delta'/\Delta) \times P(\Delta'/\Delta)$$

where g is from (34); put again $f = \beta.g$. Consider also the divisorclass

$$W = (\beta \times \beta)^* (\Xi^*)$$

Since β is an isomorphism and Ξ^* a principal polarization, W is a principal polarization on $Alb(S)$ and hence it defines an isomorphism

$$\eta' : Alb(S) \longrightarrow Pic(S).$$

Put

$$W_* = (g \times g)^*(W) = (f \times f)^*(\Xi^*)$$

By lemma 7 we have on $S \times S$

(39) $$W_* \equiv \mathcal{D} \qquad \text{(numerical equivalence)}$$

We have

$$\eta'(g(s) = \text{Class}\ \{W(g(s)) - W(g(s_o))\} = \text{Class}\ \{W_*(s) - W_*(s_o))\} \quad ,$$

where class means: with respect to linear equivalence. Using (39) we get

$$\text{Class}\ \{W_*(s) - W_*(s_o)\} = \mathcal{D}(s) - \mathcal{D}(s_o) = \eta(g(s)) \quad ,$$

hence $\eta'=\eta$. Hence η is an isomorphism.

Corollary ("theorem of Abel"): Let $\{s'_j\}$ and $\{s''_j\}$ $(j= 1,..,q)$ be two sets of (say) q-points on the Fano surface S. Then

$$\sum_j (\ell_{s'_j} - \ell_{s''_j}) \approx o \quad \text{(equivalence in the sense of } P(\Delta'/\Delta))$$

$$\Updownarrow$$

$$\sum_j (D(s'_j) - D(s''_j)) \sim o \quad \text{(linear equivalence on } S) \quad .$$

R E F E R E N C E S

[1]. M. Artin, Faisceaux constructible. Cohomologie d'une courbe algébrique; Exposé IX in S.G.A. 4, I.H.E.S. Lecture Notes in Math. no. 305, Springer 1973.

[2]. M. Artin and D. Mumford, Some elementary examples of unirational varieties which are not rational; Proc.London Math.Soc., 25, 1972.

[3]. E. Bombieri and H.P.F. Swinnerton-Dyer, On the local zeta function of a cubic threefold; Ann.Sc.Norm.Sup.Pisa, 21, 1967.

[4]. C. Chevalley, Anneaux de Chow; Séminaire Paris 1958.

[5]. W.L. Chow, On equivalence classes of cycles in an algebraic variety; Annals of Math., 64,1956.

[6]. C.H. Clemens and P.A. Griffiths, The intermediate Jacobian of the cubic threefold; Annals of Math.,95,1972.

[7]. G. Fano, Sul sistema ∞^2 di rette contenuto in una varietà cubica generale dello spazio a quattro dimensioni; Atti R.Acc.Sc. Torino, 39,1904.

[8]. A. Grothendieck, Sur quelques propriétés fondamentales en théorie des intersections; Exposé 4, Séminaire Chevalley: Anneaux de Chow, Paris 1958.

[9]. S.L. Kleiman, Algebraic cycles and Weil conjectures; Dix exposés sur la cohomologie des schémas, North-Holland, 1968.

[10]. D. Mumford, Abelian Varieties; Bombay, Oxford Univ.Press, 1970.

[11]. D. Mumford, Prym Varieties; to appear.

[12]. J.P. Murre, Algebraic equivalence modulo rational equivalence on a cubic threefold; Compositio Math.,25,1972.

[13]. J.P. Murre, Reduction of the proof of the non-rationality of a non-singular cubic threefold to a result of Mumford; Compositio Math.,27,1973.

[14]. A.N. Tyurin, Five lectures on three-dimensional varieties; Russian Math. Surveys Vol. 27, 1972.

[15]. J.L. Verdier, A duality theorem in the étale cohomology of schemes; Proc. conference on Local Fields, Springer, 1967.

STUDIES ON DEGENERATION

Yukihiko Namikawa

Introduction

In this article all algebraic varieties are defined over the complex number field $\mathbb{C}$.

As Ueno has shown in his talk (see his article in this volume), in order to study the classification of algebraic varieties, we are led to consider families of algebraic varieties.

Let $\pi : X \longrightarrow S$ be a proper morphism of non-singular algebraic varieties. There is a Zariski open dense subset S' in S such that the restriction $\pi' : X' = \pi^{-1}(S') \longrightarrow S'$ is smooth. Then our study of this family π is done in the following steps:

1. examining the smooth part $\pi' : X' \longrightarrow S'$;
2. examining the local behaviour of π near $S - S'$;
3. combining these to obtain global results.

The first step is concerned with the deformation (in sense of Kodaira-Spencer) and with moduli of algebraic varieties. We have several beautiful and deep results, in the case of curves, polarized abelian varieties, and recently K3 surfaces. Here we are mainly concerned with the second and the third steps, which we call study of degeneration (and its application).

The first systematical study in both steps was done by K. Kodaira in the case of families of elliptic curves in [9] . The method of approach we give here has mostly originated from his ideas on this subject. More recently, a seminar on degeneration was held at Princeton in 1969 - 70, where two topics were mainly studied. One was the notion of stable curves, which was studied later more systematically and extensively by P. Deligne and D. Mumford in [4] . Another was the notion of Gauss-Manin connection which was studied by P.A. Griffiths and several other mathematicians. On the latter subject we refer the reader to Schmid's article in this volume or Griffith's survey [5] . Finally we must take note of two of Mumford's recent papers on degeneration of curves and abelian varieties ([10] , [11]) .

In Chapter I we shall study the degeneration of curves to give almost complete solutions to the second step. The notion of stable curves plays an essential role there.

In Chapter II a special kind of degeneration of abelian variety is considered, namely in the case where the monodromy is unipotent. We introduce there very interesting results due to I. Nakamura. However, since we have no good theory of minimal models, we find it very difficult to reach a complete solution, and many problems are still left for the future.

In Chapter III we give a brief note to see how the methods given here should be generalized and what is known up to now.

Because of shortage of time, the author's talk at the Mannheim conference covered only Chapter I, §1 - §4 of this article.

Chapter I. Degeneration of Curves

§ 1. Statement of problems and historical survey.

(1.1) We consider a (local) family $\pi : X \longrightarrow D$ of curves of genus g which satisfies the following conditions:

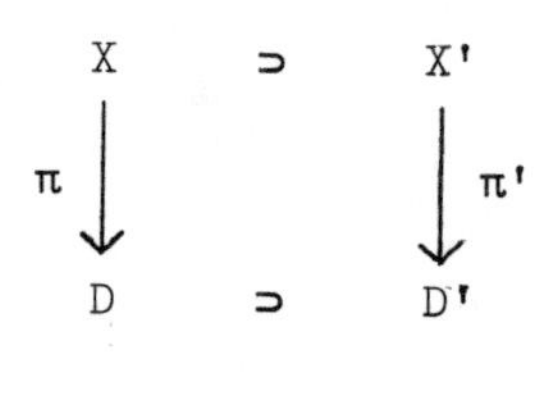

i) X is a smooth connected surface ;

ii) $D = \{t \in \mathbb{C} \; ; \; |t| < \varepsilon\}$;

iii) π is proper, surjective, and smooth over $D' = D - \{o\}$. The general fibre $X_t = \pi^{-1}(t)$, $t \in D'$, is a non-singular complete curve of genus g .

Moreover by theory of minimal models we may assume

iv) X contains no exceptional curves of the first kind(*).

Definition (1.2). If we regard π as a holomorphic function on X, then the equation $\pi = o$ defines a divisor X_o on X . We call this X_o the singular fibre (or the degenerated fibre) of π .

(*) An exceptional curve of the first kind in a surface means a non-singular rational curve C with $C^2 = -1$. If X contains an exceptional curve of the first kind C , then we can contract C to one point to obtain a new "smooth" surface X_1 . The family $\pi_1 : X_1 \longrightarrow D$ also satisfies the above conditions i) $\sim$ iii) . As π is proper, and D is a disc, exceptional curves can be contained only in the fibre over $0 \in D$. Hence after a finite number of successive contractions we have a surface free from exceptional curves.

Problems (1.3). I) What kinds of curves appear as X_o ?

II) How can we classify or characterize the family $\pi : X \longrightarrow D$ by suitable invariants?

II') Especially, how is the singular fibre X_o characterized by invariants?

III) Apply this local theory to a global study of (compact) surfaces.

Clearly Problems I), II) and II') are related to Step 2 and Problem III) to Step 3 of the introduction

In case of $g = o$ we can see easily that $X_o = \mathbb{P}_1$ (with multiplicity 1), hence there is no problem. (Note, however, that this implies that any minimal ruled surface with irregularity $q > o$ is a projective line bundle over a curve with genus g.)

In case of $g = 1$ Kodaira [9] has made deep investigations on each of the problems and given the complete solutions for Problems I), II) and II') .

For Problem I) the necessary conditions given by Kodaira for the case of $g = 1$ can be easily generalized (§ 2) and along this line Ogg and Iitaka gave all possible types of singular fibres in case of $g = 2$ ([18], [7] ; on both lists a few types were missing, cf. [15]) . Ueno and the author studied Problem II) and II') in case of $g = 2$ and gave the complete solutions on them ([16]). As a corollary we can see that all the possible types given by Ogg and Iitaka really arise as singular fibres ([15]).

On the other hand Winters has shown in general that the necessary conditions given by Kodaira for a curve to be a singular fibre are also sufficient ([25]) . Hence, theoretically, Problem I) was solved completely. It is, however, an awfully cumbersome work to give the complete list of singular fibres by these conditions even in case of $g = 3$.

For Problem II) the author succeeded to generalize the method used in case of $g = 2$, which will be explained later in § 3 . However, for Problem II') an algorithmic problem is left unsolved and we have arrived only at a weaker solution (cf. § 4) . In the algebraic category E.Viehweg has also obtained analogous results including the positive characteristic case ($\mathrm{char}(k) > 2g+1$) ([24])

Problem III) is still far from having sufficient solutions. We give here two partial results. Problem III) seems to be of much importance for a good theory of surfaces of general type.

§ 2. Problem I)

We shall first look for the necessary conditions for a divisor X_o to arise as a singular fibre.

Let $\pi : X \longrightarrow D$ be a local family of curves of genus g in (1.1) . Write

(2.1) $$X_o = \Sigma\, n_i C_i$$

where C_i is an irreducible curve in X and $n_i > o$.

Definition (2.2)

$p_i = \pi(C_i) = \dim H^1(C_i, \mathcal{O}_{C_i})$,

$c_{ij} = (C_i C_j)$ = the intersection number of C_i and C_j ,

$c_i = c_{ii}$,

$m_i = (C_i K)$, where K is the canonical divisor on X .

Condition (2.3). These invariants cannot be arbitrary. They are subject to the following:

i) $p_i \geq o$; $c_{ij} \geq o$ if $i \neq j$;

ii) $\sum_j n_j c_{ij} = o$ for $\forall i$; since $C_i X_o = C_i X_t = o$;

iii) $\sum_i n_i m_i = 2g - 2$ since $KX_o = KX_t = 2g - 2$;

iv) $c_i \leq o$, and $c_i = o$ iff $X_o = nC$. Use that X_o is connected, and ii) ;

v) $2p_i - 2 = c_i + m_i$ (adjunction formula);

vi) $(p_i, c_i, m_i) = (o, -1, -1)$ is impossible by (1.1), iv) .

Example (2.4). In case of $g = 1$ all possible types of singular fibres are listed as follows. In this case $(p, c, m) = (1, o, o)$, if the fibre has only one irreducible component, and otherwise every component has the invariants $(p, c, m) = (o, -2, o)$.

${}_nI_o$: $X_o = nC_o$, $n > o$, where C_o is a non-singular elliptic curve.

${}_nI_1$: $X_o = nC_o$, $n > o$, where C_o is a rational curve with one ordinary double point.

${}_nI_b$: $X_o = nC_1 + nC_2 + \dots + nC_b$, $n > o$, with $c_{12} = c_{23} = \dots = c_{b-1,b} = c_{1b} = 1$ $(b > 2)$ or $C_1 \cdot C_2 = p_1 + p_2$ $(b = 2)$

II : $X_o = C_o$ is a rational curve with one cusp.

III : $X_o = C_1 + C_2$ with $C_1 \cdot C_2 = 2p$.

IV : $X_o = C_1 + C_2 + C_3$ with $C_1 \cdot C_2 = C_2 \cdot C_3 = C_3 \cdot C_1 = \{p\}$.

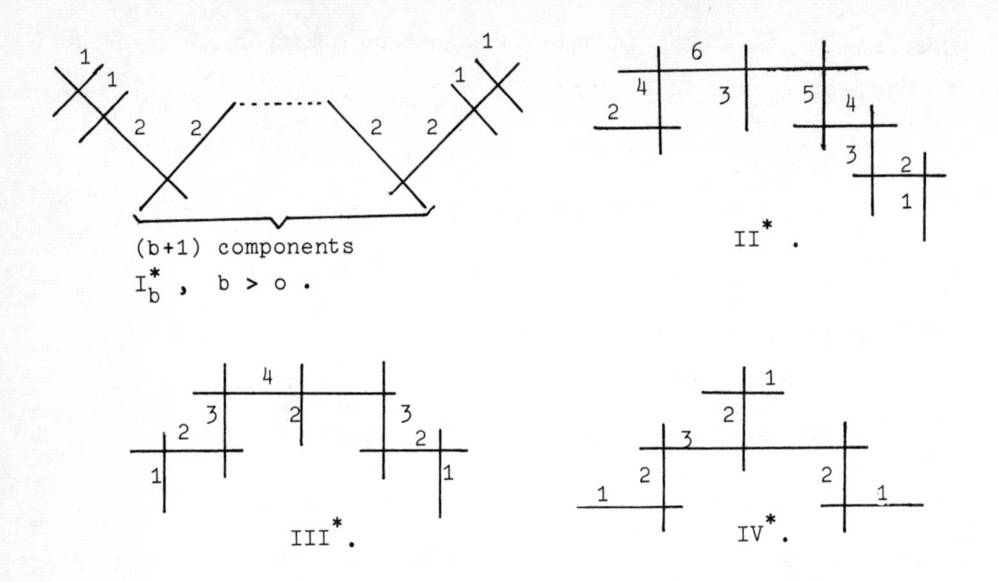

(Each integer in the figures stands for the multiplicity n_i of the corresponding component.)

The following theorem due to Winters shows that these necessary conditions are also sufficient.

Theorem (2.5) ([25]). *Suppose that the following data are given*:

i) $X_o = \sum_i n_i C_i$, *where* C_i *is an irreducible curve with* $p_i = \pi(C_i)$ *every singularity of which can be embedded in a surface* ;

ii) *non-negative divisor* $\mathfrak{r}_{ij}$ *on* C_i *and* C_j *with* $\deg(\mathfrak{r}_{ij}) = c_{ij}$ *for* $i \neq j$;

iii) *a pair of integers* (c_i, m_i) *for each* i .

If (p_i, c_{ij}, m_i) *satisfy Conditions* (2.3), *then there is a local family of curves of genus g with* X_o *as its singular fibre.*

Therefore we can classify "all" curves arising in local families of curves. However, even if we want to calculate only all the possible numerical invariants (p_i, c_{ij}, m_i) , it is already far beyond our strength in general. In case of $g = 2$ this was done by Ogg [18] and Iitaka [7] .

On the other hand Viehweg has given another characterisation by using stable reduction ([24]) .

§ 3. Problem II)

A) Characteristic map.

To solve Problem II) we must first look for "good" invariants which characterize the family. They have been already obtained by Kodaira.

Let $\pi : X \longrightarrow D$ be a local family of curves of genus g as in (1.1). We employ the notations in § 1.

Denote by M_g the coarse moduli space of smooth curves of genus g , which is a quasi-projective algebraic variety. As $\pi' : X' \longrightarrow D'$ is a family of smooth curves, there is a canonical holomorphic map $T_\pi : D' \longrightarrow M_g$ sending $t \in D'$ to the isomorphy class of X_t .

On the other hand, by transforming 1-cycles in X_t along a circle γ in D' with base point t rounding the origin once counterclockwise, we obtain an automorphism M_π of $H_1(X_t, \mathbb{Z})$, which is a free abelian group of rank 2g . Since M_π preserves the intersection form in $H_1(X_t, \mathbb{Z})$, it can be represented with an element in $Sp(g,\mathbb{Z})$. The conjugacy class of representatives is determined uniquely by π , which we denote by the same letter M_π .

<u>Definition</u> (3.1). We call T_π the <u>moduli map</u> of π and M_π the <u>monodromy</u> or the <u>Picard - Lefschetz transformation</u> of π .(*)

The pair (T_π, M_π) is called the <u>characteristic map</u> of π . This definition makes sense by Theorem (3.17) below.

(*) Kodaira called T_π the functional invariants and M_π the homological invariant in case of g = 1 .

These invariants are concerned only with the smooth part of π. The next lemma shows, however, that this is sufficient.

Lemma (3.2) ([16](5.3), [14](3.5)). Let $\pi_1 : X_1 \longrightarrow D$ and $\pi_2 : X_2 \longrightarrow D$ be two families of curves of genus $g > o$. If i is a bimeromorphic map between X_1 and X_2 with $\pi_2 \circ i = \pi_1$ which is an isomorphism over D', then i is in fact an isomorphism over D.

$$\begin{array}{ccc} X_1 & \overset{i}{\sim} & X_2 \\ \pi_1 \downarrow & & \downarrow \pi_1 \\ D & = & D \end{array}$$

B) Stable curves.

The monodromy M_π of π has a very special property. In case of families of curves we have the following theorem. (In general cases, see Chap. III.)

Theorem (3.3) (Quasi-unipotentness theorem). *The monodromy M_π of π is quasi - unipotent. More precisely to say, the conjugacy class M_π contains an element M such that*

$$M^n = \begin{pmatrix} 1_g & B \\ o & 1_g \end{pmatrix}$$

with $B \geq o$ for an integer n.

Hence a question arises naturally, "What is X_o when M_π is unipotent?" This leads to the notion of stable curves, which appear as singular fibres of families with unipotent monodromy.

Definition (3.4) (Deligne-Mumford-Mayer). A compact reduced, connected curve C is called a *stable curve of genus* $g > o$ if

i) C has only ordinary double points as singularity;

ii) each non-singular rational component T of C meets the other components in more than two points;

iii) $\dim_{\mathbb{C}} H^1(C,\mathcal{O}_C) = g$.

The condition ii) seems to be curious at first. But by virtue of this condition we can prove that the group of automorphisms of C has the same dimension as that of a non-singular curve of genus g (i.e. = 1 if $g = 1$, = o if $g > 1$), which is very plausible.

The precise study of stable curves was done by Deligne and Mumford in [4] , to which we refer the reader for details.

Example (3.5). In case of $g = 1$ the curves of types ${}_1I_o$ and ${}_1I_1$ are stable. Deligne and Mumford treat only stable curves of $g \geq 2$, but we can find in Kodaira [9] all the corresponding facts in case of $g = 1$.

Example (3.6). In case of $g = 2$ the stable curves are as follows:

1) a non-singular curve of genus 2,
2) a join of two non-singular elliptic curves meeting at one point transversally,
3) an elliptic curve with one double point,
4) a join of a non-singular elliptic curve and a rational curve with one double point meeting at one point transversally,
5) a rational curve with two double points,
6) a join of two rational curves with one double point meeting at one point transversally,
7) a join of two non-singular rational curves meeting at three points transversally.

The quasi-unipotentness theorem above, then, corresponds to the following theorem (which will be also generalized in Chap.III).

Theorem (3.7) (Stable reduction theorem) ([1], [2], [4]). Let $\pi : X \longrightarrow D$ be a local family of curves as in (1.1). There exists an $n > o$ such that for a disc $E = \{s; |s| < \varepsilon^{1/n}\}$ and a map $\mu : E \longrightarrow D$ sending s to s^n, the inverse family $\mu^*\pi : X \times_D E \longrightarrow E$ is bimeromorphic to a family of stable curves, namely, there are a flat family $\rho : Y \longrightarrow E$ of stable curves (which does not necessarily satisfy the conditions in (1.1)(*)) and a bimeromorphic mapping $i : X \times_D E \longrightarrow Y$ over E which is an isomorphism over $E' = E - \{ o \}$.

We call $\rho : Y \longrightarrow E$ a stable reduction of π. Another important fact is the existence of the coarse moduli space S_g of stable curves. Popp has shown the existence in the category of algebraic spaces ([20] , cf. his article in this volume) and I was informed that Mumford and Knudsen proved its projectivity. The stable reduction theorem already asserts the completeness of S_g. M_g is contained in S_g as a Zariski open dense subset, hence S_g gives a compactification of M_g. Also we have

Corollary (3.8). The moduli map $T_\pi : D' \longrightarrow M_g$ extends to a holomorphic map $T_\pi : D \longrightarrow S_g$.

Definition (3.9). The image $T_\pi(o)$ is called the modulus

(*) If $\rho : Y \longrightarrow E$ is a flat family of stable curves which is smooth over E', then the total space Y may have singularities at double points of the fibre Y_o over the origin. Near each double point of Y_o, ρ can be expressed in the form

$$\{(x, y, s) \; ; \; xy - s^d = o \} \longrightarrow s \, .$$

Hence if we consider $\tilde{Y}$ obtained by replacing each double point of Y_o with $d > 1$ by a chain of $(d-1)$ smooth rational curves, then the induced family $\tilde{\rho} : \tilde{Y} \longrightarrow E$ is the one considered in (1.1).

point of π and denoted by Z_{π} .

The geometrical meaning of Z_{π} is clear by the stable reduction theorem. If we consider a family $\rho : Y \longrightarrow E$ which is a stable reduction of π , then the isomorphy class of Y_o corresponds to Z_{π} in $\mathcal{S}_g$.

In order to solve Problem II) by using stable reduction, we need to investigate (flat) deformations of stable curves. Especially the existence of universal local deformation space plays an essential role. (For details see [14].)

Theorem (3.10) Let C be a stable curve. Then there exists a universal local deformation space of C . That is, there is a flat family $\varpi : \mathcal{Z} \longrightarrow \mathcal{U}$ of stable curves with the properties:

i) for a point u_o in $\mathcal{U}$, $\mathcal{Z}_{u_o} = \varpi^{-1}(u_o)$ is isomorphic to C ;

ii) for any flat family $f : \mathcal{X} \longrightarrow S$ with a point s such that $f^{-1}(s) = C$, there are a neighbourhood V of s in S and a holomorphic map $\varphi : V \longrightarrow \mathcal{U}$ with $\varphi(s) = u_o$ such that $f_{/V} : \mathcal{X}_{/V} \longrightarrow V$ is isomorphic to $\varphi^*(\tilde{\omega}) : \mathcal{Z} \times_{\mathcal{U}} V \longrightarrow V$;

iii) Aut(C) acts on $\tilde{\omega} : \mathcal{Z} \longrightarrow \mathcal{U}$ and the above φ is determined uniquely up to Aut(C) .

Moreover, we know

Proposition (3.11). i) $\mathcal{U}$ is smooth.

ii) The discriminant $\mathcal{D}$ of $\tilde{\omega}$ in $\mathcal{U}$ is a divisor with only normally crossings near u_o . The components of $\mathcal{D}$ have one-to-one correspondence with the double points on C .

iii) The canonical map $p : \mathcal{U} \longrightarrow \mathcal{S}_g$ is quasi-finite and the induced map $\mathcal{U}/\mathrm{Aut}(C) \longrightarrow \mathcal{S}_g$ is an open immersion.

Finally we state another essential theorem due to Mumford.

Theorem (3.12). For a stable curve C the canonical homomorphism

$$i : \mathrm{Aut}(C) \longrightarrow \mathrm{Aut}(H_1(C, \mathbb{Z}))$$

is injective.

C) Solution of Problem II).

First we note (This should have been noted in the previous paragraph.)

Proposition (3.13). A local family of curves in (1.1) comes from a flat family of stable curves (as in the footnote of (3.7)) if and only if the monodromy is unipotent.

Now let us consider in general. Let $\pi : X \longrightarrow D$ be a family of curves of genus g in (1.1).

Take a stable reduction $\rho : Y \longrightarrow E$ of π which exists by (3.7). Next we use (3.10). Denote by $\tilde{\omega} : \mathcal{Z} \longrightarrow \mathcal{U}$ a universal deformation space of Y_o. Then there is a holomorphic map $\varphi : E \longrightarrow \mathcal{U}$ (we replace D by a smaller disc if necessary) such that $\rho : Y \longrightarrow E$ is isomorphic to $\varphi^*(\tilde{\omega}) : \mathcal{Z} \times_{\mathcal{U}} E \longrightarrow E$.

We have obtained a commutative diagramm:

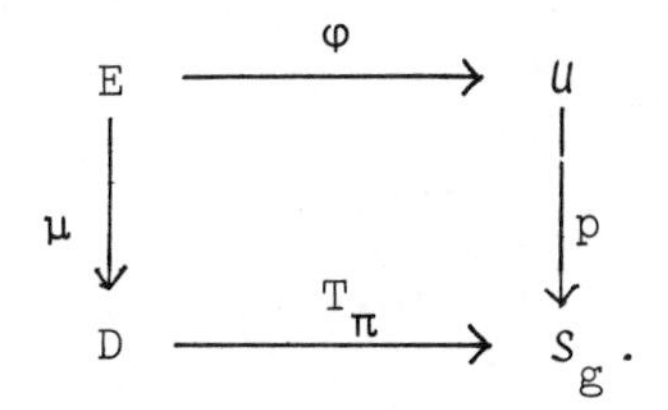

On the other hand, on E there is an automorphism g by sending s to $e_n s$ where $e_n = \exp(2\pi\sqrt{-1}/n)$. It can be lifted naturally to an automorphism of $X \times_D E$, hence induces a bimeromorphic map $\mathfrak{g}$ of Y onto itself, and we can see that $\mathfrak{g}$ is a

fortiori an isomorphism (cf. (3.2) and the footnote of (3.7)).
Let $\mathcal{g}$ (resp G) denote the finite cyclic group generated by $\mathcal{g}$ (resp G). Take a note on the fact that $Y/\mathcal{g} \longrightarrow E/G$ is bimeromorphic to $\pi : X \longrightarrow D$ and isomorphic over D'.

As $\mathcal{g}$ is a lift of g, it induces an automorphism $\mathcal{g}_o$ of Y_o. An easy but important observation is

<u>Lemma</u> (3.15). $\mathcal{g}$ <u>is induced through</u> φ <u>from the action of</u> $\mathcal{g}_o$ <u>on</u> Z.

Hence this lemma and the observation before it show with help of (3.2) that:

(3.16) $\pi : X \longrightarrow D$ is determined by $\varphi : Y \longrightarrow E$ and $\mathcal{g} \in \mathrm{Aut}(Y_o)$.

Now it is clear for us how to prove the next theorem, which solves Problem II).

<u>Theorem</u> (3.17) <u>Any family</u> $\pi : X \longrightarrow D$ <u>of curves of genus</u> $g \geq 2$ in (1.1) <u>is uniquely determined by its characteristic map</u> (T_π, M_π).

<u>Remark</u> (3.18) In case of $g = 1$ the solution of Problem II) is more complicated because of the existence of multiple fibres (${}_nI_b$ with $n > 1$). If the singular fibre is not multiple, then the same assertion is valid also in this case ([9] Th. 9.1).

<u>Outline of proof of</u> (3.17) First we note that we have only to prove the uniqueness near the origin. Hence we can consider the situation in (3.16).

However, $\rho : Y \longrightarrow E$ is determined by $\varphi : E \longrightarrow \mathcal{U}$ and φ is determined by T_π up to Aut(C) (cf.(3.14) and (3.11) iii)).

On the other hand we can see with geometric observations that M_π induces naturally an automorphism of $H_1(Y_o, \mathbb{Z})$ which coincides with the one induced by $\mathcal{g}_o$.

By virtue of Theorem (3.12) $\mathcal{g}_o$ is determined by M_π (up to conjugates).

Hence the conclusion follows from (3.16). For more precise proof we refer the reader to [14] .

Also we can give a necessary and sufficient condition for a pair (T,M) of a map $T : D' \longrightarrow M_g$ and a conjugacy class M in $Sp(g,\mathbb{Z})$ to be the characteristic map of a family of curves, but it cannot be stated shortly, hence we omit it (cf. [14]).

§ 4. Problem II').

In case of $g = 1$ Kodaira has proved the following.

Theorem (4.1). For a local family of curves of genus 1, the singular fibre X_o is determined uniquely by the monodromy M_π and the modulus point Z_π if X_o is not a multiple fibre.

We show the explicit correspondence between the singular fibres and the invariants in the following list (cf.(2.4)).

Type	Z_π (*)	M_π	Type	Z_π	M_π
$_1I_o$	τ	$\begin{pmatrix} 1 & o \\ o & 1 \end{pmatrix}$,	II	ω	$\begin{pmatrix} 1 & 1 \\ -1 & o \end{pmatrix}$,
$_1I_o^*$	τ	$\begin{pmatrix} -1 & o \\ o & -1 \end{pmatrix}$,	II*	ω	$\begin{pmatrix} o & -1 \\ 1 & 1 \end{pmatrix}$,
$_1I_n$, n>o,	∞	$\begin{pmatrix} 1 & n \\ o & 1 \end{pmatrix}$,	III	i	$\begin{pmatrix} o & 1 \\ -1 & o \end{pmatrix}$,
$_1I_n^*$, n>o,	∞	$\begin{pmatrix} -1 & -n \\ o & -1 \end{pmatrix}$,	III*	i	$\begin{pmatrix} o & -1 \\ 1 & o \end{pmatrix}$,
			IV	ω	$\begin{pmatrix} o & 1 \\ -1 & -1 \end{pmatrix}$,
			IV*	ω	$\begin{pmatrix} -1 & -1 \\ 1 & o \end{pmatrix}$.

(*) see the next page

In case of $g = 2$ we need one more invariant.

Consider the case when Z_π corresponds to a stable curve C of type 2), 4) or 6) in (3.6). C has two irreducible components C_1 and C_2. Take the least positive n such that M_π^n is unipotent. Then for this n, $\bar{T}_\pi$ can be lifted to $\varphi : E \longrightarrow \mathcal{U}$ as in (3.14), where $E = \{s ; |s| < \varepsilon\}$ and $\mathcal{U}$ is a local universal deformation space of C. By (3.11) ii) there is an irreducible component of the discriminant in $\mathcal{U}$ of $\tilde{\omega} : \mathcal{Z} \longrightarrow \mathcal{U}$ which corresponds to the double point $C_1 \cap C_2$. Let $f = o$ be the minimal defining equation of the component in $\mathcal{U}$.

Definition (4.2). $\deg \pi$ = the order of zero of $f \circ \varphi$ at $s = o$. We call it the degree of π.

Then we can answer Problem II') in case of $g = 2$.

Theorem (4.3) ([16]). For a family of curves of genus 2 the singular fibre X_o is determined by M_π, Z_π and $\deg \pi$.

The complete classification of the singular fibres by these invariants is given in [15]. We note here only two phenomena which don't occur in case of $g = 1$.

(4.4) 1) The third invariant "degree" is essential. Suppose X_o is composed of two elliptic curves joined with a series of (d-1) non-singular rational curves. In this case $\deg \pi = d$ and $M_\pi = 1_{2g}$.

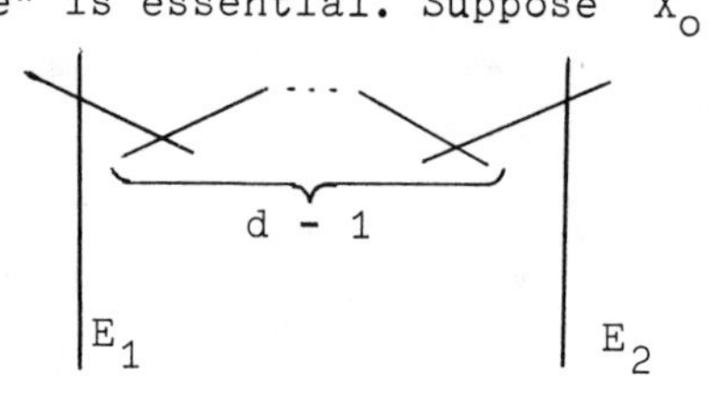

2) The same singular fibre can appear in completely different families. For example a singular fibre shown in the figure arises in the families with the following invariants:

(*) $M_1 \xrightarrow{\sim} H/SL(2,\mathbb{Z})$ where $H = \{\tau \in \mathbb{C} ; \operatorname{Im} \tau > o\}$, and τ mod $SL(2,\mathbb{Z})$ corresponds to the isomorphy class of $\mathbb{C}/(1,\tau)\mathbb{Z}^2$. $H/SL(2,\mathbb{Z})$ is isomorphic to $\mathbb{C}$ via j-function and $S_1 = M_1 \cup \{\infty\}$. $\omega = \exp(2\pi\sqrt{-1}/3)$ $i = \sqrt{-1}$.

i) $Z_\pi = \begin{pmatrix} z & o \\ o & z \end{pmatrix}$, (type 2))

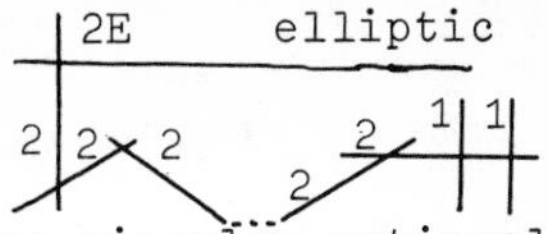

2E elliptic

2 2 2 2 2 2 1 1

non singular rational

$M_\pi = \begin{pmatrix} o & 1 & o & o \\ 1 & o & o & o \\ o & o & o & 1 \\ o & o & 1 & o \end{pmatrix}$

$\deg \pi = 2m$, $m > o$;

ii) $Z_\pi = \begin{pmatrix} z & z/2 \\ z/2 & \infty \end{pmatrix}$, (type 3))

$M_\pi = \begin{pmatrix} 1 & o & o & o \\ 1 & -1 & o & -m \\ o & o & 1 & 1 \\ o & o & o & -1 \end{pmatrix}$,

$\deg \pi = o$.

(Here m is equal to the number of non-singular rational components with multiplicity 2.)

In general cases a weaker solution can be given. Let C be the stable curve corresponding to Z_π and $z_1, \ldots, z_r$ the double points of C . Then for each double point z_i we can define i-degree of π in the same ways as (4.2). Then we can prove in a similar way as (4.3)

Theorem (4.4). *The singular fibre* X_o *is determind by* M_π , Z_π *and i-degrees.*

This statement is weak because some i-degrees are already determined by M_π . For example, if C is of type 6) in (3.6) , then we have two degrees other than degree of π . However, if $M_\pi = \begin{pmatrix} 1_2 & B \\ o & 1_2 \end{pmatrix}$ with $B = \begin{pmatrix} p & o \\ o & q \end{pmatrix}$, then they are p and q .

Viehweg has shown that Theorem (4.4) holds also in algebraic case ([24]) .

§ 5. Problem III

For this problem we have not yet satisfactory solutions. In case of $g = 1$ Kodaira made a very precise study on elliptic surfaces in [9], which is the only systematic research on this problem up to now.

Here we shall give a few applications of the previous local theory.

A) Global characterization of families of curves

(5.1) Let S be a non-singular compact surface and suppose that it admits a structure of a fibre space $\pi : S \longrightarrow C$ with conditions

i) C is a non-singular curve ;

ii) S contains no exceptional curves of the first kind in fibres ;

iii) for any point t in the complement C' of a finite subset D of C the fibre $F_t = \pi^{-1}(t)$ is a non-singular curve of genus $g \geq 2$.

Definition (5.2) i) The canonical map $T_\pi : C' \longrightarrow M_g$ is called the moduli map of π (which can be extended to $\bar{T}_\pi : C \longrightarrow S_g$).

ii) The canonically defined representation

$$M_\pi : \pi_1(C' , o) \longrightarrow GL(H_1(F_o , \mathbb{Z}))$$

is called the monodromy of π . M_π can be expressed as a representation into $Sp(g,\mathbb{Z})$ as before.

iii) We call the pair (T_π , M_π) the characteristic map of π . Then as a corollary of (3.17) we can prove

Theorem (5.3). *A family of curves of genus $g \geq 2$ in (5.1) is uniquely determined by the characteristic map.*

B) Calculation of Chern numbers.

This result is due to Ueno.

We consider again a surface S with a fibre structure $\pi : S \longrightarrow C$ as in (5.1). Let p be the genus of C. Then we can calculate the Chern numbers c_1^2 and c_2 of S with this fibre structure.

The calculation of c_2(= Euler characteristic) is standard.

$$(5.4)\quad c_2 = (2 - 2g)(2 - 2p) + \sum_{t \in D} (\chi(F_t) + (2g - 2))$$

$$= 4(g - 1)(p - 1) + \sum_{t \in D} (\chi(F_t) + (2g - 2)) .$$

Let us calculate c_1^2.

First we note that the Leray spectral sequence

$$H^q(C , R^p\pi_* \mathcal{O}_S) \Rightarrow H^{p+q}(S , \mathcal{O}_S)$$

yields the equality

$$(5.5)\quad \sum_i (-1)^i \dim H^i(S , \mathcal{O}_S) = \sum_{pq} (-1)^{p+q} \dim H^q(C , R^p\pi_* \mathcal{O}_S)$$

$$= \dim H^0(C , \pi_* \mathcal{O}_S) - \dim H^1(C , \pi_* \mathcal{O}_S)$$

$$- \dim H^0(C , R^1\pi_* \mathcal{O}_S) + \dim H^1(C , R^1\pi_* \mathcal{O}_S) .$$

Since every fibre is connected, $\pi_* \mathcal{O}_S = \mathcal{O}_C$ and $\dim H^1(F_t, \mathcal{O}_{F_t}) = g$ hence $R^1\pi_* \mathcal{O}_S$ is a locally free sheaf of rank g. Let $L = \overset{g}{\wedge} R^1\pi_* \mathcal{O}_S$. Then we have

$$(5.6)\quad \dim H^0(C , \pi_* \mathcal{O}_S) = \dim H^0(C , \mathcal{O}_S) = 1 ,$$

$$\dim H^1(C , \pi_* \mathcal{O}_S) = \dim H^1(C , \mathcal{O}_C) = p ,$$

$$\dim H^0(C , R^1\pi_* \mathcal{O}_S) - \dim H^1(C , R^1\pi_* \mathcal{O}_S)$$

$$= \deg R^1\pi_* \mathcal{O}_S + g(1 - p)$$

$$= \deg L + g(1 - p) .$$

Proposition (5.7) Let $\rho : J \longrightarrow C$ be the family of generalized Jacobian varieties associated with π (or, equivalently, the connected component of the zero section of $\mathrm{Pic}(S/C)$) and

$o : C \longrightarrow J$ <u>the zero section</u>. Then we have a canonical isomorphism

$$R^1\pi_* \mathcal{O}_S \xrightarrow{\sim} N_{o(C)/J}$$

<u>where</u> $N_{o(C)/J}$ <u>denotes the normal bundle of</u> $o(C)$ <u>in</u> J .

Now let us consider the smooth part $\pi' : S' \longrightarrow C'$.

Then $J' = J_{/C'}$ can be constructed explicitly as follows.

Take a fibre $F_t = \pi^{-1}(t)$ for $t \in C'$. Let $\{\omega_1, \ldots, \omega_g\}$ be a basis of the space of holomorphic forms on F_t, $H^0(F_t, \Omega_{F_t})$ and $\{\alpha_1, \ldots, \alpha_g, \beta_1, \ldots, \beta_g\}$ a basis of $H_1(F_t, \mathbf{Z})$ subject to

$$(\alpha_i , \alpha_j) = (\beta_i , \beta_j) = o ,$$

$$(\alpha_i , \beta_j) = \begin{cases} 1 , & i = j , \\ o , & i \neq j . \end{cases}$$

Then

$$\tilde{T}_\pi(t) = (\textstyle\int_{\beta_k} \omega_l)\ (\int_{\alpha_i} \omega_j)^{-1}$$

is symmetric and has the positive imaginary part. Denote by $\mathfrak{S}_g$ the set of symmetric matrices of degree g with positive imaginary part, which is called the Siegel upper half plane of degree g . Hence we can define a multiple-valued map

$$\begin{array}{cccc} \tilde{T}_\pi : & C' & \longrightarrow & \mathfrak{S}_g \\ & \cup\!\!\!\!\shortmid & & \cup\!\!\!\!\shortmid \\ & t & \longrightarrow & \tilde{T}_\pi(t), \end{array}$$

which is seen to be holomorphic ([13] or [16]) .

<u>Definition</u> (5.8) $\tilde{T}_\pi$ is called the <u>period map</u> of π .

On the other hand we have the monodromy

$$M_\pi : \pi_1(C') \longrightarrow Sp(g , \mathbf{Z}) .$$

If we choose a representation of M_π suitably, then for any loop γ with base point t on C' , the analytic continuation $\tilde{T}_\pi(\gamma t)$

of $\tilde{T}_\pi(t)$ along γ is subject to

$$\tilde{T}_\pi(\gamma t) = M_\pi([\gamma]) \cdot \tilde{T}_\pi(t)$$
$$= (A_\gamma \tilde{T}_\pi(t) + B_\gamma)(C_\gamma \tilde{T}_\pi(t) + D_\gamma)^{-1}$$

where $[\gamma]$ denotes the homotopy class of γ and

$$M_\pi([\gamma]) = \begin{pmatrix} A_\gamma & B_\gamma \\ C_\gamma & D_\gamma \end{pmatrix} \in Sp(g, \mathbb{Z}) .$$

Let $\tilde{C}'$ be the universal covering space of C', and consider $\pi_1(C')$ as the covering transformation group of $\tilde{C}'$ and $\tilde{T}_\pi$ as a single-valued holomorphic function on $\tilde{C}'$.

First construct

$$\tilde{J}' = \tilde{C}' \times \mathbb{C}^g / (1_g , \tilde{T}_\pi(\tilde{t})) \mathbb{Z}^{2g}$$

over $\tilde{C}'$, and define the operation of $\gamma \in \pi_1(C')$ on $\tilde{J}'$ as

$$\begin{array}{ccc} \gamma : \tilde{J}' & \longrightarrow & \tilde{J}' \\ \cup & & \cup \\ (\tilde{t}, [\zeta_1,\ldots,\zeta_g]) & \longrightarrow & (\gamma(\tilde{t}), [(\zeta_1,\ldots,\zeta_g) \\ & & \times (C_\gamma \tilde{T}_\pi(\tilde{t}) + D_\gamma)^{-1}]) . \end{array}$$

Then we have

$$J' = \tilde{J}' / \pi_1(C') .$$

Now for each point p_i in D consider a small circle γ_i rounding p_i and let m_i be the smallest n such that $M_\pi(\gamma_i)^n$ is unipotent. Let $m = \mathrm{LCM}\{m_i\}$.

Take a cusp form χ of degree nm. Then

$$\omega = \chi(T_\pi(t)) \frac{(dt \wedge d\zeta_1 \wedge \ldots \wedge d\zeta_g)^{nm}}{(dt)^{nm}}$$
$$\in H^0(C', \mathcal{O}(-nmL)) .$$

**Proposition* (5.9).* ω' extends to

$$\omega \in H^0(C, \mathcal{O}(-nmL)) .$$

This was proved only in case of $g = 2$, but it seems to hold in general.

Then we have

$$-\,nm \deg L = \Sigma \text{ order of zero of } \omega = \mu .$$

By Riemann-Roch's theorem

$$\Sigma\, (-1)^i \dim H^i(S\, , \mathcal{O}_S) = \frac{1}{12}(c_1{}^2 + c_2) .$$

Hence combining this with (5.4)~(5.6), we have

<u>Theorem</u> (5.10)

$$c_1{}^2 = 8(g - 1)(p - 1) + \frac{12}{nm}\,\mu - \sum_{t \in D} (\chi(F_t) + 2g - 2) .$$

In case of $g = 2$ we can take $nm = 120$.

Also we shall take $\chi = (\chi_{10})^{12}$ where χ_{10} is the discriminant of binary sextics (cf.[6]). χ does not vanish at any points in $\mathfrak{S}_2$ corresponding to the periods of non-singular curves of genus 2, hence the above ω has zeros only on D. (This fact holds no more if $g > 2$.) Hence we have

<u>Corollary</u> (5.11) <u>In case of</u> $g = 2$, $\deg L \leq o$, <u>and</u> $\deg L = o$, <u>if and only if</u> π <u>is a holomorphic fibre bundle.</u>

<u>Corollary</u> (5.12) <u>For</u> $t \in D$, <u>let</u> $12\gamma_t$ <u>be the order of zero of</u> ω <u>at</u> t. <u>Then we have</u>

$$c_1{}^2 = 8(p - 1) + \frac{2}{10}\,\gamma_t + \sum_{t \in D} (\gamma_t - (\chi(F_t) + 2)) .$$

As the calculation of γ_t is local, we may do it with the classification table of X_o by the invariants in (4.3).

In [18] Ogg conjectured that $\gamma_t = \chi(F_t) + 2$ ($= n_t + \varepsilon_t - 1$ in his notation). But this is not the case if $\deg \pi > o$ because of the zeros of χ_{10} at $\begin{pmatrix} \tau_1 & o \\ o & \tau_2 \end{pmatrix}$.

Due to Ueno γ_t is subject to

$$\left[\frac{\deg \pi_t}{n_t}\right] \leq \gamma_t - (\chi(F_t) + 2) \leq \left[\frac{\deg \pi_t}{n_t}\right] + 1$$

where $\deg \pi_t$ denotes the degree of π at t (cf.(4.2)) , n_t is the least positive integer such that ${M_{\pi_t}}^n$ is unipotent and [ ] denotes the Gauss symbol. In almost all cases

$\left[\frac{\deg \pi_t}{n_t}\right] = \gamma_t - (\chi(F_t)+2)$. The local families given in (4.4)2) give, however, pathological examples, namely

$\gamma_t - (\chi(F_t)+2) = \left[\frac{\deg \pi_t}{n_t}\right] + 1$.

Addendum. After having written this manuscript, the author obtained the following generalization of (5.3).

Theorem. *The "bimeromorphic" type of families of curves of genus $g \geq 2$ (over arbitrary dimensional parameter spaces) is uniquely determined by the characteristic map*.

The proof will appear in [14] .

Chapter II. Degeneration of abelian varieties

§ 1. Period map and monodromy

Let us consider a family $\pi' : A' \longrightarrow D'$ of abelian varieties over a punctured disc $D' = \{t : o < |t| < \varepsilon\}$. Assume, moreover, that $A' \longrightarrow D'$ is polarized in the sense of Ueno ([23]), that is, we are given a non-degenerate bilinear form

$$(1.1) \quad R^1\pi'_*\mathbb{Z}_{A'} \times R^1\pi'_*\mathbb{Z}_{A'} \longrightarrow \mathbb{Z}_{D'} .$$

Over each $t \in D'$ the above form can be expressed with a suitable basis of $H^1(A_t, \mathbb{Z})$ in the form

$$P = \begin{pmatrix} o & \Delta \\ \Delta & o \end{pmatrix}, \quad \Delta = \begin{pmatrix} d_1 & & & o \\ & d_2 & & \\ & & \ddots & \\ o & & & d_g \end{pmatrix}$$

where d_i are positive integer with $d_1|d_2|\ldots|d_g$.

Let $\mathfrak{S}_g$ denote the Siegel upper-half plane of degree g . In a similar way as a family of curves in Chap.I, § 5, we can define a <u>period map</u>

$$(1.2) \quad T_\pi : D' \longrightarrow \mathfrak{S}_g$$

of π' which is a multiple-valued holomorphic map. <u>The monodromy</u>

$$(1.3) \quad M_\pi = \begin{pmatrix} A & B \\ C & D \end{pmatrix} \in Sp(P, \mathbb{Z})$$

(where $Sp(P, \mathbb{Z})$ is the symplectic group with respect to P above) can be defined also, and the analytic continuation $T_\pi(\gamma t)$ of $T_\pi(t)$ along a circle γ, in D' rounding $\{o\}$ once counter-clockwise is subject to

$$(1.4) \quad T_\pi(\gamma t) = M_\pi \cdot T_\pi(t)$$
$$= (AT_\pi(t) + B\Delta)(CT_\pi(t) + D\Delta)^{-1}\Delta .$$

<u>Assume further that π' has a section</u> $s : D' \longrightarrow A'$.
<u>Then we can construct</u> π' <u>from</u> T_π <u>and</u> M_π <u>as follows</u>.

Let $\tilde{D}' = \{\tilde{t}\ ,\ \mathrm{Im}\tilde{t} > -\frac{1}{2\pi}\log \varepsilon\}$ be the universal covering of D' with the covering map $p : \tilde{t} \longrightarrow t = \exp(2\pi\sqrt{-1}\ \tilde{t})$. $\tilde{T}_\pi = T_\pi \circ p$ is a single-valued holomorphic map. For $\nu \in \mathbb{Z}^{2g}$ define an automorphism g_ν of $\tilde{D}' \times \mathbb{C}^g$ by

$$g_\nu : (t, (\xi_1,\ldots,\xi_g)) \longrightarrow (\tilde{t}, (\xi_1,\ldots,\xi_g) + \nu \begin{pmatrix}\Delta \\ \tilde{T}_\pi(\tilde{t})\end{pmatrix}) .$$

Then $\tilde{A}' = \tilde{D}' \times \mathbb{C}^g / \{g_\nu\}_{\nu\in\mathbb{Z}^{2g}}$ is a family of abelian varieties. For $(\tilde{t}, (\zeta_1,\ldots,\zeta_g))$ denote the corresponding point in A' by $(\tilde{t}, [\zeta_1,\ldots,\zeta_g])$. Also write

$$M_\pi{}^n = \begin{pmatrix} A_n & B_n \\ C_n & D_n \end{pmatrix} \in Sp(P\ ,\ \mathbb{Z})$$

for $n \in \mathbb{Z}$. Let $\pi_1(D') = \mathbb{Z}$ act on A' as

$$g_n : (\tilde{t}, [\zeta_1,\ldots,\zeta_n]) \longrightarrow (\tilde{t} + n, [(\zeta_1,\ldots,\zeta_n)\ f_n(\tilde{t})])$$

where

$$f_n(\tilde{t}) = (C_n\ \tilde{T}_\pi(\tilde{t}) + D_n\ \Delta)^{-1}\Delta .$$

Then it is easily seen that

$$A' \xrightarrow{\ \sim\ } \tilde{A}' / \pi_1(D') .$$

§ 2. Problems

First of all

<u>Problem</u> I). <u>How can one construct a singular fibre</u> A_o <u>over</u> o <u>such that</u> $\pi : A = A' \cup A_o \longrightarrow D$ <u>is proper</u>?

As we have no good theory of minimal models, there are many choices of A_o . Hence there arises

<u>Problem</u> II). <u>What should be a "good" or "natural" singular fibre</u>?

One can expect that for a good singular fibre the polarization on π' extends to π .

In this case also the quasi-unipotentness theorem on monodromy does hold. We say that π' is a <u>stable family</u> if the monodromy is unipotent. If we follow the method of construction in case of curves, it will be the following.

Let n be a positive integer such that $M_\pi{}^n$ is unipotent. Define $\mu : E \longrightarrow D$ by $s \longrightarrow s^n$. Then $\rho' : \mathcal{B}' = \mathcal{A}' \times_D E' \longrightarrow E'$ is a stable family of abelian varieties. We shall construct a singular fibre B_o to obtain $\rho : \mathcal{B} \longrightarrow E$. The covering transformation group G of E' is naturally lifted to a group $\mathcal{G}$ of automorphisms of $\mathcal{B}'$. We construct $\mathcal{B}$ so that the action of $\mathcal{G}$ extends to $\mathcal{B}$. Then $\pi : \mathcal{B}/\mathcal{G} \longrightarrow E/G = D$ (or a bimeromorphic model of $\mathcal{B}/\mathcal{G}$) is the desired family.

Hence the problems have become more precise.

<u>Problem</u> I_s). <u>Construct a singular fibre for stable families of abelian varieties.</u>

<u>Problem</u> II_s). i) <u>For which singular fibre can the polarization be extended</u>?

ii) <u>For which singular fibre can the automorphism of</u> $\mathcal{B}'$ <u>be extended to</u> $\mathcal{B}$?

In case M_g is of finite order, they are easily solved. Namely, if M_g is of finite order and unipotent, then $M_g = 1_{2g}$. Hence T_π is already single-valued and we can construct a family of abelian varieties over the whole D . In case of $g = 2$ Ueno has constructed singular fibres concretely along the method mentioned above ([23]) .

Mumford has studied these problems in [11] (especially § 6), and shown a quite general method of construction and a criterion for the extendability of the polarization (ibid. (6.7))

Here we shall introduce another kind of construction due to Nakamura. This method is closely related with Mumford's, but his construction is very explicit. A remarkable property of his model is that the total space B is non-singular. Instead his model does not necessarily fit to Problem II_s). For this purpose we must modify his model.

On the other hand in case π' is a family of Jacobian varieties associated with a local family of curves in Chap.I.(1.1) induced by a flat family of stable curves, Seshadri and Oda have constructed the singular fibre which automatically solves Problem II_s) i) ([22]). Their method of construction is so to say intrinsic, namely, they construct B_o only by using the singular fibre Y_o.

We may expect further good fruits in this direction for the future.

§ 3. Neron models of stable families of abelian varieties.

The results in this section and the next are due to Nakamura [12] . The method of construction is, however, slightly different from his original one.

We consider here only a stable family of abelian varieties. For simplicity we shall moreover restrict ourselves to consider the principal polarization, namely, $\Delta = 1_g$ in (1.1) . By the observation in § 5 we may start from a period map $T_{\pi'} : D' \longrightarrow \mathfrak{S}_g$.

Since $M_\pi \in Sp(g,\mathbb{Z})$ is assumed to be unipotent, it is subject to (up to conjugate)

$$(3.1) \qquad M_\pi = \begin{pmatrix} 1_g & B \\ o & 1_g \end{pmatrix} , \quad B = \begin{pmatrix} o & o \\ o & B_o \end{pmatrix}$$

where $B_o \in GL(g'', \mathbb{Z})$, ${}^tB_o = B_o$ and $B_o > o$. Put $g' = g - g''$. We keep employing the notations in § 1 . By (3.1) $\tilde{T}_\pi : \tilde{D}' \longrightarrow \mathfrak{S}_g$ is rewritten as

$$(3.2) \quad \tilde{T}_\pi(\tilde{t}) = \tilde{t}B + S(t)$$

where $S(t)$ is single-valued and bounded on D' . Note that the first g' column vectors $\Sigma_i = \Sigma_i(t)$, $i = 1, \ldots, g'$, of $\tilde{T}_\pi(\tilde{t})$ are, then, a single-valued holomorphic function on the whole D .

Put $G = \{g_\nu\}_{\nu \in \mathbb{Z}^{2g}}$, and $N = \{g_\nu\}_{\nu = \binom{\nu_1}{o}, \nu_1 \in \mathbb{Z}^g}$ in § 2 . Since the automorphic factor $f_n(\tilde{t})$ is trivial, we have

$$\begin{array}{ccc} \hat{A}' = \tilde{D}' \times \mathbb{C}^g / \pi_1(D') \times N & \xrightarrow{\sim} & D' \times (\mathbb{C}^*)^g \\ (\tilde{t}, [\zeta_1, \ldots, \zeta_g]) & & (t, (w_1, \ldots, w_g)) \end{array}$$

with $t = e(\tilde{t}) = \exp(2\pi\sqrt{-1}(\tilde{t}))$, $w_i = e(\zeta_i)$. Define a group $H = \{h_{\nu_2}\}_{\nu_2 \in \mathbb{Z}^g}$ of automorphisms of $D \times (\mathbb{C}^*)^g$ by

$$(3.3) \quad \begin{array}{cccc} h_{e_i} : & D \times (\mathbb{C}^*)^g & \longrightarrow & D \times (\mathbb{C}^*)^g \\ & (t, (w_j)) & \longrightarrow & (t, (t^{b_{ij}} \sigma_{ij}(t)\, w_j)) \end{array}$$

where $e_i = {}^t(o, \ldots, 0, \overset{i}{1}, 0, \ldots, 0) \in \mathbb{Z}^g$, $B = (b_{ij})$ and $\sigma_{ij}(t) = e(S_{ij}(t))$ except for $h_{e_{g'+1}}, \ldots, h_{e_g}$ over $t = o$. Then

$$A^o = D \times (\mathbb{C}^*)^g / H$$

is a family of abelian Lie groups over D and isomorphic to A' over D' . The fibre A^o_o over o is isomorphic to $\mathbb{C}^g / (1_g, \Sigma_1(o), \ldots, \Sigma_{g'}(o))\, \mathbb{Z}^{g+g'}$ which is an extension of an abelian variety by $(\mathbb{C}^*)^{g''}$,

$$(3.4) \quad o \longrightarrow (\mathbb{C}^*)^{g''} \longrightarrow A^o_o \longrightarrow A_o \longrightarrow o$$

where the last surjection is induced from the projection of $(\mathbb{C}^*)^g$

onto the first g' components (*).

However, the singular fibre of Neron minimal model ([17]) is in general not connected. Therefore, in order to construct it, we must patch together a finite number of copies of A^o .

For that purpose we shall go back a little. Let $H_1 = \{h_\nu \in H ; \nu = \binom{\nu'}{o} , \nu' \in \mathbb{Z}^{g'}\}$, and $B = D \times (\mathbb{C}^*)^g/H_1$. B is a family of extensions of g'-dimensional abelian varieties by $(\mathbb{C}^*)^{g''}$, and the fibre over the origin is A^o_o . Put

$$\tilde{B} = \bigcup_{\mu \in \mathbb{Z}^{g''}} B_\mu$$

where each B_μ is a copy of B and

$$\underset{B_{\mu(1)}}{(t^{(1)}, [w_i^{(1)}])_{\mu(1)}} = \underset{B_{\mu(2)}}{(t^{(2)}, [w_i^{(2)}])_{\mu(2)}}$$

if and only if

$$(t^{(1)} , [t^{\mu_i^{(1)}} w_i^{(1)}]) = (t^{(2)} , [t^{\mu_i^{(2)}} w_i^{(2)}])$$

in B , and $t^{(1)} = t^{(2)} \neq o$. Hence, over D' , $\tilde{B}$ and B_o are the same, and we identify them. The natural induced topology on $\tilde{B}$ satisfies the condition that, for $\mu = (\mu_i) \in \mathbb{Z}^{g''}$,

$$\begin{aligned} \lim_{t \to o} (t, [w_1, \ldots , w_{g'} , t^{-\mu_1} w_{g'+1}, \ldots , t^{-\mu_{g''}} w_g]) \\ = (o, [w_1, \ldots , w_{g'} , w_{g'+1}, \ldots , w_g]_\mu \in (B_\mu)_o , \end{aligned}$$

(*) If $\tilde{T}_\pi$ comes from a family of stable curves as Chap. I § 5 B), then A^o is nothing but the associated family of generalised Jacobian varieties. The abelian variety A_o is the Jacobian variety of the normalisation of the singular fibre.

and we can see that $\tilde{B}$ admits a structure of a complex manifold. On $\tilde{B}$ we can extend the operation of h_{e_i}, $i = g' + 1, \ldots, g$, (which was not defined over the origin) as

$$(3.3)^{bis} \quad h_{e_i} : \tilde{B} \longrightarrow \tilde{B}$$
$$\quad\quad (t,[w_j]_\mu) \mapsto (t,[\sigma_{ij}(t)w_j]_{\mu+B_o e_i})$$

Then $Z = H/H_1 (\xrightarrow{\sim} \mathbb{Z}^{g''})$ can be seen to operate on $\tilde{B}$ freely and totally discontinuously. Put

$$A = \tilde{B} / Z .$$

<u>Definition</u> (3.5). We call the family $\pi : A \longrightarrow D$ the <u>"Neron" model</u> of A'.

We don't know whether it really coincides with the algebraic Neron minimal model, but it satisfies the following universal property.

<u>Theorem</u> (3.6) <u>Let</u> $\rho : X \longrightarrow D$ <u>be a family of complex Lie groups whose fibres over</u> D' <u>are abelian varieties and whose fibre over the origin is an extension of abelian varieties by a split torus. Denote by</u> $\rho' : X' \longrightarrow D'$ <u>the restriction of</u> ρ <u>over</u> D'. <u>If there is a morphism</u> $\alpha' : X' \longrightarrow A'$ <u>of families of abelian varieties, then</u> α' <u>extends to a morphism</u> $\alpha : X \longrightarrow A$ <u>of families of Lie groups</u>.

Moreover we can see

<u>Proposition</u> (3.7) i) π <u>is smooth</u>.

ii) <u>The fibre over the origin is an extension of a finite group</u> N <u>by</u> A_o^o . N <u>is isomorphic to</u> $\mathbb{Z}^{g''} / B_o \mathbb{Z}^{g''}$.

§ 4. Nakamura's compactification of Neron models.

The construction is done in four steps.
First we construct a special compactification of $(\mathbb{C}^*)^{g''}$.
Let $K = K_{g''}$ be the image of a rational map,

$$(4.1) \quad (\mathbb{P}_1)^{g''} \longrightarrow (\mathbb{P}_1)^N$$
$$(w_i) \longmapsto (w_i,\ w_{ij} = w_i/w_j)_{1 \leq i < j \leq g''}$$

where w_i , w_{ij} are inhomogeneous coordinates of $\mathbb{P}_1$ and $N = g''(g''+1)/2$.

This is a compactification of $(\mathbb{C}^*)^{g''}$ corresponding to a decomposition of $\mathbb{R}^{g''} = \{(x_1, \ldots, x_{g''})\}$ by $x_i = o$, $x_i = x_j$ ([8] Chap.I) . The boundary is a union of $2(2^{g''}-1)$ divisors $W(e)$ each of which is defined for e = non-zero g''-vector with coefficients o or 1 as

$$(4.2) \quad W(e) = \{(w_i, w_{ij}) \in K \subset (\mathbb{P}_1)^N ;$$
$$w_i = o \text{ or } \infty \text{ according as } e_i = 1 \text{ or } -1 ,$$
$$w_{ij} = o \text{ or } \infty \text{ according as } e_i - e_j = 1 \text{ or } -1\}.$$

Secondly we compactify A_o^o .

By (3.4) A_o^o is a principal bundle over A_o with fibre $(\mathbb{C}^*)^{g''}$. We compactify each fibre by $K_{g''}$ to obtain a $K_{g''}$ - bundle $K(A)$ over A_o . It is constructed explicitly as follows. Take $(\mathbb{C}^*)^{g'} \times K = \{(w_i', w_j'', w_{kl}'') ; (w_i') \in (\mathbb{C}^*)^{g'}, (w_j'', w_{kl}'') \in K\}$. It contains $(\mathbb{C}^*)^g$ and the action of H_1 defined by $\{(h_\nu)_{t=o} ; h_\nu \in H_1\}$ (3.3) on $(\mathbb{C}^*)^g$ extends to $(\mathbb{C}^*)^{g'} \times K$. Then we have $K(A) = (\mathbb{C}^*)^{g'} \times K / H_1$. For $(w_i', w_j'', w_{kl}'') \in (\mathbb{C}^*)^{g'} \times K$ we denote by $[w_i', w_j'', w_{kl}'']$ the corresponding point in $K(A)$.

Thirdly we compactify $\tilde{B}_o$. Note that $\tilde{B}_o = \coprod_{\mu \in \mathbb{Z}^{g''}} (B_o)_\mu$ with $B_o = A_o^o$. We construct its compactification $K(\tilde{B})$ as

$$(4.3) \qquad K(\tilde{B}) = \bigcup_{\mu \in Z^{g''}} K(A)_\mu$$

where $[w_i'^{(1)}, w_j''^{(1)}, w_{kl}''^{(1)}]_{\mu(1)} \in W_{\mu(1)}(e^{(1)}) \subset K(A)_{\mu(1)}$

and $[w_i^{\prime(2)}, w_j^{\prime\prime(2)}, w_{kl}^{\prime\prime(2)}]_{\mu^{(2)}} \in W_{\mu^{(2)}}(e^{(2)}) \subset K(A)_{\mu^{(2)}}$

are the same point in $K(\tilde{B})$ if and only if $e^{(1)}+e^{(2)} = o$, $\mu^{(1)}+ e^{(1)}= \mu^{(2)}$, $w_i^{\prime(1)}= w_i^{\prime(2)}$, $w_j^{\prime\prime(1)} = w_j^{\prime\prime(2)}$ if $e_j^{(1)}= o$, and $w_{kl}^{\prime\prime(1)} = w_{kl}^{\prime\prime(2)}$ if $e_k^{(1)}= e_l^{(1)}$.

A remarkable fact is

<u>Proposition</u> (4.4) <u>The union</u>

(4.5) $\quad \bar{B} = \tilde{B} \amalg K(\tilde{B})$

<u>admits a structure of a complex manifold.</u>

We shall show the easiest case $g' = o$, $g'' = 1$. In general it is essentially the same though far more complicated.

Then $B = D \times \mathbb{C}^*$, $K = K(A) = \mathbb{P}_1 = \{(w)\}$, and $K(\tilde{B}) = \bigcup_{\mu\in\mathbb{Z}} K_\mu$ where $(w^{(1)})_{\mu^{(1)}} = (w^{(2)})_{\mu^{(2)}}$ if and only if $\mu^{(1)} - \mu^{(2)}= \pm 1$, and $(w^{(1)}, w^{(2)}) = (\infty, o)$ or (o, ∞) according as $\mu^{(1)}= \mu^{(2)}+ 1$ or $\mu^{(1)}= \mu^{(2)}- 1$.

It is enough to consider near $p = (o)_o = (\infty)_1$.
On a neighbourhood U of $\tilde{B} = \bigcup B_\mu$ which is a union of $U_1 = \{[t, w]_o \in B_o ; |t| < \varepsilon , |w| < \varepsilon\}$ and $U_2 = \{[t,w]_1 \in B_1 ; |t| < \varepsilon , |w|^{-1} < \varepsilon\}$ let us define a map $i : U \longrightarrow \mathbb{C}^2$ by $[t,w]_o \longrightarrow (tw^{-1},w)$ on U_1 and $[t,w]_1 \longrightarrow (w^{-1},tw)$ on U_2. Then for a sufficiently small ε, i is an open immersion to $\mathbb{C}^2 - \{o\}$.

On the other hand on a neighbourhood $V = V_1 \cup V_2$ of $K(\tilde{B})$ with $V_1 = \{(w)_o ; |w| < \varepsilon\}\subset K_o$ and $V_2 = \{(w)_1; |w|^{-1} < \varepsilon\}\subset K_1$, let us define a map $j : V\longrightarrow \mathbb{C}^2$ by $(w)_o \longrightarrow (o,w)$ and $(w)_1 \longrightarrow (w^{-1},o)$. Then j is a closed immersion of V into $\mathbb{C}^2$.

It is clear that $i \cup j : U \cup V \longrightarrow \mathbb{C}^2$ is an isomorphism of a neighbourhood of p in $\bar{B}$ to an open subset in $\mathbb{C}^2$, which defines the structure of a complex manifold on $\bar{B}$ near p.(Q.E.D.)

The last step is already evident. We see that h_{e_i} in $(3.3)^{bis}$ extends to an automorphism $\bar{h}_{e_i}$ of $\bar{B}$ and thus Z acts on $\bar{B}$. As the action is properly discontinuous and fixed point free, we obtain a complex manifold

$$\bar{A} = \bar{B} / Z .$$

Summing up, we have obtained

<u>Theorem</u> (4.6) <u>The Neron model</u> $\pi : A \longrightarrow D$ <u>has a properification</u> $\bar{\pi} : \bar{A} \longrightarrow D$ <u>such that</u>

i) $\bar{A}$ <u>is a complex manifold</u> ;

ii) <u>each irreducible component of the singular fibre is a fibre bundle over an abelian variety</u> A_o <u>with fibre</u> $K_{g''}$.

§ 5. Examples and comments.

The study on Nakamura's compactification is still on the way. We shall show explicit structures of singular fibres by his construction in case of $g = 2$, and make relation to some of their properties, especially with respect to Problem II_s). Results by further investigation will appear in his article [12] .

<u>Example</u> (5.1)

$$\tilde{T}(\tilde{t}) = \begin{pmatrix} o & o \\ o & p \end{pmatrix} \tilde{t} + (S_{ij}(t)) , \quad p > o .$$

Put $s = S_{11}(o)$ and $e = S_{12}(o)$. We have by (3.4)

$$o \longrightarrow \mathbb{C}^* \longrightarrow A_o^o \longrightarrow A_o \longrightarrow o$$

where $A_o = \mathbb{C} / (1, s)\mathbb{Z}^2$. The point $[e]$ in A_o corresponding to $e \in \mathbb{C}$ gives the extension class of this exact sequence via the canonical isomorphism : $Ext^1(A_o, \mathbb{C}^*) \xrightarrow{\sim} Pic(A_o) \xrightarrow{\sim} A_o$([13] , [21]) .

By this exact sequence A_o^o is a principal $\mathbb{C}^*$- bundle over A_o . Regard $\mathbb{C}^*$ as a subgroup of $PGL(1)$ by $w \longrightarrow \begin{pmatrix} w & o \\ o & 1 \end{pmatrix}$. Then $K(A)$ is the $\mathbb{P}_1$-bundle associated with A_o^o. It has two sections, o-section $o(A_o)$ and ∞ -section $\infty(A_o)$. The singular fibre A_o is then constructed as follows. Take p-copies of $K(A)$, $K(A)_o$, $K(A)_1$, ... , $K(A)_{p-1}$. For $o \leq i < p - 1$ we identify $o(K(A)_i)$ and $\infty(K(A)_{i+1})$ trivially, and identify $o(K(A)_{p-1})$ and $\infty(K(A)_o)$ <u>with the translation by</u> $[e]$

$$+ [e] : \infty(K(A)_o) \longrightarrow o(K(A)_{p-1})$$
$$\qquad\qquad\qquad a \longrightarrow a + [e] .$$

A_o is, hence, no more a fibre space over A_o . Instead the principal polarization on the general fibre extends naturally to $\bar{A}$. Let L' be the relatively ample line bundle on A' corresponding to the polarization. Then there are p-different extensions of L', $L, tL, \ldots , t^{p-1}L$ (with a vague notation), and $\bigotimes_{i=o}^{p-1} t^iL$ is a relatively ample bundle on $\bar{A}$. This phenomenon is more precisely explained in the next example.

<u>Example</u> (5.2)

$$\tilde{T}(\tilde{t}) = \begin{pmatrix} p & o \\ o & q \end{pmatrix} \tilde{t} + (S_{ij}(t)) , \quad p \geq q > o .$$

Put $\lambda = e(S_{12}(o))$. $A_o^o = (\mathbb{C}^*)^2 = \{(w_1, w_2)\}$ and $K(A)$ is isomorphic to a blowing up of $\mathbb{P}_1 \times \mathbb{P}_1$ at two points (o,o) and (∞,∞). The boundary $K(A) - A_o^o$ is a cycle of

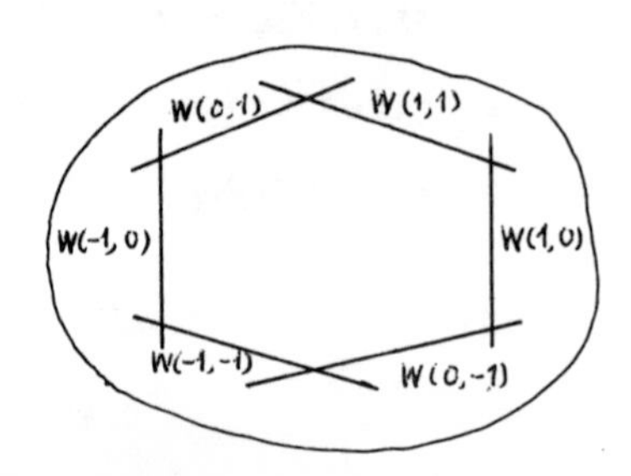

6 projective lines,

$$W(1,o) = \{(w)_{(1,o)} \longrightarrow (o,w) \in (\mathbb{P}_1)^2\} \quad ,$$

$$W(-1,o) = \{(w)_{(-1,o)} \longrightarrow (\infty,w) \in (\mathbb{P}_1)^2\} \quad ,$$

$$W(o,1) = \{(w)_{(o,1)} \longrightarrow (w,o) \in (\mathbb{P}_1)^2\} \quad ,$$

$$W(o,-1) = \{(w)_{(o,-1)} \longrightarrow (w,\infty) \in (\mathbb{P}_1)^2\} \quad ,$$

$$W(1,1) = \{(w = w_1/w_2)_{(1,1)} \longrightarrow (o,o) \in (\mathbb{P}_1)^2\} \quad ,$$

$$W(-1,-1) = \{(w=w_1/w_2)_{(-1,-1)} \longrightarrow (\infty,\infty) \in (\mathbb{P}_1)^2\} \quad .$$

Then the singular fibre A_o is

$$A_o = \bigcup_{\substack{o \le i \le p-1 \\ o \le j \le q-1}} K(A)_{(i,j)}$$

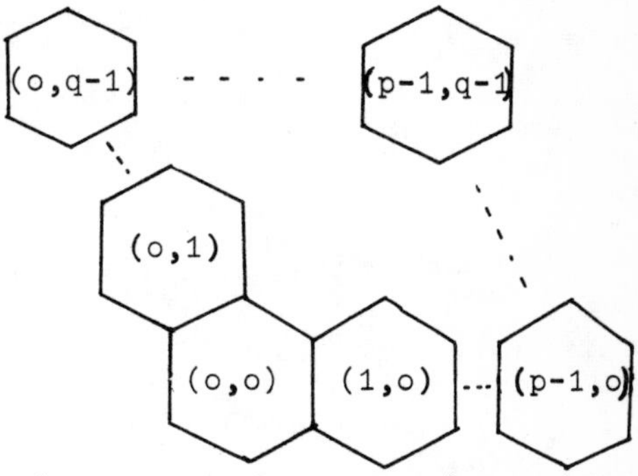

where $K(A)_{(i,j)}$ are glued together as indicated in the figure and through the isomorphisms

$$\begin{array}{ccccc} W(-1,o)_{(o,j)} & \xrightarrow{\sim} & W(1,o)_{(p-1,j)} & \text{for} & o \le j \le q-1 \ , \\ \cup & & \cup & & \\ (w)_{(1.o)} & \longrightarrow & (\lambda w)_{(1,o)} & & \end{array}$$

$$\begin{array}{ccccc} W(-1,-1)_{(o,j)} & \xrightarrow{\sim} & W(1,1)_{(p-1,j)} & \text{for} & o \le j \le q-1 \ , \\ \cup & & \cup & & \\ (w)_{(-1,-1)} & & (w)_{(1,1)} & & \end{array}$$

$$\begin{array}{ccccc} W(o,-1)_{(i,o)} & \xrightarrow{\sim} & W(o,1)_{(i,q-1)} & \text{for} & o \le i \le p-1 \ , \\ \cup & & \cup & & \\ (w)_{(o,-1)} & \longrightarrow & (\lambda w)_{(o,1)} & & \end{array}$$

$$\begin{array}{ccccc} W(-1,-1)_{(i,o)} & \xrightarrow{\sim} & W(1,1)_{(i,q-1)} & \text{for} & o \le i \le p-1 \ . \\ \cup & & \cup & & \\ (w)_{(-1,-1)} & \longrightarrow & (w)_{(1,1)} & & \end{array}$$

Now we consider Problem II_s i).

To the line bundle L' on A' corresponds a theta divisor Θ'

on A', defined by

$$\vartheta(\tilde{t},\zeta) = \sum_{m\in\mathbf{Z}^2} e\left(\frac{1}{2}\,{}^t m\,\tilde{T}(\tilde{t})m + {}^t m(\zeta + b\tilde{t})\right) = 0$$

where $e(\ ) = \exp(2\pi\sqrt{-1}(\ \ ))$ and $b = {}^t(p/2, q/2)$. Then the closure of Θ' in $\bar{A}$ becomes a divisor Θ whose fibre over o can be described in each component $K(A)_{(i,j)}$ in the form

$$K(A)_{(o,o)} \ :\ (w_1^{-1}+1)(w_2^{-1}+1) = \begin{cases} o & \text{if } \lambda = 1 \\ 1 & \text{if } \lambda \neq 1, \end{cases}$$

$$K(A)_{(i,o)} \ :\ w_2^{-1} + 1 = o ,$$

$$K(A)_{(o,j)} \ :\ w_1^{-1} + 1 = o ,$$

otherwise $K(A)_{(i,j)} \cap \Theta = \emptyset$.

This Θ determines a line bundle L which is an extension of L' .

For an integral vector $a = {}^t(i,j)$, $o \leq i \leq p-1$, $o \leq j \leq q-1$, let $\Theta'^{(a)}$ be the theta divisor on A' defined by

$$\vartheta^{(a)}(\tilde{t},\zeta) = \sum_{m\in\mathbf{Z}^2} e\left(\frac{1}{2}\,{}^t m\,\tilde{T}(\tilde{t})\,m + {}^t m(\zeta+(a+b)\tilde{t})\right) = o .$$

Then $\Theta'^{(a)}$ defines the same line bundle L' on A' , but the closure $\Theta^{(a)}$ of $\Theta'^{(a)}$ in $\bar{A}$ is subject to

$$\Theta^{(a)} \cap K(A)_{(k,l)} = \Theta \cap K(A)_{(k+i,l+j)}$$

(Consider $k+i$ and $l+j$ modulo p and q respectively.)
Hence the extensions $L^{(a)}$ of L' defined by $\Theta^{(a)}$ are different.

$\otimes L^{(a)}$ is not yet relatively ample, because no $\Theta^{(a)}$ hits the components $W(1,1)_{(k,l)}$ and $W(-1,-1)_{(k,l)}$ generally. Now we contract these $W(1,1)_{(k,l)}$ and $W(-1,-1)_{(k,l)}$ to points to obtain a new model $\bar{A}^{(1)}$, where $W(1,1)$ and $W(-1,-1)$ correspond to ordinary double points. The bundle induced by $\bigotimes_a L^{(a)}$ is then relatively ample on $\bar{A}^{(1)}$.

Hence <u>concerning Problem</u> II_s i), $\bar{A}^{(1)}$ <u>seems to be a better model</u>(*).

With respect to Problem II_s ii) we come to the same conclusion. An automorphism σ on A' which induces the monodromy $\begin{pmatrix} 0 & 1 & 0 & 0 \\ -1 & 0 & 0 & 0 \\ 0 & 0 & 0 & 1 \\ 0 & 0 & -1 & 0 \end{pmatrix}$ cannot be extended to $\bar{A}$, but can be to $\bar{A}^{(1)}$. This comes from a general fact that Nakamura's compactification depends on the choice of representatives of monodromy, namely, the matrix B_o in (3.1) (which can move under transformations $B_o \longrightarrow {}^tUB_oU$ for $U \in GL(g'', \mathbb{Z})$) . If we don't choose B_o suitably, the polarizing divisor does not necessarily extend. In our case there are essentially two kinds of "good" representatives, hence two kinds of models(**). The above automorphism σ extends <u>to an isomorphism between these two models</u>

<u>Example</u> (5.3)

$$\tilde{\tilde{T}}(\tilde{t}) = \begin{pmatrix} p+r & r \\ r & q+r \end{pmatrix} \tilde{t} + (S_{ij}(t)) , \quad p \geq q \geq r > o .$$

(*) This phenomenon is related with Mumford's Proposition (6.7) in [11] .

(**) Another "good" model can be obtained easily using $\bar{A}^{(1)}$. It is well known that the ordinary double points in $\mathbb{C}^3$ have two kinds of resolutions which replace the double point by $\mathbb{P}_1$. (The blowing-up at the point has $\mathbb{P}_1 \times \mathbb{P}_1$ as its inverse image and both fibrations can be contracted.) Hence the other resolution of $\bar{A}^{(1)}$ is the desired one.

A_o^o and $K(A)$ are as in (5.2). Then the singular fibre A_o is a union of $K(A)_{(i,j)}$ naturally patched together according as the following figure. ("Naturally" means that the boundaries are identified without twisting.)

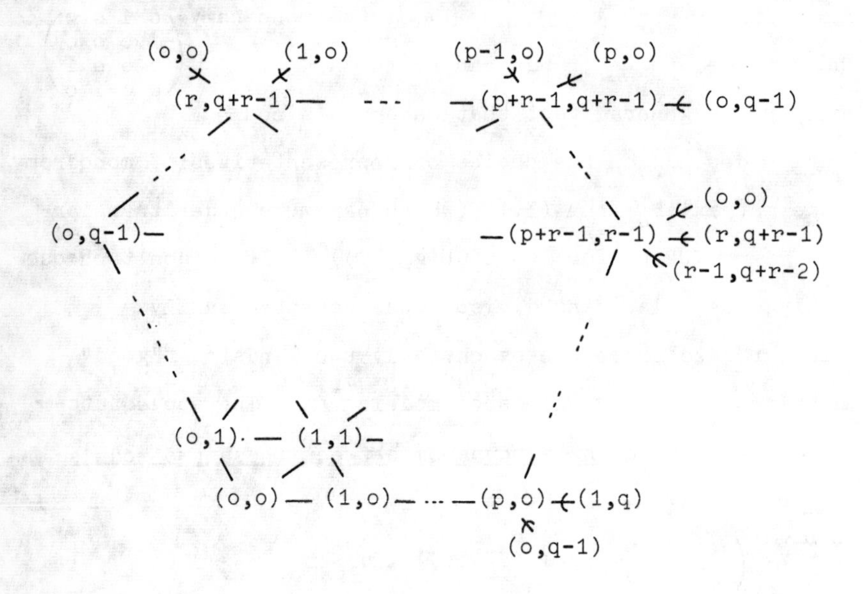

In a similar way as before we can define a divisor Θ on A whose fibre over o can be expressed in the form ;

$\Theta \cap K(A)_{(o,o)} : w_1^{-1} + w_2^{-1} + 1 = o,$

(which hits $W(1,1)$, $W(-1,o)$, $W(o,-1)$ once for each)

$\Theta \cap K(A)_{(r,r)} : w_1 + w_2 + 1 = o,$

$\Theta \cap K(A)_{(i,i)} : w_1 + w_2 = o ,$

$(o < i < r)$

$$\Theta \cap K(A)_{(i,r)} \ : \ w_2 + 1 = o \ , \quad (r < i < p+r)$$

$$\Theta \cap K(A)_{(r,j)} \ : \ w_1 + 1 = o \ , \quad (r < j < q+r)$$

otherwise $\Theta \cap K(A)_{(i,j)} = \emptyset$.

We can see that this model works well for both Problems II_s i) and ii) .

We note that by virtue of the theory of quadratic forms these examples cover all the possible cases of $g = 2$ (cf.[15]) .

Addendum

The author has found very recently another good candidate of singular fibres in stable families of abelian varieties. It fits to Problem II_s i). A remarkable fact is that all these fibres form a global family over a new compactification of $\mathfrak{S}_g/\Gamma_g(n)$ ($n \geq 3$), which Mumford named as Delony compactification. However this singular fibre is not a compactification of our "Neron" model. Nakamura has already constructed this family in case of $g = 2$ and 3 over the Igusa compactification (cf. [13]) which coincides with the Delony compactification in these cases, and has shown that it has a universal property on deformation. The detail will be discussed elsewhere.

Chapter III. General methods to study degeneration

One of key steps of our study was to reduce the problem to the "stable" case. This method has two generalisations corresponding to Theorems (3.3) and (3.7).

Theorem 1. (Quasi-unipotentness theorem, or monodromy theorem.)

Consider a proper algebraic family $\pi : X \longrightarrow D$ over a disc D which is smooth over $D' = D - \{o\}$. Then the monodromy M_q on $H^q(X_t, \mathbb{C})$, $t \in D'$, is subject to

$$(M_q^N - 1)^{q+1} = o$$

for an $N > o$.

For references see Schmid's article in this volume or [5].

Theorem 2. (Semi-stable reduction theorem) ([8]).

Consider the same algebraic family $\pi : X \longrightarrow D$ as above. For $n > o$ let $\mu : E \longrightarrow D$ be a map of discs sending s to s^n. Then for a suitable $n > o$ there is an algebraic family $\rho : Y \longrightarrow E$ which is bimeromorphic to $\mu^* \pi : X \times_D E \longrightarrow E$ through an epimorphism $p : Y \longrightarrow X \times_D E$ which is an isomorphism over E', and whose fibre over the origin is reduced and with non-singular components crossing normally.

Next step was to study the degeneration of "stable" families. It was done so far by studying the behaviour of the "period" map near the discriminant. In this direction Schmid's result (cf. his article in this volume) is the first important step. Also we should make use of the theory of mixed Hodge structures due to Deligne ([3]).

However, the author is not acquainted with these topics, so more developed explanation will be left to experts in them.

However, for example, by virtue of Pjatetsuki-Šapiro and Šafarevič's work [19] we can expect to attack

Problem 3. Study the degeneration of K3 surfaces.

"Now we see only puzzling reflections in a mirror, our knowledge is partial. But then it will be whole, like God's knowledge of me".

REFERENCES

[1] M. Artin and G. Winters: Degenerate fibres and stable reduction of curves, Topology, Vol.10 (1971), 373 - 383.

[2] C.H. Clemens et al.: Seminar on degeneration of algebraic varieties, Institute for Advanced Study, Princeton, 1969 - 1970.

[3] P. Deligne: Théorie de Hodge, II, Publ. math. IHES, t.40(1972), 5-57; III, mimeographed notes, IHES, 1972.

[4] P. Deligne and D. Mumford; The irreducibility of the space of curves given genus, Publ. math. IHES, t. 36 (1969), 75 - 110.

[5] P.A. Griffiths: Periods of integrals on algebraic manifolds: summary of main results and discussion of open problems, Bull. AMS, Vol. 76 (1970, 228 - 296.

[6] I. Igusa: On Siegel modular forms of genus two. I. Amer.Math., Vol. 84(1962), 175 - 200.

[7] S. Iitaka: On the degenerates of a normally polarized abelian variety of dimension 2 and an algebraic curve of genus 2, (in Japanese), Master degree thesis, University of Tokyo, 1967.

[8] G. Kempf et al.: Troidal embeddings I, Lecture Notes in Math., Springer, 1973.

[9] K. Kodaira: On compact analytic surfaces, II - III, Ann. of Math., Vols. 77 and 78 (1963), 563 - 626 and 1-40.

[10] D. Mumford: An analytic construction of degenerating curves over complete local rings, Compositio Math., Vol. 26 (1972, 129 - 174

[11] D. Mumford: An analytic construction of degenerating abelian varieties over complete rings, Compositio Math., Vol. 24 (1972, 239 - 272.

[12] I. Nakamura: On degeneration of abelian varieties,to appear.

[13] Y. Namikawa: On the canonical holomorphic map from the moduli space of stable curves to the Igusa monoidal transform, Nagoya Math.J., Vol. 52 (1973), 197 - 259.

[14] Y. Namikawa: On families of curves of genus $g > 2$, to appear.

[15] Y. Namikawa and K. Ueno: The complete classification of fibres in pencils of curves of genus two, Manuscripta Math., Vol. 9 (1973, 163 - 186.

[16] Y. Namikawa and K. Ueno: On fibres in families of curves of genus two. I., Number theory, algebraic geometry, and commutative algebra, in honor of Y. Akizuki, Kinokuniya, Tokyo, 1973, 297 - 371; II., to appear.

[17] A.Néron: Modèles minimaux des variétés abéliennes sur les corps locaux et globaux, Publ. math. IHES, t.21(1964, 5 - 128.

[18] A.P. Ogg: On pencils of curves of genus two, Topology, Vol.5 (1966), 355 - 362.

[19] I.I. Pjateckii - Šapiro and I.R. Šafarevič: A Torelli theorem for algebraic surfaces of type K3, Izv. Akad. Nauk SSSR, Tom 35 (1971);English translation, Math. USSR Izvestija, Vol. 5 (1971), 547 - 588.

[20] H. Popp: On moduli of algebraic varieties, II, Compositio Math., to appear.

[21] J.P. Serre: Groupes algébriques et corps de classes, Hermann, Paris, 1959.

[22] C.S. Seshadri and T. Oda: Compactification of the generalized Jacobian variety, to appear.

[23] K. Ueno: On fibre spaces of normally polarized abelian varieties of dimension 2, I - II, J. Fac. Sc. Univ. Tokyo, Vols. 18 and 19 (1971 - 1972), 33 - 95 and 163 - 199.

[24] E. Viehweg: Invarianten degenerierter Fasern in lokalen Familien von Kurven, to appear.

[25] G.B. Winters: On the existence of certain families of curves, to appear.

HYPERELLIPTIC CURVES OVER NUMBER FIELDS

Frans OORT (Amsterdam)

In this note we give a proof for the Shafarevich-Parshin theorem (cf. [11]; [10], 1, 4; [8], p. 79), which states that there are only a finite number of hyperelliptic curves of given genus over an algebraic number field having smooth reduction outside a given finite set of discrete valuations; this proof is not very much different from the one given by Parshin.

I thank H.W. Lenstra jr. for drawing my attention to [1], and I thank K. Lønsted for conversation on this topic.

1. Sums of powers of prime numbers.

Consider a finite number of (rational) prime numbers, e.g. $S = \{2, 3, 5\}$ and try to solve the equation

$$2^x \pm 3^y \pm 5^z = 0 ;$$

the number of such solutions turns out to be finite. More generally: let $S = \{p_1, \ldots, p_n\}$, $A \in \mathbb{Z}^n$, $A = (a_1, \ldots, a_n)$, then we write symbolically:

$$p^A = p_1^{a_1} \times \ldots \times p_n^{a_n} .$$

With these notations, the finiteness statement can be formulated:

Theorem (1.1).

$$|\{(A,B) \mid A,B \in \mathbb{Z}^n,\ p^A \pm p^B \pm 1 = 0\}| < \infty .$$

In case $n = 2$, the proof of this fact is not difficult, however for $n \geq 3$ it seems a deep fact. It follows from a more general statement: let L be an algebraic number field, i.e. $[L : \mathbb{Q}] < \infty$, and let T be a finite set of discrete valuations on L; denote by $\mathcal{O}_{(T)}$ the ring of elements of L integral outside, T, i.e.

$$\mathcal{O}_{(T)} = \bigcap_{v \notin T} \mathcal{O}_v$$

(if $T = \{2, 3, 5\}$, $L = \mathbb{Q}$, then $\mathcal{O}_{(T)} = \mathbb{Z}[\frac{1}{30}]$). We denote:

$$J_{L,T} := \{\lambda \mid \lambda \in \mathcal{O}^*_{(T)} \text{ and } (\lambda - 1) \in \mathcal{O}^*_{(T)}\}$$

where the star denotes the group of units. The following theorem seems to be known under the name "conjecture of Julia Robinson" (cf. [1]):

Theorem (1.2). (cf. [4], VII.4). Let $[L : \mathbb{Q}] < \infty$ and $|T| < \infty$ as above, then

$$|J_{L,T}| < \infty .$$

Note that (1.2) implies (1.1) : if $\pm p^A - 1 = \pm p^B$, then $\lambda := \pm p^A$ is in $O^*_{(T)}$ and also $(\lambda-1) \in O^*_{(T)}$ thus any solution (A,B) yields an element in $J_{\mathbb{Q},T}$. Note that we allow T to be non-empty; it is not so clear that the methods as in [1] can be used to prove (1.2) in case $|T| \geq 3$. However a proof of (1.2) is known, using the Siegel-Mahler theorem. Note that (1.2) can be generalized to the case L is finitely generated over $\mathbb{Q}$, and $O_{(T)}$ replaced by a subring R of L which is of finite type over $\mathbb{Z}$, and which has a finitely generated group R^* of units (cf. [4], VII.2 Th. 4 and page 134).

Sometimes one can give a bound for $|J_{L,T}|$, e.g. if $L = \mathbb{Q}, T = \{p,q\}$ using methods of Baker and Coates one can show that $p < q$, $p^a + 1 = q^b$, $3\alpha \leq a \leq 3\alpha + 2$, $3\beta \leq b \leq 3\beta + 2$, then

$$p^\alpha, q^\beta < \exp \exp \exp (2q^2)10^{3^{10}},$$

which however clearly is much too big: $\alpha \leq 1$ and $\beta \leq 1$.

2. Properties of good reduction.

Let R be a Dedekind domain, K its field of fractions, and C a complete, smooth, absolutely irreducible algebraic curve over K.

Definition. We say C has good and irreducible reduction, or smooth reduction, at all places of R if there exists a smooth curve $\underline{C} \to \mathrm{Spec}(R)$ with $\underline{C} \otimes_R K \cong C$.

Note that in this case the geometric fibres are irreducible (being connected by Zariski's connectedness theorem).

Remark. One can define C to have good reduction whenever its Jacobian variety $J = \mathrm{Jac}(C)$ has good reduction, i.e. if there exists an abelian scheme $\underline{J} \to \mathrm{Spec}(R)$ with $\underline{J} \otimes_R K \cong J$. If this is the case, and $C(K) \neq \emptyset$, the flat extension of $C \hookrightarrow J$ into $\underline{J}$ is a curve $\underline{C} \to \mathrm{Spec}\ R$, $\underline{C} \subset \underline{J}$,

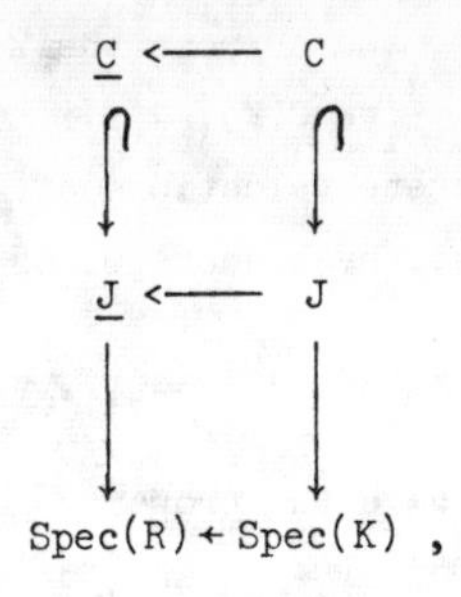

and C has good and irreducible reduction if and only if the fibres of $\underline{C} \to \mathrm{Spec}(R)$ are smooth.

Definition. A complete, smooth, absolutely irreducible algebraic curve C of genus $g \geq 2$ is called hyperelliptic if there exists a two-fold covering $C \otimes_K k \to \mathbb{P}^1_k$, where k is some field containing K; the ramification points of this covering are called the (hyperelliptic) Weierstrass points of the hyperelliptic curve C.

Suppose $\mathrm{char}(K) \neq 2$, and let k be an algebraically closed field containing K, and let $P_1, \ldots, P_w$ be the ramification points of the two-fold covering $C \otimes k \to \mathbb{P}^1_k$; then $|2P|$ is a non-trivial linear series if $P = P_i$, and the gap sequence (cf. [2], p. 216) at P is $(1, 3, \ldots, 2g-1)$. Because the degree of the covering is not divisable by $\mathrm{char}(k)$, there is no wild ramification, the Zeuthen-Hurwitz formula (e.g. cf. [2], p. 215) can be applied and we conclude $w = 2g + 2$; moreover

$$\sum_{j=1}^{g} (n_j(P)-j) = 0 + 1 + \ldots + (g-1) = \frac{1}{2} g(g-1),$$

and because

$$\sum_P \sum_j (n_j(P) - j) = (g-1)g(g+1)$$

(cf. [2], p. 217), we conclude that a hyperelliptic curve C has no other Weierstrass points except $P_1, \ldots, P_{2g+2}$. From this it follows that the points P_i, $1 \leq i \leq 2g + 2$, are rational over an algebraic extension of K, the two-fold covering $C \otimes k \to \mathbb{P}^1_k$ is essentially unique, and if K is perfect the divisor $D = P_1 + \ldots + P_{2g+2}$ is rational over K because for any $\sigma \in \mathrm{Gal}(\overline{K}/K)$ we see $\sigma D = D$.

Lemma (2.1). Let R be a Dedekind domain, suppose its class number is one (i.e. R is a principal ideal domain), and suppose all residue characteristics to be different from two. Let C be a hyperelliptic curve over the field of fractions K of R, and suppose the Weierstrass points of C to be rational over K, i.e. $P_i \in C(K)$, $1 \le i \le 2g+2$. Suppose C has smooth reduction for all discrete valuations of R. Then one can choose coordinates on $\mathbb{P}^1_K$ such that the branch points

$$\pi(P_1) = e_1, \ldots, \pi(P_{2g+1}) = e_{2g+1}, \pi(P_{2g+2}) = \infty$$

of the covering $\pi : C \to \mathbb{P}^1_K$ have the property $e_i \in R$, $1 \le i \le 2g+1$, and e_i not congruent with e_j for $i \ne j$ and for any discrete valuation of R.

__Proof__: Let $J = \mathrm{Jac}(C)$, and choose $C \hookrightarrow J$ such that one of the points P_i is mapped onto $0 \in J$; then the involution $\tau : C \to C$ which interchanges the sheets of π is induced by the map $-\mathrm{id} : J \to J$, and the Weierstrass points $P_1, \ldots, P_{2g+2}$ are mapped onto points of order two of J, or on $0 \in J$. Because C has good reduction, J extends to an abelian scheme $\underline{J} \to \mathrm{Spec}(R)$, and τ extends to involution $\tau : \underline{C} \to \underline{C}$ as the restriction of $-\mathrm{id} : \underline{J} \to \underline{J}$ (with $\underline{C}$ as in the remark above). Because no residue characteristic of R equals two, the group scheme ${}_2\underline{J}$ of 2-torsion points on $\underline{J}$ is etale over $\mathrm{Spec}(R)$, and the same holds for the flat extension $\underline{P}_1 \cup \ldots \cup \underline{P}_{2g+2}$ of $P_1 \cup \ldots \cup P_{2g+2}$ inside $\underline{C}$; thus for any $x \in \mathrm{Spec}(R)$, the fibre $C_x = \underline{C} \otimes k(x)$ is a hyperelliptic curve : $\underline{C}$ has absolutely irreducible fibres over $\mathrm{Spec}(R)$, and $\tau | C_x$ is an involution with $2g+2$ fixed points, and $\mathrm{genus}(C_x) = g$, thus $C_x/\tau \cong \mathbb{P}^1_{k(x)}$. Let P be one of the points P_i, say $P = P_{2g+2}$, and $\underline{P} \subset \underline{C}$ its flat extension. The divisorial sheaf $\underline{O}(2\underline{P})$ on $\underline{C}$ has the property that for any $x \in \mathrm{Spec}(R)$,

$$\dim_{k(x)} \Gamma(C_x, \underline{O}(2\underline{P}) \otimes k(x)) = 2$$

(because P_x is a Weierstrass point on C_x), thus $F := \Gamma(\underline{C}, \underline{O}(2\underline{P}))$ is a projective module of rank 2 over $\Gamma(\mathrm{Spec}\, R, \underline{O}) = R$; because R has class number equal to one, F is free over R, and we see there exists $f \in F$ with $F = R.1 \oplus R.f$. The function f defines a morphism $f : \underline{C} \to \mathbb{P}^1_{\mathrm{Spec}(R)}$, which makes every fibre C_x of $\underline{C} \to \mathrm{Spec}(R)$ a hyperelliptic curve (this is because the linear series $|2P|$ is the canonical map $D \to \mathbb{P}^1$ if P is a Weierstrass point on a hyperelliptic curve D); the sections

$$\underline{P}_1, \ldots, \underline{P}_{2g+2} : \mathrm{Spec}(R) \to \underline{C}$$

yield sections

$$E_i := f.\underline{P}_i : \mathrm{Spec}(R) \to \mathbb{P}^1_{\mathrm{Spec}(R)}$$

and because $P_i \otimes k(x)$ are the $2g+2$ Weierstrass points of C_x (which are distinct), the proof of the lemma is concluded.

Suppose K is a perfect field with $\mathrm{char}(K) \neq 2$, and let C be a hyperelliptic curve over K. The divisor $P_1 + \ldots + P_{2g+2}$ of all Weierstrass points determines a K-rational reduced subscheme which we denote by $W(C) \subset C$. Let

$$L := K(W(C))$$

be the smallest extension of K over which the points P_i are rational, i.e. let H be the subgroup of $G = \mathrm{Gal}(\overline{K}/K)$ consisting of those elements which operate trivially on $W(C) = \{P_1, \ldots, P_{2g+2}\}$, then $L = \overline{K}^H$.

Lemma (2.2). Let C be a hyperelliptic curve over K, with $[K : \mathbb{Q}] < \infty$, let $L = K(W(C))$, let v be a discrete valuation of K which does not divide 2, and suppose C has smooth reduction at v. Then L/K is unramified at all places of L dividing v.

Proof: Because C has smooth reduction at v, its $J = \mathrm{Jac}(C)$ can be extended to an abelian scheme $\underline{J} \to \mathrm{Spec}(R_v)$; let w be a discrete valuation of L dividing v, and $\overline{w}$ an extension to $\overline{L}$; let $\underline{C} \to \mathrm{Spec}(R_v)$ be a smooth curve extending C, and let $\underline{W}$ be the flat extension of $W(C) \subset C$. Note that $\underline{W}$ is etale over $\mathrm{Spec}(R_v)$: because the points P_i are rational over L, we can choose an embedding $C \otimes L \hookrightarrow J \otimes L$ which identifies $W(C) \otimes L$ and the scheme ${}_2J \otimes L$ of points of order two; uniqueness of minimal models ensures that $C \otimes L \hookrightarrow J \otimes L$ extends to an embedding $\underline{C} \otimes R_w \hookrightarrow \underline{J} \otimes R_w$; thus $\underline{W} \otimes R_w$ and ${}_2\underline{J} \otimes R_w$ are isomorphic; because v does not divide 2, the group scheme ${}_2\underline{J}$ is etale over $\mathrm{Spec}(R_v)$ and we conclude $\underline{W}$ to be etale. Let $I(\overline{w})$ be the inertia group; it acts trivially on $\underline{W} \otimes k(w)$, thus it acts trivially on $W(C)$ because $\underline{W}$ is etale over $\mathrm{Spec}(R_v)$, thus $I(\overline{w}) \subset H$, i.e. L/K is unramified at w, and the lemma is proved.

Remark: The lemma (2.2) stays correct if we only assume C has good reduction at v (instead of smooth reduction at v).

3. Finiteness.

For $[K : \mathbb{Q}] < \infty$, S a set of discrete valuations of K, and g an integer, $g \geq 2$, we denote by $\mathrm{Sh}_{g,K,S}$ the set of K-isomorphism classes of (complete, smooth, absolutely irreducible) hyperelliptic algebraic curves over K which have smooth reduction outside S. In case $g = 1$, we write $Sh_{1,K,S}$ for the set of K-isomorphism classes of abelian curves (elliptic curves with at least one K-rational point), etc.

Theorem (3.1). If $[K : \mathbb{Q}] < \infty$ and $|S| < \infty$, $g \geq 1$, then

$$|Sh_{g,K,S}| < \infty .$$

Note that in case $g = 1$ the theorem is due to Shafarevich (cf [10], IV.1.4); a proof for $g = 2$, and a remark for the case $g > 2$ can be found in [8], end of section 1. A general philosophy (cf. [11]) suggests that (3.1) is a very particular case of a much more general analogon to a theorem by Hermite: fix discrete invariants for a function field Γ over an algebraic number field K (e.g. transcendence degree, genus, etc.), insist on properties of "good reduction", corresponding to properties of L/K being unramified in case $[\Gamma = L : K] < \infty$, and try to show finiteness of the numbre of K-isomorphism classes of such objects (cf [11], section 4). However even for function fields in one variable over algebraic number fields such theorems seem to be unproven, e.g. the case of (3.1) with $g \geq 3$, and the word "hyperelliptic" in the definition of Sh omitted has not yet been established.

Proof. For any element C of $Sh_{1,K,S}$ we choose one point $0 \in C(K)$, and denote by $W(C)$ the two-torsion points on the abelian curve C with 0 as zero element. Fix $g \geq 1$, and let L/K be the smallest Galois extension containing $K(W(C))$ for all $C \in Sh_{g,K,S}$; we claim $[L : K] < \infty$. In fact L is the compositum of extensions of degree at most $2g + 2$ which are unramified outside $S' = S \cup \{\text{all primes dividing } 2\}$: for $g \geq 2$ this follows from (2.2), and for $g = 1$ the same arguments as in the proof of (2.2) hold. Thus by the theorem of Hermite (cf [3], p. 595), we conclude $[L : K] < \infty$. Choose a finite set T of discrete valuations of L such that all primes dividing primes in S, and all primes dividing 2 are in T and such that $\mathcal{O}_{(T)}$ has class number one (this is possible : let S'' be the set of all extensions of elements of S' to L ; the class number of $\mathcal{O}_{(S'')}$ is finite, consider a finite set of ideals representing the class group of $\mathcal{O}_{(S'')}$ and obtain T from S'' by adding all primes dividing these ideals). We denote by $\underline{L}_{g,L,T}$ the set of elements of $Sh_{g,L,T}$ which can be given over L by an equation

$$Y^2 = (X - e_1) \times \ldots \times (X - e_{2g+1}), \; e_i \in \mathcal{O}_{(T)}.$$

Note that L and T are constructed in such a way that $\alpha(C) := C \otimes_K L$,

$$\alpha : Sh_{g,K,S} \to \mathcal{L}_{g,L,T} ;$$

by the choice of L, for any $C \in Sh_{g,K,S}$ we have $W(C) \subset C(L)$, thus by lemma (2.1) the curve C can be choosen as a twofold covering of $\mathbb{P}^1$ branching at points e_i, $1 \le i \le 2g+1$ and at ∞. Thus such a curve C can be given by an equation

$$e\eta^2 = \prod_{i=1}^{2g+1} (\xi - e_i\zeta), \; e_i \in \mathcal{O}_{(T)},$$

and a substitution $\eta = e^gY$, $\xi = eX$, $\zeta = eZ$ yields the desired equation. Note that the fibres of α are finite sets: let D be a curve of genus $g \ge 2$, respectively an elliptic curve with $0 \in D(L)$ distinguished, over L, then $A = \mathrm{Aut}(D)$, resp. $A = \mathrm{Aut}(D,0)$, is finite; the fibre $\alpha^{-1}(D)$ corresponds bijectively with $H^1(\mathrm{Gal}(L/K),A)$ (cf [9] , III.1.3, Prop. 5), which certainly is a finite group, $[L : K]$ and A being finite. Thus in order to prove finiteness of $Sh_{g,K,S}$ it suffices to show the same for $\mathcal{L}_{g,L,T}$. For any $D \in \mathcal{L}_{g,L,T}$ choose en equation as indicated, i.e. choose an ordering for the set $W(D)$ and define

$$\beta : \mathcal{L}_{g,L,T} \to (J_{L,T})^{2g-1}$$

as follows: if $D \in \mathcal{L}_{g,L,T}$ and we have choosen $e_1, e_2, \ldots, e_{2g+1}, \infty$, then

$$\beta D = (\ldots, cr(e_i, e_2, \infty, e_1), \ldots), \; 3 \le i \le 2g+1,$$

where cr denotes the cross ratio; one can normalize such that $e_1 = 0$, $e_2 = 1$, and

$$\beta D = (e_3, \ldots, e_{2g+1}).$$

By lemma (2.1) in fact the coordinates of β land in $J_{L,T}$. Clearly β is injective, and application of (1.2) ends the proof of (3.1).

As Tate has proved, $Sh_{1,\mathbb{Q},\emptyset} = \emptyset$, i.e. every elliptic curve over $\mathbb{Q}$ has bad reduction somewhere (e.g. cf. [5] , pp. 144/145), and this would follow if a deep conjecture of Weil could be verified (cf. [6] , p. 205). Such general principles seem to be unknown, and in particular it seems unknown whether $Sh_{g,\mathbb{Q},\emptyset} = \emptyset$ for $g \ge 2$ (cf. [7] , C2).

References

[1] Chowla, S., Proof of a conjecture of Julia Robinson. Norske Vid. Selsk.Forh. (Trondheim) 34 (1961), 100-101.

[2] Fulton, W., Algebraic curves. Benjamin, 1969.

[3] Hasse, H., Zahlentheorie. Akad. Verlag, Berlin, 1963.

[4] Lang, S., Diophantine geometry. Intersc. Publ., 1962.

[5] Ogg, A.P., Abelian curves of 2-power conductor. Proc. Camb. Phil. Soc. 62 (1966), 143-148.

[6] Ogg, A. P., Abelian curves of small conductor. Journ. r. angew. Math. 226 (1967), 204-215.

[7] Parshin, A.N., Quelques conjectures de finitude en gémétrie diophantienne. Actes, Congrès intern. math., 1970, 1, 467-471.

[8] Parshin, A.N., Minimal models of curves of genus 2 and homomorphisms of abelian varieties defined over a field of finite characteristic. Izv. Akad. Nauk SSSR 36 (1972) (Math. USSR Izvestija, 6 (1972), 65 - 108).

[9] Serre, J.-P., Cohomologie Galoisienne. Lect. N. Math. 5, Springer Verlag, 1964.

[10] Serre, J.-P., Abelian l-adic representations and elliptic curves (McGill University lecture notes). Benjamin, 1968.

[11] Shafarevich, I.R., Algebraic number fields. Proc. ICM, Stockholm 1962, 163-176 (Amer. Math. Soc. Translat. 31 (1963),25 - 39).

Modulräume algebraischer Mannigfaltigkeiten

Herbert Popp

Für algebraische Mannigfaltigkeiten[1] X der Dimension n, definiert über dem komplexen Zahlkörper $\mathbb{C}$, ergibt sich mit Hilfe des Begriffs der Kodaira-Dimension $\kappa(X)$ und des Iitaka'schen Struktursatzes für m-kanonische Abbildungen (vgl. Ueno's Beitrag wegen der genauen Formulierung) die folgende Klasseneinteilung.

1. Algebraische Mannigfaltigkeiten X mit $\kappa(X) = \dim X$. Man nennt sie <u>Mannigfaltigkeiten allgemeinen Typs</u>.
2. Algebraische Mannigfaltigkeiten X mit $\dim X > \kappa(X) \geqslant 1$.
3. Algebraische Mannigfaltigkeiten X mit $\kappa(X) = 0$.
4. Algebraische Mannigfaltigkeiten X mit $\kappa(X) = -\infty$.

Es besteht die Aufgabe eine Übersicht über die einzelnen Klassen algebraischer Mannigfaltigkeiten zunächst im Sinne der birationalen Geometrie herzustellen.

Zur Untersuchung der Klasse 2. sind Faserraummethoden natürlich. Denn nach dem Iitaka'schen Satz gibt es zu jeder algebraischen Mannigfaltigkeit X dieser Klasse eine dazu birational äquivalente Mannigfaltigkeit X^*, die die Struktur eines Faserraumes $X^* \xrightarrow{f} W$ (eindeutig im birationalen Sinne) über einer algebraischen Mannigfaltigkeit W hat. Dabei ist $\dim W = \kappa(X)$ und die allgemeine Faser von f eine algebraische Mannigfaltigkeit mit Kodaira Dimension 0. Die Faserräume $X^* \longrightarrow W$ gilt es zu beschreiben, in Analogie zu den Ergebnissen von Kodaira [8], Kawai [7] und Ueno [20] über Faserräume des obigen Typs mit einer elliptischen Kurve als allgemeine Faser.

[1] Algebraische Mannigfaltigkeiten ohne nähere Spezifizierung sollen immer irreduzible, projektive und glatte $\mathbb{C}$-Schemata sein.

Dazu werden für algebraische Mannigfaltigkeiten der Dimension > 1 und Kodaira Dimension 0 universelle Familien mit kompakten Basisräumen benötigt, auf die sich die Beschreibung der Faserräume beziehen kann. (Vgl. in diesem Zusammenhang die Ausführungen über feine Modulräume auf Seite 15 ff.)
Leider kennt man solche universellen Familien nach dem jetzigen Stand nur für Kurven.

Für die Klasse 3. zeigt Ueno's Arbeit [20], dass diese mit Hilfe der Albanese Abbildungen weiter zu unterteilen ist. Iitaka hat über die Struktur der Albanese Abbildung α einer algebraischen Mannigfaltigkeit X mit Kodaira Dimension 0 die folgenden Vermutungen aufgestellt, die inzwischen von Ueno für Kummer Mannigfaltigkeiten in [20] bewiesen worden.

1. $\dim \mathrm{Alb}(X) = q \leq \dim X$ und α ist surjektiv. ($\mathrm{Alb}(X)$ = Albanese Torus von X.)
2. Die allgemeine Faser von α ist zusammenhängend und hat Kodaira-Dimension 0.

Falls diese Vermutungen auch allgemein richtig sind, können für die Mannigfaltigkeiten der Klasse 3., für welche die Irregularität q positiv, aber kleiner als die Dimension ist, erneut Faserraummethoden angewandt werden. Die noch übrigbleibenden Mannigfaltigkeiten der Klasse 3. sind dann gerade diejenigen mit Irregularität 0 oder Irregularität = Dimension. Letztere sind birational isomorph zu abelschen Mannigfaltigkeiten. Beide Typen aber können mit Hilfe der Riemannschen Theorie der Moduln behandelt werden, wie wir weiter unten ausführen.

Über die Klasse 4. ist fast nichts bekannt, wenn die Dimension der Mannigfaltigkeiten > 2 ist. Die Regelmannigfaltigkeiten gehören dazu und im Falle von Flächen besteht die Klasse 4. genau aus den Regelflächen.

Was ist jedoch mit den Mannigfaltigkeiten allgemeinen Typs ? Die Albanese Abbildung ist dort für eine weitere Unterteilung zunächst nicht zu gebrauchen, da die auftretenden Fasern und Basen keine Gemeinsamkeiten erkennen lassen. Aber es bietet sich die durch Riemann für Kurven bekannte Modultheorie an.

Riemann's Standpunkt ist der folgende: Um eine Übersicht über die Mannigfaltigkeiten allgemeinen Typs, die über $\mathbb{C}$ definiert sind, zu erhalten, sollte man versuchen die Menge der Isomorphieklassen dieser Mannigfaltigkeiten in natürlicher Weise zu einem geometrischen Objekt, d.h. analytischen Raum, algebraischen Raum, $\mathbb{C}$-Schema zu machen, je nachdem was möglich ist. Das Objekt zusammen mit seiner Geometrie ergibt die gewünschte Übersicht.

Diese Methode ist für Riemann'sche Flächen oder irreduzible, glatte, projektive Kurven eines festen Geschlechts g von Riemann in [18] formuliert worden.

Wir zeigen hier, dass für gewisse Typen algebraischer Mannigfaltigkeiten, wie z.B. Mannigfaltigkeiten mit ampler, kanonischer Garbe, oder Flächen allgemeinen Typs usw., in der Tat die Isomorphieklassen der über $\mathbb{C}$ definierten Mannigfaltigkeiten (des jeweiligen Typs) in natürlicher Weise zu einem algebraischen Raum von endlichem Typ über $\mathbb{C}$ gemacht werden können. Um dies zu präzisieren sind Vorbereitungen notwendig.

X sei eine algebraische Mannigfaltigkeit mit ampler kanonischer Garbe ω_X. Für eine natürliche Zahl $t \geqslant 1$ sei $h_X(t) = \chi(\omega_X^{\otimes t})$ die Euler Charakteristik der Garbe $\omega_X^{\otimes t}$. Dann ist $h_X(t)$ ein Polynom in t, welches man das <u>Hilbertpolynom</u> von X nennt.

$\mathfrak{M}_h$ sei die Menge der Isomorphieklassen über $\mathbb{C}$ definierter algebraischer Mannigfaltigkeiten mit ampler kanonischer Garbe und

Hilbertpolynom h[2]. Riemann's Modulproblem kann wie folgt formuliert werden, falls man wie wir in der Kategorie der algebraischen Räume arbeitet. (Vgl. S.14.)

Finde einen algebraischen Raum M_h von endlichem Typ über $\mathbb{C}$ mit folgenden Eigenschaften.

1. Es gibt eine 1-1 deutige Abbildung φ der Menge $\mathfrak{M}_h$ auf die $\mathbb{C}$-wertigen Punkte von M_h, so dass für Familien die Bedingung 2. erfüllt ist.

2. Zu jeder eigentlichen, glatten Familie $V \xrightarrow{\rho} S$ von Mannigfaltigkeiten aus $\mathfrak{M}_h$ mit reduzierter Basis S(V,S sind algebraische Räume von endlichem Typ über $\mathbb{C}$; ρ ist eine eigentliche und flache Abbildung, die geometrischen Fasern von ρ sind Mannigfaltigkeiten aus $\mathfrak{M}_h$) ist die im folgenden definierte Abbildung $f: S \longrightarrow M_h$ ein Morphismus algebraischer Räume. Definition von f: Sei $P \in S$ ein $\mathbb{C}$-wertiger Punkt und V_P die Faser über P in $V \longrightarrow S$. V_P bestimmt eine Klasse in $\mathfrak{M}_h$ und bezüglich φ einen $\mathbb{C}$-wertigen Punkt $\varphi(V_P)$ von M_h. Wir setzen $f(P) = \varphi(V_P)$.

<u>Definition 1.</u> Ein algebraischer Raum mit diesen Eigenschaften heisst <u>Modulraum</u> für $\mathfrak{M}_h$. Ist X eine Mannigfaltigkeit einer Isomorphieklasse $\tilde{X}$ aus $\mathfrak{M}_h$, so heisst $\varphi(\tilde{X}) \in M_h$ <u>Modulpunkt</u> von X.

Diese "naive" Formulierung des Riemann'schen Modulproblems ist zu vage. Z.B. ist nicht klar ob der Modulraum M_h, falls er existiert, eindeutig bestimmt ist. Es gibt eine funktorielle präzise Formulierung des Problems in Mumford's Buch [12].

Darauf gehen wir kurz ein.

[2] Für Kurven hängt h(x) nur von dem geometrischen Geschlecht ab. Bei Flächen allgemeinen Typs ist h durch die Selbstschnittzahl K^2 der kanonischen Garbe und durch das arithmetische Geschlecht p_a bestimmt.

X sei eine algebraische Mannigfaltigkeit aus $\mathfrak{M}_h$ [3]. Matsusaka [10] hat gezeigt, dass es eine Zahl $c > 0$ gibt, die nur von h abhängt und für die gilt: die Garbe $\omega_X^{\otimes t}$ ist sehr ample und die Kohomologiegruppen $H^i(X,\omega_X^{\otimes t})$ sind 0, für alle $i \geqslant 1$, $t \geqslant c$. Dies impliziert

$$\chi(\omega_X^{\otimes t}) = \dim H^o(X,\omega_X^{\otimes t}) = h(t), \text{ für } t \geqslant c.$$

Nun sei $t \geqslant c$ fest gewählt.

$$\phi_t : X \longrightarrow P^N$$
$$P \longrightarrow (f_o(P),\ldots,f_N(P))$$

sei der durch die Basis $f_o,\ldots,f_N$ von $H^o(X,\omega_X^{\otimes t})$ bestimmte Morphismus.

Wir nennen die Mannigfaltigkeit $\phi_t(X) \subset P^N$ eine <u>t-kanonische Einbettung</u> von $X \in \mathfrak{M}_h$. Für eine Mannigfaltigkeit X ist $\phi_t(X)$ bis auf eine projektive Transformation eindeutig bestimmt. Die Mannigfaltigkeiten $\phi_t(X) \subset P^N$, $X \in \mathfrak{M}_h$ haben alle $g(x) = h(t\cdot x)$ als Hilbertpolynom und liegen in demselben P^N. Dies legt es nahe, das Hilbertschema $H^g_{P^N}$ zusammen mit der universellen Familie $\Gamma \longrightarrow H^g_{P^N}$ zu betrachten, welche die eigentlichen, flachen Familien V/S des P^N/S mit g als Hilbertpolynom parametrisiert. Ein Standardargument (vgl. [12] ,S.71) zeigt: Es gibt ein lokal abgeschlossenes Teilschema H von $H^g_{P^N}$, so dass die Pullback-Familie $\Gamma_H \longrightarrow H$ von $\Gamma \longrightarrow H^g_{P^N}$ nach H universell ist bezüglich eigentlicher, glatter Familien $V \longrightarrow S$ des P^N/S mit g als Hilbertpolynom, die t-kanonisch eingebettet sind.

$\Gamma_H \longrightarrow H$ nennen wir die <u>universelle Familie t-kanonisch eingebetteter Mannigfaltigkeiten</u> aus $\mathfrak{M}_h$.

Der Grauert'sche Kohärenzsatz impliziert, dass jede eigentliche, glatte Familie algebraischer Mannigfaltigkeiten $V \longrightarrow S$ aus $\mathfrak{M}_h$ über

[3] Präzise müsste es heissen, X sei eine Mannigfaltigkeit einer Isomorphieklasse aus $\mathfrak{M}_h$. Wir sagen dafür im folgenden kurz, X sei aus $\mathfrak{M}_h$.

einem noetherschen algebraischen $\mathbb{C}$-Raum S lokal Pullback von $\Gamma_H \longrightarrow H$ ist. Weiter ergibt sich aus den universellen Eigenschaften von H^g_pN, dass PGL(N) auf H^g_{pN} und damit auch auf H operiert.

Proposition 1. (Vgl. [12], S. 101 oder [15], S. 28.) Sei S ein noethersches $\mathbb{C}$-Schema. V_1/S und V_2/S seien t-kanonisch eingebettete Familien des P^N/S, welche Pullbacks von $\Gamma_H \longrightarrow H$ bezüglich der Morphismen $f_i : S \longrightarrow H$, $i = 1,2$, sind. Dann sind V_1/S und V_2/S genau dann als Familien isomorph, wenn die S-wertigen Punkte f_1 und f_2 von H in derselben Bahn bezüglich PGL(N) liegen.

Diese Proposition führt dazu den Quotientenfunktor

$$\mathcal{H}(S) = H(S)/PGL(N)(S), \text{ S ein noethersches } \mathbb{C}\text{-Schema,}$$

zu betrachten, wobei $\mathcal{H}(S)$ die Menge der Isomorphieklassen eigentlicher, glatter, t-kanonisch eingebetteter Familien von Mannigfaltigkeiten aus $\mathfrak{M}_h$ mit Basis S ist.

Es stellt sich die Frage nach der Repräsentierbarkeit von $\mathcal{H}(S)$.

Da wir in der Kategorie der algebraischen Räume arbeiten werden, ist $\mathcal{H}$ abzuändern, damit diese Frage Sinn hat.

Wir betrachten die Kategorie $\mathfrak{S}$ der noetherschen $\mathbb{C}$-Schemata mit der etalen Topologie. $\widetilde{\mathcal{H}}$ sei die durch $\mathcal{H}$ auf dieser Kategorie bestimmte Garbe und $\mathcal{M}_h$ sei die eindeutige Erweiterung von $\widetilde{\mathcal{H}}$ zu einer Garbe (Garbe bezüglich der etalen Topologie) auf die Kategorie der noetherschen algebraischen $\mathbb{C}$-Räume. $\mathcal{M}_h$ ist dann ein Funktor auf der Kategorie der noetherschen algebraischen Räume, den wir den Funktor der Familien von Mannigfaltigkeiten aus $\mathfrak{M}_h$ nennen. Man sieht leicht, dass $\mathcal{M}_h$ unabhängig von den t-kanonischen Einbettungen ist, die bei seiner Definition benutzt wurden.

Es ist sinnvoll nach der Repräsentierbarkeit von $\mathcal{M}_h$ zu fragen, also zu fragen, ob es einen (noetherschen) algebraischen Raum M_h gibt, so dass die Funktoren $\mathrm{Hom}(-,M_h)$ und $\mathcal{M}_h$ übereinstimmen.

Leider ist $\mathcal{M}_h$ fast nie repräsentierbar; die nichttrivialen Automorphismen der algebraischen Mannigfaltigkeiten aus $\mathfrak{M}_h$ verhindern dies. Man muss etwas weniger fordern als Repräsentierbarkeit. Das geschieht gerade im Begriff des groben Modulraumes.

Definition 2. Ein algebraischer Raum M_h über $\mathbb{C}$ und ein Morphismus $\phi: \mathcal{M}_h(S) \longrightarrow M_h(S)$ von Funktoren (S sei ein noetherscher algebraischer $\mathbb{C}$-Raum, $M_h(S) = \mathrm{Hom}(S,M_h)$) heisst grober Modulraum für $\mathfrak{M}_h$, wenn folgendes gilt:

1. $\phi : \mathcal{M}_h(\mathbb{C}) \longrightarrow M_h(\mathbb{C})$ ist eineindeutig und surjektiv.
2. Ist $\mathcal{M}_h \xrightarrow{\Psi} N$ ein Morphismus von Funktoren, wobei N ein algebraischer $\mathbb{C}$-Raum ist, so gibt es genau einen Morphismus $\chi: M_h \longrightarrow N$ algebraischer Räume, so dass

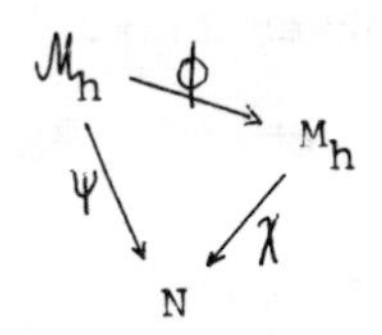

kommutativ ist.

Wir bemerken, dass ein Modulraum, der Definition 2 erfüllt, eindeutig bestimmt ist, falls er existiert, und auch Modulraum nach Definition 1 ist.

Die Vermutung liegt nahe, dass der Quotient $\overline{H}$ von H nach PGL(N), wenn er in "einem guten Sinne" existiert, ein grober Modulraum für $\mathfrak{M}_h$ ist. Dies ist in der Tat richtig und ist im Mumford'schen Buch [12] ausgeführt. (Vgl. auch [15].) Es gilt:

Proposition 2. Der geometrische Quotient $\overline{H}$ von H nach PGL(N) in der Kategorie der algebraischen Räume ist, falls er existiert, grober Modulraum für $\mathfrak{M}_h$.

Die Frage nach groben Modulräumen ist damit reduziert auf die Frage der Existenz geometrischer Quotienten in der Kategorie der algebraischen Räume.

Wir wiederholen zunächst die Definition dieses Begriffes.

Definition 3. Sei X ein algebraischer Raum über $\mathbb{C}$ und G eine algebraische Gruppe (definiert über $\mathbb{C}$), welche auf X operiert. Ein algebraischer Raum $\overline{X}$ über C und ein Morphismus $\varphi : X \longrightarrow \overline{X}$ heisst geometrischer Quotient von X nach G, wenn:

1. $\varphi : X \longrightarrow \overline{X}$ kategorieller Quotient (bezüglich der Kategorie der algebraischen Räume über $\mathbb{C}$) ist.
2. Die Funktionen auf $\overline{X}$ sind genau diejenigen Funktionen auf X, die bei G invariant sind.
3. Die Bahnen $\mathbb{C}$-wertiger Punkte von X werden durch φ eineindeutig auf die $\mathbb{C}$-wertigen Punkte von $\overline{X}$ abgebildet.

Es gilt der Satz (vgl. [16]).

Satz 1. Ist X ein quasiprojektives $\mathbb{C}$-Schema und G eine algebraische Gruppe über $\mathbb{C}$, die auf X eigentlich und mit endlichen Stabilisatoren operiert, so existiert der geometrische Quotient $\overline{X}$ von X nach G in der Kategorie der algebraischen Räume über $\mathbb{C}$. $\overline{X}$ ist von endlichem Typ über $\mathbb{C}$[4)].

Wir beschreiben in groben Zügen den Beweis.

Man zeigt, dass der analytische Quotient von X nach G, der nach Holmann [4] und Kaup [6] existiert, ein algebraischer Raum über $\mathbb{C}$ mit den gewünschten funktoriellen Eigenschaften ist. (Beachte, Holmann behandelt in seiner Arbeit den Fall, dass X, betrachtet als analytischer Raum, reduziert ist. Kaup hat in [6] Holmann's Methoden auf den nichtreduzierten Fall erweitert.) $\overline{X}_{an}$ sei der analytische

[4)] Man kann allgemeiner für X einen separierten algebraischen Raum von endlichem Typ über $\mathbb{C}$ nehmen.

Quotient von X nach G im Sinne von Holmann. Zum Beweis, dass $\overline{X}_{an}$ ein algebraischer Raum ist, benötigt man die Existenz eines affinen $\mathbb{C}$-Schemas U von endlichem Typ über $\mathbb{C}$ und einer etalen Abbildung $U \xrightarrow{\overline{h}} \overline{X}_{an}$, so dass der analytische Teilraum $U \times_{\overline{X}_{an}} U$ des affinen Raumes $U \times U$ ein $\mathbb{C}$-Schema ist. $U \times_{\overline{X}_{an}} U$ definiert dann eine etale Äquivalenzrelation auf U und das Diagramm $U \times_{\overline{X}_{an}} U \rightrightarrows U$ einen algebraischen Raum über $\mathbb{C}$, für welchen der assoziierte analytische Raum gerade $\overline{X}_{an}$ ist. Wir konstruieren U zuerst lokal.

$\overline{P} \in \overline{X}_{an}$ sei ein Punkt und $P \in X$ ein $\mathbb{C}$-wertiger Punkt mit $\varphi(P) = \overline{P}$; $\varphi : X \longrightarrow \overline{X}_{an}$ bezeichnet dabei die analytische Quotientenabbildung. O_P sei die Bahn von P und $I = I_P$ der Stabilisator von P bezüglich G. Sei W_P eine I-stabile affine Umgebung von P auf X. (W_P existiert, da X quasiprojektiv und I endlich ist.) Dann kann durch Cartan's Methode eine Einbettung $W_P \xrightarrow{\rho} \mathbb{C}^N$ von W_P in einen geeigneten affinen Raum $\mathbb{C}^N$ und eine Untergruppe I^* der linearen Gruppe GL(N) gefunden werden, so dass gilt:

1. I^* ist isomorph zu I.
2. I und I^* induzieren dieselben Operationen auf $\rho(W_P)$.

Mit anderen Worten, man linearisiert die Operation von I auf W_P. I^* operiert auf der glatten Mannigfaltigkeit $\rho(O_P \cap W_P) = O_P^*$ und lässt den Punkt $P^* = \rho(P) \in O_P^*$ fest. Dann operiert I^* auch auf den Tangentialraum T_{P^*} von O_P^* im Punkte P^*.

Sei L ein linearer Teilraum des $\mathbb{C}^N$, auf welchem I^* operiert, so dass $L \cap T_{P^*} = P^*$ und L und T_{P^*} den Raum $\mathbb{C}^N$ aufspannen. (Ein solches L findet sich leicht, wenn man eine Hermit'sche Metrik des $\mathbb{C}^N$ benutzt, die I^*-invariant ist.)

Das affine Teilschema $\rho^{-1}(L \cap \rho(W_P)) = U_P'$ von X ist dann transversal zu O_P im Punkte P und I-invariant. Man zeigt, dass ein geeignetes offenes, affines Teilschema U_P von U_P' existiert, so dass gilt: I operiert auf U_P und für alle Punkte $Q \in U_P$ ist der Stabilisator I_Q

von Q bezüglich G in der Gruppe I enthalten. Weiter ist U_P in den Punkten $Q \in U_P$ transversal zu der Bahn von Q.
Sei U_P^I der Quotient von U_P nach I und $U_P^I \xrightarrow{\bar{h}} \bar{X}_{an}$ die natürliche Abbildung. Dann zeigt man, dass $\bar{h}$ etal ist und dass für endlich viele geeignet gewählte Punkte $P_1,\ldots,P_n \in X$ gilt: Sind U_{P_i} affine Mannigfaltigkeiten die nach der obigen Vorschrift zu den Punkten $\varphi(P_i)=\bar{P}_i$ gehören, so ist die natürliche Abbildung $U = \coprod_i U_{P_i}^{I_{P_i}} \longrightarrow X$ etal und surjektiv. ($\coprod_i U_{P_i}^{I_{P_i}}$ = direkte Summe.) Dass für ein solches U das Faserprodukt $U \times_{\bar{X}_{an}} U$ Teilschema von $U \times U$ ist und das Diagram $U \times_{\bar{X}_{an}} U \rightrightarrows U$ geometrischer Quotient von X nach G ist, findet sich in [16] ausgeführt.

Um Satz 1 auf die Operation von PGL(N) auf H anzuwenden, muss gezeigt werden, dass PGL(N) auf H eigentlich und mit endlichen Stabilisatoren operiert. Dabei bedeutet eigentlich, dass die Graphabbildung

$$\begin{array}{ccc} PGL(N) \times H & \xrightarrow{\Psi} & H \times H \\ (g,x) & \longrightarrow & (x,g(x)) \end{array}$$

eigentlich ist. Dies folgt aber mit Hilfe des Bewertungskriteriums für eigentliche Abbildungen sofort aus [11], S. 672, Corollar 1. (Vgl. auch [16], S. 75.)

Um die Endlichkeit der Stabilisatoren nachzuweisen zeigt man, dass für einen Punkt $P \in H$ die Stabilisatorgruppe bezüglich PGL(N) isomorph zur Automorphismengruppe der Faser Γ_P der Familie $\Gamma_H \longrightarrow H$ ist. Diese ist aber nach [9] endlich.

Zusammengefasst ergibt sich:

<u>Satz 2.</u> Es gibt einen algebraischen Raum M_h von endlichem Typ über $\mathbb{C}$ der grober Modulraum für $\mathfrak{M}_h$ ist.

Die eben beschriebene Methode führt über $\mathbb{C}$ auch für andere Typen algebraischer Mannigfaltigkeiten zu groben Modulräumen. So zeigen wir in [16], dass über $\mathbb{C}$ die Menge $\mathfrak{M}_{p_a}^{K^2}$ der Isomorphieklassen algebraischer Flächen allgemeinen Typs mit arithmetischem Geschlecht p_a und Selbstschnittzahl K^2 der kanonischen Garbe einen groben Modulraum besitzt, der ein algebraischer Raum von endlichem Typ über $\mathbb{C}$ ist. Dabei wird nicht gefordert, dass die kanonische Garbe der betrachteten Flächen ample ist.

Für polarisierte, algebraische Mannigfaltigkeiten über $\mathbb{C}$ mit Irregularität 0 ergeben sich mit dieser Methode ebenfalls grobe Modulräume in der Kategorie der noetherschen, algebraischen Räume über $\mathbb{C}$, wenn gewisse Regelmannigfaltigkeiten ausgeschlossen werden. Insbesondere ergeben sich (vgl. [17]) Modulräume für polarisierte K-3 Flächen und Enriquesflächen.

Für polarisierte abelsche Mannigfaltigkeiten, die über $\mathbb{C}$ definiert sind, ergibt die Methode die Existenz grober Modulräume als algebraische Räume. Allerdings wird die bekannte Tatsache, (vgl. [12]) dass für glatte Kurven und polarisierte abelsche Mannigfaltigkeiten die Modulräume quasiprojektive Mannigfaltigkeiten sind, nicht erhalten. Wichtig ist zu bemerken, dass über $\mathbb{C}$ die beschriebene Methode für stabile Kurven vom Geschlecht $g > 1$ (vgl. [1] wegen der Definition dieses Begriffes) zu Modulräumen $\overline{M}_g$ führt, welche algebraische Räume von endlichem Typ über $\mathbb{C}$ und darüberhinaus eigentlich (kompakt) über $\mathbb{C}$ sind. Wegen der Einzelheiten vergleiche man [16] und Seite 16 ff. $\overline{M}_g$ enthält die grobe Modulmannigfaltigkeit M_g für glatte Kurven vom Geschlecht g als dichten Zariski-offenen Teilraum. $\overline{M}_g$ ist also eine Kompaktifizierung von M_g, wobei die Randpunkte den Isomorphieklassen singulärer stabiler Kurven vom Geschlecht g entsprechen. Die Existenz von $\overline{M}_g$ ist auch implizit in [1] enthalten. Allerdings sind die Betrachtungen dort von anderer Art. Nach Mumford und Knudsen ist $\overline{M}_g$ sogar projektiv.

Entscheidend für die Existenz grober Modulräume als algebraische Räume ist der Satz über die Existenz geometrischer Quotienten in der Kategorie der algebraischen Räume. Klar ist, dass wir mit denselben Überlegungen Modulräume erhalten, die Schemata oder sogar quasiprojektive Schemata sind, falls der geometrische Quotient von H nach PGL(N) in der entsprechenden Kategorie existiert. Es scheint eine Bemerkung über geometrische Quotienten, insbesondere in der Kategorie der Schemata angebracht.

Mumford's Buch [12] ist für diese Frage zuständig. Dort wird die Operation einer reduktiven algebraischen Gruppe G auf einem Schema X betrachtet und u.a. gezeigt:

1. Ist X = Spec(A) ein affines $\mathbb{C}$-Schema (statt $\mathbb{C}$ kann ein beliebiger Körper genommen werden), so ist $\overline{X} = \mathrm{Spec}(A^G)$, A^G = Fixring von A nach G, zusammen mit der kanonischen Abbildung $X \longrightarrow \overline{X}$ ein geometrischer Quotient, falls die Operation auf X eigentlich ist. Ist X von endlichem Typ über $\mathbb{C}$, so ist auch $\overline{X}$ von endlichem Typ über $\mathbb{C}$. (Vgl. [12], S. 27 ff.)
2. Ist X ein $\mathbb{C}$-Schema (für $\mathbb{C}$ kann wieder ein beliebig algebraisch abgeschlossener Körper stehen) auf dem die reduktive algebraische Gruppe G operiert. L sei eine G-lineare invertierbare Garbe auf X und X^s die Menge der bezüglich L stabilen Punkte, dann operiert G auf X^s und der geometrische Quotient $\overline{X^s}$ von X^s nach G existiert und ist ein quasiprojektives algebraisches $\mathbb{C}$-Schema. (Vgl. [12], S. 38 und [23] für einen analogen Satz in Charakteristik $p > 0$.)

Der formulierte Quotientensatz für Schemata wird von Mumford in [12] auf die Operation von PGL(N)/Spec($\mathbb{Q}$) auf H_g/Spec($\mathbb{Q}$) angewandt. Dabei ist H_g/Spec($\mathbb{Q}$) das $\mathbb{Q}$-Schema, das in Charakteristik 0 die 3-kanonisch eingebetteten glatten Kurven vom Geschlecht g parametrisiert. Man erhält so in Charakteristik 0 für glatte Kurven vom Geschlecht g die

Existenz grober Modulräume, die über $\mathbb{Q}$ definierte quasiprojektive Schemata sind. Schwierigkeiten ergeben sich beim Nachweis, dass die Punkte aus H_g stabile Punkte bezüglich der gegebenen Operation sind. Diese Schwierigkeiten entfallen, wenn man in der Kategorie der algebraischen Räume arbeitet, man erhält aber auch schwächere Ergebnisse.

In Mumford's Buch findet sich ein zweiter Quotientensatz (vgl. [12], S. 76). Dieser Satz besagt, dass, unabhängig von der Charakteristik, der geometrische Quotient des Teilschemas der stabilen Punkte eines endlichen Produkts $(P^N)^m$ des projektiven Raumes P^N bezüglich der natürlichen Operation von PGL(N) als quasiprojektives Schema, existiert.

Dieser Quotientensatz hat wichtige Anwendungen. So hat Mumford bei der Konstruktion der Modulschemata für polarisierte abelsche Mannigfaltigkeiten diesen Satz entscheidend benutzt zusammen mit der Tatsache, dass eine abelsche Mannigfaltigkeit A, die im P^N eingebettet ist, durch die Menge der n-Teilungspunkte bestimmt ist, falls n genügend gross ist. Die n-Teilungspunkte von A bestimmen einen Punkt im $(P^N)^{n^{2g}}$!

Narasimhan und Sheshadri [14] und andere haben diesen Satz erfolgreich auf die Modultheorie für Vektorraumbündel über Kurven angewandt.

Noch eine Bemerkung über Modulfragen in der Kategorie der analytischen Räume. Dort sind, wie oben ausgeführt, gute Quotientenkriterien durch Holmann und Kaup verfügbar. Man erhält daraus unmittelbar für die oben betrachteten Typen algebraischer Mannigfaltigkeiten die Existenz von Modulräumen als analytische Räume. Unsere Ergebnisse besagen, dass diese Modulräume, deren Existenz als analytische Räume durch die vorhandene Literatur implizit bewiesen wird, sogar algebraische Räume sind.

Weshalb ist es für die Klassifikationstheorie von Bedeutung, dass Modulräume algebraische Räume und nicht nur analytische Räume sind? Wir führen zwei Gründe an.

1. Für algebraische Räume gilt das Bewertungskriterium für eigentliche (proper) Abbildungen. Dieses Kriterium ist für algebraische Räume von endlichem Typ über $\mathbb{C}$ ein guter Test für Kompaktheit, denn "eigentlich" und kompakt sind dort äquivalent. Für analytische Räume ist das Bewertungskriterium nicht richtig, wie der in O punktierte offene Einheitskreis zeigt. Das besagte Kriterium kann wie folgt formuliert werden.

   <u>Bewertungskriterium</u>. Sei X ein (separierter) algebraischer Raum über $\mathbb{C}$ und U eine Zariski-offene dichte Teilmenge von X. R sei ein kompletter, diskreter Bewertungsring vom Rang 1, der $\mathbb{C}$ enthält und der K als Quotientenkörper hat. X ist eigentlich (proper) über $\mathbb{C}$ genau dann, wenn zu jedem Morphismus $f:\mathrm{Spec}(K) \longrightarrow U$ eine endliche Körpererweiterung K'/K und ein Morphismus $\mathrm{Spec}(R') \longrightarrow X$, R' = ganzer Abschluss von R in K', existieren, derart, dass das Diagram

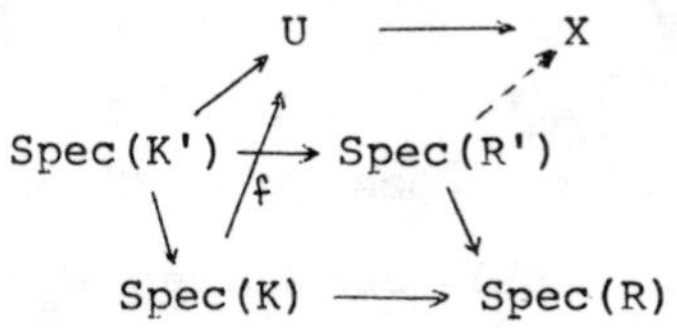

$$\begin{array}{ccccc} & & U & \longrightarrow & X \\ & \nearrow & \uparrow f & & \nearrow \\ \mathrm{Spec}(K') & \longrightarrow & & \mathrm{Spec}(R') & \\ & \searrow & & \searrow & \\ & & \mathrm{Spec}(K) & \longrightarrow & \mathrm{Spec}(R) \end{array}$$

   kommutativ ist.

   Dieses Kriterium wurde in [16] angewandt um zu zeigen, dass der Modulraum $\overline{M}_g$ für stabile Kurven vom Geschlecht $g \geqslant 2$ über $\mathbb{C}$ eigentlich über $\mathbb{C}$ ist. In der Tat kann man zeigen, dass das stabile Reduktionstheorem für glatte Kurven (vgl. [1]) vom Geschlecht g äquivalent ist mit dem Bewertungskriterium für die groben Modulräume stabiler Kurven.

2. Die Existenz der Modulräume für algebraische Mannigfaltigkeiten allgemeinen Typs als algebraische Räume ist wichtig für die Klassifikation kompakter, komplexer Mannigfaltigkeiten ohne meromorphe Funktionen. Z.B. sei X eine solche Mannigfaltigkeit der Dimension n und sei $\alpha : X \longrightarrow T$ die Albanese Abbildung von X. Man kann zeigen, dass α surjektiv ist und zusammenhängende Fasern hat. Ist Dimension T = n-2, so ist die allgemeine Faser von α eine Fläche. Für diese Fläche folgt aus der Existenz der Modulräume für Flächen allgemeinen Typs als algebraische Räume, dass sie nicht von allgemeinem Typ ist. Benutzt man noch die Ergebnisse von Iitaka [5], so ergibt sich, dass die allgemeine Faser von α eine Fläche von Kodaira Dimension ≤ 0 ist. Mehr dazu findet sich in Ueno's Beitrag in diesem Buch oder in [21].

Bisher haben wir nur über grobe Modulräume gesprochen. Der Nachteil dieser Räume ist, dass über ihnen keine "guten" Familien existieren. Für glatte Kurven vom Geschlecht g zum Beispiel ist wohlbekannt, dass es über der groben Modulmannigfaltigkeit M_g keine glatte, eigentliche Familie $\Gamma_g \longrightarrow M_g$ von Kurven vom Geschlecht g gibt derart, dass für jeden Punkt $P \in M_g$ die Faser Γ_P den Punkt P als Modulpunkt hat. Die Automorphismen der Kurven oder -äquivalent dazu- das nicht fixpunktfreie Operieren von PGL(N) auf H sind für die Nichtexistenz dieser Familien verantwortlich (vgl. Seite17). Dieses Phänomen tritt bei den anderen Typen algebraischer Mannigfaltigkeiten entsprechend auf. Die Klassifikationstheorie algebraischer Mannigfaltigkeiten benötigt aber universelle Familien. Weshalb, soll kurz erläutert werden.
Will man etwa alle Faserräume, die für Mannigfaltigkeiten der Klasse 2 auftreten, beschreiben, so kann man daran denken, nach universellen Faserräumen oder universellen Familien zu suchen (endlich viele, wenn eine Polarisation und die Dimension der Mannigfaltigkeiten fixiert

wird), so dass alle anderen Faserräume modulo gewisser Korrekturen, wie Auflösen von Singularitäten, Pullbacks dieser universellen Faserräume sind. Da die betrachteten Faserräume im allgemeinen eine kompakte Basis besitzen, sollten die universellen Faserräume oder Familien ebenfalls über einem kompakten algebraischen Raum als Basis definiert sein. Natürlich sind die gewüschten universellen Familien im allgemeinen nicht glatt.

Nach dem jetzigen Stand der Klassifikationstheorie kennt man die Existenz universeller Familien für glatte, polarisierte abelsche Mannigfaltigkeiten mit n-Teilungspunktstruktur, glatte polarisierte K-3 Flächen mit n-Teilungspunktstruktur, glatte algebraische Mannigfaltigkeiten mit sehr ampler kanonischer Garbe und natürlich für glatte Kurven alles mit n-Teilungspunktstruktur. (Vgl. dazu [17].) Entscheidend bei allen diesen Typen algebraischer Mannigfaltigkeiten ist, dass die Automorphismen treu auf der ganzzahligen Homologie beziehungsweise Kohomologie operieren und dass als Folge davon die Automorphismen dieser Mannigfaltigkeiten durch n-Teilungspunkte eliminiert werden können. (Eine n-Teilungspunktstruktur von X ist, grob gesagt, eine Basis der Homologie oder Kohomologie von X mit Koeffizienten in $\mathbb{Z}/n$) Mannigfaltigkeiten des obigen Typs mit n-Teilungspunktstruktur haben, wenn n genügend gross ist, keine Automorphismen. Die für die angegebenen Typen von Mannigfaltigkeiten existierenden universellen Familien sind glatt, haben aber den Nachteil, dass ihre Basen nicht kompakt sind. Man muss die Familien kompaktifizieren. Das allerdings ist mit Schwierigkeiten verbunden und von uns bisher nur für Kurven befriedigend gelöst. Dort kann man für stabile Kurven vom Geschlecht g den Begriff der n-Teilungspunktstruktur (level n-structure) einführen und dann folgendes zeigen.

1. Ist $\Gamma^{(n)} \longrightarrow M_g^{(n)}$ die Familie der glatten Kurven vom Geschlecht g mit n-Teilungspunktstruktur, so kann die Basismannigfaltigkeit $M_g^{(n)}$

durch Hinzunahme von Punkten, welche eineindeutig den singulären, stabilen Kurven mit n-Teilungspunktstruktur entsprechen, zu einem algebraischen Raum $\overline{M}_g^{(n)}$ kompaktifiziert werden.

2. Die Familie $\Gamma^{(n)} \longrightarrow M_g^{(n)}$ kann in natürlicher Weise zu einer Familie stabiler Kurven $\overline{\Gamma}^{(n)} \longrightarrow \overline{M}_g^{(n)}$ erweitert werden, falls $n \geqslant 3$ ist, und diese Familie ist unversell.

Wir beschreiben kurz die Konstruktion von $\overline{\Gamma}^{(n)} \longrightarrow \overline{M}_g^{(n)}$.

Es sei $\Gamma_H \xrightarrow{f} H$ die universelle Familie 3-kanonisch eingebetteter stabiler Kurven vom Geschlecht g. Die Gruppe PGL(5g-6) operiert auf H und Γ_H; f ist mit diesen Operationen verträglich. Der geometrische Quotient $\overline{H}$ von H nach PGL(5g-6) ist ein grober Modulraum für die stabilen Kurven vom Geschlecht g. Man kann zeigen (vgl. [17]): Operiert PGL(5g-6) fixpunktfrei auf H und damit auch auf Γ_H, so ist der geometrische Quotient $\overline{\Gamma}_H$ von Γ_H nach PGL(5g-6) zusammen mit der durch f induzierten Abbildung $\overline{\Gamma}_H \xrightarrow{\overline{f}} \overline{H}$ eine Familie stabiler Kurven vom Geschlecht g, welche universell für stabile Kurven vom Geschlecht g ist. Aber PGL(5g-6) operiert gerade nicht fixpunktfrei auf H, z.B. ist für Kurven vom Geschlecht $g > 2$ für jeden Punkt $P \in H$ der Stabilisator bezüglich PGL(N) isomorph zur Automorphismengruppe der Faser Γ_P von $\Gamma_H \longrightarrow H$. Allerdings kann man zeigen, es gibt endliche (verzweigte) Überlagerungen $H^{(n)}$ von H, die man mit Hilfe der ganzzahligen Homologie der Fasern der Familie $\Gamma_H \longrightarrow H$ konstruiert auf welchen PGL(5g-6) fixpunktfrei operiert. Wir erläutern die Konstruktion und betrachten zuerst glatte Kurven.

$\Gamma \longrightarrow S$ sei eine glatte Familie von Kurven vom Geschlecht g, wobei S ein $\mathbb{C}$-Schema endlichen Typs ist. Es sei $P \in S$ und Γ_P die Faser. $H_1(\Gamma_P, \mathbb{Z})$ bezeichnet die ganzzahlige erste Homologiegruppe der glatten Kurve Γ_P. ($H_1(\Gamma_P,\mathbb{Z})$ ist ein freier $\mathbb{Z}$-Modul vom Rang 2g.) $\Lambda_P = \Lambda(H_1(\Gamma_P,\mathbb{Z}))$ sei die Menge der Basen von $H_1(\Gamma_P,\mathbb{Z})$. Sei $T = \bigcup_{P \in S} \Lambda_P$ die disjunkte

Vereinigung und $\alpha: T = \bigcup_{P\in S} \Lambda_P \longrightarrow S$ die natürliche Abbildung, die jedem Punkt $Q \in \Lambda_P$ den Punkt P zuordnet. Da die Familie $\Gamma \longrightarrow S$ lokal bezüglich der komplexen Topologie ein Produkt ist, kann auf T eine Topologie so definiert werden, dass $T \longrightarrow S$ eine unverzweigte topologische Überlagerung ist. Überträgt man noch die komplexe Struktur von S auf T, so wird $T \longrightarrow S$ zu einer komplexen unverzweigten Überlagerung. Die Automorphismengruppe $GL(2g,\mathbb{Z})$ von $H_1(\Gamma_P,\mathbb{Z})$, Γ_P eine beliebige Faser von $\Gamma \longrightarrow S$, operiert auf $T \longrightarrow S$ als Decktransformationsgruppe. Sei $GL(2g,\mathbb{Z})^n$ die Kongruenzuntergruppe zu der natürlichen Zahl n, d.h. $GL(2g,\mathbb{Z})^n = \{A \in GL(2g,\mathbb{Z});\ A \equiv I_{2g} \text{ modulo } n\}$, und $P^{(n)}(\Gamma/S)$ der analytische Quotient von T nach $GL(2g,\mathbb{Z})^n$ mit $P^{(n)}(\Gamma/S) \xrightarrow{\alpha} S$ als natürliche Abbildung. Dann ist $P^{(n)}(\Gamma/S)$ eine endliche etale galoissche, komplexe Überlagerung von S, die nach Grauert/Remmert [2] auch algebraisch, d.h. ein Schema ist.

<u>Definition 4.</u> Ein Schnitt von $P^{(n)}(\Gamma/S) \longrightarrow S$ über S heisst eine <u>n-Teilungspunktstruktur</u> der Familie Γ/S.

Nun sei $\Gamma_{H_o} \longrightarrow H_o$ die universelle Familie 3-kanonisch eingebetteter, glatter Kurven und $H_o^{(n)} = P^{(n)}(\Gamma_{H_o}/H_o) \xrightarrow{\alpha} H_o$ zugehörige etale Überlagerung. Die Konstruktion von $P^n(\Gamma/S)$ für eine glatte Familie Γ/S über einem Schema S von endlichem Typ zeigt, dass $P^{(n)}(\Gamma/S)$ funktoriell ist bezüglich Faserprodukten und bezüglich Isomorphismen von Familien. Diese Funktorialität erlaubt es, eine Operation von $PGL(5g-6)$ auf $H_o^{(n)}$ zu definieren, so dass $H_o^{(n)} \xrightarrow{\alpha} H_o$ ein $PGL(5g-6)$-Morphismus ist. Ist $\Gamma_o^{(n)} = \Gamma_{H_o} \times_{H_o} H_o^{(n)} \xrightarrow{f^{(n)}} H_o^{(n)}$ die Pullback-Familie, so operiert $PGL(5g-6)$ auf $\Gamma_o^{(n)}$, und $f^{(n)}$ ist ein $PGL(5g-6)$-Morphismus. Man zeigt dann, dass $PGL(5g-6)$ auf $H_o^{(n)}$ fixpunktfrei operiert, falls $n \geq 3$ ist, indem man nachweist, dass ein Punkt $P \in H_o^{(n)}$ in natürlicher Weise eine Basis von $H_1(\Gamma_P,\mathbb{Z}/n)$ bestimmt und ein $\sigma \in PGL(5g-6)$, welches P als Fixpunkt hat, durch einen Automorphismus von Γ_P

induziert ist, der dann trivial auf $H_1(\Gamma_P,\mathbb{Z}/n)$ operiert. Ein solches σ ist aber wegen [19], S.12, die Identität, falls $n \geq 3$ ist. PGL(5g-6) operiert dann auch fixpunktfrei auf $\Gamma_o^{(n)}$. Nimmt man die geometrischen Quotienten als algebraische Räume, so ergibt sich eine Familie $\overline{\Gamma}_o^{(n)} \longrightarrow \overline{H}_o^{(n)}$ von glatten Kurven vom Geschlecht g mit n-Teilungspunktstruktur, welche für glatte Kurven vom Geschlecht g mit n-Teilungspunktstruktur universell ist. $\overline{\Gamma}_o^{(n)} \longrightarrow \overline{H}_o^{(n)}$ ist oben mit $\Gamma^{(n)} \longrightarrow M_g^{(n)}$ bezeichnet worden.

Nun zu beliebigen Kurven vom Geschlecht g.

$\Gamma_H \longrightarrow H$ sei wieder die universelle Familie 3-kanonisch eingebetteter, stabiler Kurven vom Geschlecht g. Dann ist H_o Zariski offenes, dichtes Teilschema von H. Sei $H_o^{(n)} \xrightarrow{\alpha} H_o$ die oben konstruierte etale Überlagerung von H_o und $H^{(n)}$ die Normalisierung von H im Ring der rationalen Funktionen von $H_o^{(n)}$. $H^{(n)} \xrightarrow{\alpha} H$ sei die natürliche Abbildung. Man zeigt, vgl. [17], dass die Operation von PGL(5g-6) auf $H_o^{(n)}$ in eindeutiger Weise auf $H^{(n)}$ erweitert werden kann, derart, dass $\alpha: H^{(n)} \longrightarrow H$ eine PGL(5g-6)-Abbildung ist. Ist $\Gamma^{(n)} = \Gamma_H \times_H H^{(n)} \longrightarrow H^{(n)}$ die Pullbackfamilie von $\Gamma_H \longrightarrow H$, so operiert PGL(N) auch auf $\Gamma^{(n)}$. Entscheidend ist es, zu zeigen, dass PGL(5g-6) fixpunktfrei auf $H^{(n)}$ (und dann auch auf $\Gamma_H^{(n)}$) operiert, falls $n \geq 3$. Dies sieht man, wenn mit Hilfe der Monodromie eine natürliche Beziehung zwischen den Punkten $P' \in H^{(n)}$ und der Homologie $H_1(\Gamma_{\alpha(P')},\mathbb{Z}/n)$ hergestellt wird. ($\Gamma_{\alpha(P')}$ bezeichnet die Faser über $\alpha(P') \in H$ der Familie $\Gamma_H \longrightarrow H$.) Wir führen dies aus und beschränken uns auf Punkte $P' \in H^{(n)}$, für welche $P = \alpha(P') \in H - H_o$, also die zugehörigen stabilen Kurven singulär sind. Es sei $P \in H - H_o = D$. D ist ein Divisor in H mit normalen Schnitten als Singularitäten; die Anzahl d der im analytischen Sinne irreduziblen Komponenten von D durch P ist gleich der Anzahl der singulären Punkte der Faser Γ_P der Familien $\Gamma_H \longrightarrow H$. (Vgl. [1].) Da H glatt ist und D normale Schnitte als Singularitäten hat, gibt es in $P \in H$ lokale Koordinate $(t_i)_{1 \le i \le N}$ auf H und eine Umgebung U von P auf H, so dass gilt

1. $U = \{(t_1, \ldots, t_N);\ |t_i| < \varepsilon\}$
2. $U-D = (E')^d \times E^{N-d}$, wobei $E = \{t \in \mathbb{C};\ |t| < \varepsilon\}$, $E' = \{t \in \mathbb{C};\ 0 < |t| < \varepsilon\}$.

Sei $\pi_1(U-D) = \gamma_1\mathbb{Z} \oplus \cdots \oplus \gamma_d\mathbb{Z}$ (γ_i ist die Homotopieklasse eines Kreises, welcher $t_i = 0$ umschliesst, $i \leq d$) die Fundamentalgruppe von U-D und Γ_Q die Faser der Familie $\Gamma_H \longrightarrow H$ über einem beliebigen Punkt $Q \in U-D$. Dann operiert $\pi_1(U-D)$ als Monodromiegruppe auf $H_1(\Gamma_Q, \mathbb{Z})$ und auch auf $\alpha^{-1}(Q) = \Lambda(H_1(\Gamma_Q, \mathbb{Z}))/GL(2g,\mathbb{Z})^n$, da $GL(2g,\mathbb{Z})^n$ Normalteiler in $GL(2g,\mathbb{Z})$ ist. Es gilt dann, unabhängig von Q und der Umgebung U, $\alpha^{-1}(P) = \alpha^{-1}(Q)/\pi_1(U-D)$ = Quotientenmenge von $\alpha^{-1}(Q)$ nach $\pi_1(U-D)$. $\alpha_{Q,P} : \alpha^{-1}(Q) \longrightarrow \alpha^{-1}(P) = \alpha^{-1}(Q)/\pi_1(U-D)$ bezeichnet im folgenden die Quotientenabbildung.

<u>Behauptung.</u> Jeder Punkt $P' \in \alpha^{-1}(P)$, $P \in H$ bestimmt in natürlicher Weise 2g Vektoren aus $H_1(\Gamma_P, \mathbb{Z}/n)$, welche $H_1(\Gamma_P, \mathbb{Z}/n)$ aufspannen.

Nur der Fall $P \in D$ ist interessant. Sei $P' \in \alpha^{-1}(P)$ und $(a_1, \ldots, a_{2g})$ ein 2g Tupel von Vektoren aus $H_1(\Gamma_Q, \mathbb{Z})$, welches einen Punkt $Q' \in \alpha^{-1}(Q)$ bestimmt, für den $\alpha_{Q,P}(Q') = P'$ gilt. $A: H_1(\Gamma_Q, \mathbb{Z}) \longrightarrow H_1(\Gamma_Q, \mathbb{Z}/n)$ sei die natürliche Abbildung und $(\bar{a}_1, \ldots, \bar{a}_{2g})$ das Bild von $(a_1, \ldots, a_{2g})$ bei A. Dann spannen $(\bar{a}_i)$ den Raum $H_1(\Gamma_Q, \mathbb{Z}/n)$ auf. Wir bewegen nun Q entlang eines Weges, der ganz in U verläuft, nach P und erhalten eine Abbildung $\beta_{Q,P} : H_1(\Gamma_Q, \mathbb{Z}/n) \longrightarrow H_1(\Gamma_P, \mathbb{Z}/n)$, von der man mit Hilfe der Theorie von Picard-Lefschetz [3] zeigt, dass sie unabhängig vom Weg ist. Durch die Abbildung $\beta_{Q,P}$ erhält man eine Abbildung $\beta : \alpha^{-1}(P) \longrightarrow (H_1(\Gamma_P, \mathbb{Z}/n)$ wie folgt:
Sei $P' \in \alpha^{-1}(P)$ und $Q' \in \alpha^{-1}(Q)$, $Q \in U-D$ mit $\alpha_{Q,P}(Q') = P'$. $(a_1, \ldots, a_{2g})$ sei eine Basis von $H_1(\Gamma_Q, \mathbb{Z})$, welche Q' bestimmt. Man definiert

$$\beta(P') = \beta_{Q,P}(A(a_i)).$$

$\beta(P')$ ist ein 2g-Tupel von Elementen aus $H_1(\Gamma_P, \mathbb{Z}/n)$, das $H_1(\Gamma_P, \mathbb{Z}/n)$ aufspannt. Man muss zeigen, dass β wohldefiniert ist und dass ein Element $\sigma \in PGL(N)$, welches den Punkt $P' \in H^{(n)}$ fest lässt, von einem Automorphismus von $\Gamma_{\alpha(P')}$ induziert ist, der die Vektoren $\beta(P')$ fest lässt, falls $n \geq 3$ ist. Ein solcher Morphismus ist die Identität, wie in [17] gezeigt wird.

Der geometrische Quotient $\bar{\Gamma}^{(n)} \longrightarrow \bar{H}^{(n)}$ von $\Gamma^{(n)} \longrightarrow H^{(n)}$ ist die gesuchte universelle Familie stabiler Kurven vom Geschlecht g mit n-Teilungspunktstruktur, welche oben durch $\bar{\Gamma}^{(n)} \longrightarrow \bar{M}_g^{(n)}$ bezeichnet ist.

Die Idee dieser Konstruktion ist es, die Monodromie, welche eine wichtige Invariante für Familien von Kurven darstellt[5], zu benutzen, um die Familien $\Gamma^{(n)} \longrightarrow M_g^{(n)}$ zu kompaktifizieren. Dass dies möglich ist, ist bemerkenswert und sollte auch in anderen Situationen, z.B. bei Abelschen Mannigfaltigkeiten, Anwendung finden.

Wir wollen abschliessend auf die Bedeutung der universellen Familien $\bar{\Gamma}^{(n)} \longrightarrow \bar{M}_g^{(n)}$ hinweisen.

Die Methode der Lefschetzbüschel stellt eine Beziehung zwischen Flächen und eindimensionalen Familien (Büscheln) von Kurven her, die für ein tieferes Studium der Flächen allgemeinen Typs von Bedeutung ist. Die m-kanonischen Abbildungen bringen elliptische Flächen allgemeinen Typs mit Büscheln von elliptischen Kurven in Beziehung. Beide Tatsachen legen es nahe, Familien von Kurven und insbesondere Büschel von Kurven zu klassifizieren. Die Idee ist wie folgt.

Zur Vereinfachung nehmen wir an, dass $X \xrightarrow{\pi} S$ eine eindimensionale Familie von Kurven ist, d.h. X und S sind glatte, zusammenhängende Schemata oder zusammenhängende, komplexe Mannigfaltigkeiten der Dimension 2 bzw. 1 und π ist ein eigentlicher, flacher Morphismus. Weiter

[5] Ist $X \longrightarrow S$ eine Familie von Kurven (mit 1-dimensionaler Basis), so bestimmt der glatte Teil $X_o \longrightarrow S_o$ von $X \longrightarrow S$ zusammen mit der Monodromieoperation von $\pi_1(S_o)$ auf der ganzzahligen Homologie von $X_o \longrightarrow S_o$ im Wesentlichen die Familie $X \longrightarrow S$. Man vergleiche Namikawa's Beitrag.

sei X minimal. Dann ist π auf $\pi^{-1}(S-\{P_1,\dots,P_n\})$ eingeschränkt glatt, wenn $P_1,\dots,P_n$ geeignet gewählte Punkte aus S sind. Das stabile Reduktionstheorem für Kurven führt zur Existenz einer galoisschen Überlagerung $S' \xrightarrow{f} S$ mit Galoisgruppe G, welche in den Punkten P_i und unter Umständen in noch weiteren Punkten $Q_1,\dots,Q_s$ von S verzweigt ist, sowie zu einer Familie $X' \xrightarrow{\pi'} S'$ von stabilen Kurven, so dass gilt:

1. Ist $U = S - \{P_1,\dots,P_n,Q_1,\dots,Q_n\}$ und $U' = f^{-1}(U)$, so sind die Familien $X \times_S U' \longrightarrow U'$ und $X' \times_{S'} U' \longrightarrow U'$ isomorph.
2. Die Gruppe G operiert auf X', und π' ist ein G-Morphismus.
3. Der Quotient $X'^G \longrightarrow S'^G = S$ ist ein Faserraum über S, der auf U eingeschränkt zu $X \times_S U \longrightarrow U$ isomorph ist. Aus $X'^G \longrightarrow S$ erhält man $X \longrightarrow S$ durch Auflösung von Singularitäten und Zusammenblasen von Divisoren.

Man sieht, dass das Problem Büschel $X \longrightarrow S$ von Kurven zu klassifizieren in zwei Teile zerfällt, nämlich 1) Büschel von stabilen Kurven zu klassifizieren und 2) die Auflösung der bei Quotientenbildung nach G entstehenden Singularitäten durch Invarianten zu beschreiben. (Für Familien von Kurven mit höherdimensionalen Basen zerfällt das Problem entsprechend.) Die Klassifikation der Familien stabiler Kurven kann vorgenommen werden, indem man sich auf die universellen Familien $\overline{\Gamma}^{(n)} \longrightarrow \overline{M}_g^{(n)}$ bezieht. Das zweite Problem ist z.B. von Namikawa/Ueno [13] und Viehweg [22] bearbeitet worden und wird in Namikawa's Beitrag diskutiert. Die von Viehweg angegebenen Invarianten, die über das Hilbertschema 3-kanonisch eingebetteter Kurven definiert sind, können leicht auf $\overline{M}_g^{(n)}$ umgeschrieben werden.

Literatur

[1] P. Deligne and D. Mumford, The irreducibility of the space of curves of given genus. Public. Math. 36(1969), 75-109.

[2] H. Grauert und R. Remmert, Komplexe Räume. Math. Annalen 136 (1958), 245-318.

[3] P.A. Griffiths, Seminar on degeneration of algebraic varieties. Institute for Advanced Study, Princeton 1969/70.

[4] H. Holmann, Quotienten komplexer Räume. Math. Annalen 142 (1961), 407-440.

[5] S. Iitaka, Deformations of compact complex surfaces III. J. Math. Soc. Japan 22(1970), 247-261.

[6] B. Kaup, Äquivalenzrelationen auf allgemeinen komplexen Räumen. Schriftenreihe Math. Institut Univ. Münster. Heft 39(1968).

[7] S. Kawai, Elliptic fibre spaces over compact surfaces. Comment. Math. Univ. St. Paul 15 (1967), 119-138.

[8] K. Kodaira, On compact analytic surfaces II, III. Ann. of Math. 77 and 78 (1963), 563-626 and 1-40.

[9] H. Matsumura, On algebraic groups of birational transformations. Lincei-Rend. Sc. mat.e.nat. XXXIV (1963), 151-154.

[10] T. Matsusaka, On canonically polarized varieties, II. Am. Journal of Math. 92 (1970), 283-292.

[11] T. Matsusaka and D. Mumford, Two fundamental theorems on deformations of polarized varieties. Am. Journal of Math. 86 (1964), 668-684.

[12] D. Mumford, Geometric invariant theory. Ergebnisse der Math. 34. Springer Verlag 1965.

[13] Y. Namikawa and K. Ueno, The complete classification of fibres of pencils of curves of genus 2. Manuscripta Math. 9 (1973), 143-186.

[14] M.S. Narasimhan and C.S. Seshadri, Stable and unitary vector bundles on a compact Riemann surface. Ann. of Math. 82 (1965), 540-567.

[15] H. Popp, On moduli of algebraic varieties I. Invent. Math. 22 (1973), 1-40.

[16] H. Popp, On moduli of algebraic varieties II. Compositio Math. 28 (1974), 51-81.

[17] H. Popp, On moduli of algebraic varieties III. To appear.

[18] B. Riemann, Theorie der abelschen Funktionen. Journal f.d.r.u.a. Math. 54 (1857).

[19] Séminaire Henri Cartan, 1960/61. II. Auflage. Exposé 17.

[20] K. Ueno, Classification of algebraic varieties, I. Compositio Math. 27 (1973), 277-342.

[21] K. Ueno, Lectures on classification of algebraic varieties and compact complex manifolds. To appear in Springer Lecture Notes.

[22] E. Viehweg, Invarianten degenerierter Fasern in lokalen Familien von Kurven.

[23] C.S. Seshadri, Quotient spaces modulo reductive algebraic groups. Annals of Math. 95 (1972), 511-556.

Abbildungen in arithmetische Quotienten hermitesch symmetrischer Räume

Wilfried Schmid*)

In den Arbeiten [4,5,6] hat Phillip Griffiths den Versuch begonnen, die klassische Theorie der Moduln Riemannscher Flächen für beliebige projektive Mannigfaltigkeiten zu verallgemeinern. Insbesondere konstruiert er zu jeder Familie von projektiven Mannigfaltigkeiten $\mathcal{V} \to T$, die von einer komplexen Mannigfaltigkeit T parametrisiert wird, eine Periodenabbildung $\Phi : T \longrightarrow \Gamma\backslash D$; dabei ist D ein Raum, der die möglichen Hodge Zerlegungen der Kohomologie der Fasern klassifiziert, und Γ ist eine arithmetisch definierte Gruppe von Automorphismen des Raumes D. Falls die Familie $\mathcal{V} \longrightarrow T$ auch singuläre Fasern enthält, so ist Φ nur auf dem Komplement derjenigen analytischen Menge S erklärt, die den singulären Fasern entspricht. In diesem Fall spiegeln die Singularitäten der Abbildung Φ entlang S gewisse geometrische Eigenschaften der Familie $\mathcal{V} \longrightarrow T$ wider.

Für einparametrige Familien habe ich in [11] die möglichen Singularitäten der Periodenabbildungen beschrieben. Diese Beschreibung, wie auch ihre Herleitung, ist relativ kompliziert. Wenn D ein hermitesch symmetrischer Raum ist -und das passiert zum Beispiel bei Familien von Riemannschen Flächen, abelschen Mannigfaltigkeiten, oder K3-Flächen-, dann kommt man mit wesentlich weniger Aufwand aus. Es ist Ziel der vorliegenden Arbeit, die Hauptaussagen von [11] in diesem Spezialfall auf einfachere Weise herzuleiten.

Ich betrachte also die folgende Situation. Es sei D ein hermitesch symmetrischer Raum des nichtkompakten Typs. Die grösste zusammenhängende

*) Diese Arbeit ist während eines Gastaufenthaltes am SFB 40 für reine Mathematik an der Universität Bonn entstanden.

Gruppe von Automorphismen[1] bezeichne ich mit $G_{\mathbb{R}}$. Dann ist $G_{\mathbb{R}}$ eine halbeinfache Lie Gruppe, alle einfachen Faktoren von $G_{\mathbb{R}}$ sind nichtkompakt, und $G_{\mathbb{R}}$ operiert transitiv auf D mit kompakten Isotropiegruppen. Ferner sei $\Gamma \subset G_{\mathbb{R}}$ eine arithmetisch definierbare[2] Untergruppe. Als diskrete Untergruppe von $G_{\mathbb{R}}$ operiert Γ eigentlich diskontinuierlich auf D, und der Quotient $\Gamma\backslash D$ hat die Struktur eines komplexen Raumes. Als typische und wichtigste Beispiele für D und Γ sollen die Siegelsche obere Halbebene H_g und die Siegelsche Modulgruppe $Sp(g,\mathbb{Z})$ genannt sein. In diesem Fall ist $\Gamma\backslash D$ der Modulraum der abelschen Mannigfaltigkeiten mit einer Hauptpolarisierung.

Gegeben sei eine Abbildung

$$\Phi : T - S \longrightarrow \Gamma\backslash D, \tag{1}$$

die auf dem Komplement einer analytischen Menge S in einer komplexen Mannigfaltigkeit T definiert ist. Ich nehme an

(2) Φ ist holomorph, und Φ kann lokal durch die Quotientenabbildung $D \longrightarrow \Gamma\backslash D$ faktorisiert werden.

Wenn man Φ mit der universellen Überlagerung $(\widetilde{T - S}) \longrightarrow T - S$ zusammenschaltet, kann man die so erhaltene Abbildung wegen (2) global durch $D \longrightarrow \Gamma\backslash D$ faktorisieren. Es existiert also ein kommutatives Diagramm

$$\begin{array}{ccc} (\widetilde{T - S}) & \xrightarrow{\tilde{\Phi}} & D \\ \downarrow & & \downarrow \\ T - S & \xrightarrow{\Phi} & \Gamma\backslash D. \end{array} \tag{3}$$

Für jedes $\sigma \in \pi_1(T - S)$ gibt es ein (möglicherweise nicht eindeutig bestimmtes) Element $\gamma(\sigma) \in \Gamma$, so dass

---

[1]) oder eine endliche Überlagerung derselben

[2]) d.h. $G_{\mathbb{R}}$ kann so als Gruppe der reellen Punkte in einer halbeinfachen Matrizengruppe über $\mathbb{Q}$ realisiert werden, dass Γ und $G_{\mathbb{Z}}$ kommensurabel sind; vgl. [2] .

(4) $\tilde{\Phi}(\sigma t) = \gamma(\sigma)\circ\tilde{\Phi}(t)$, für alle $t \in T - S$;

dabei wird $\pi_1(T - S)$ als Transformationsgruppe für $(\widetilde{T - S})$ aufgefasst. Wie ich nun zusätzlich voraussetze, sollen die Elemente $\gamma(\sigma)$ so gewählt werden können, dass gilt:

(5) $\sigma \longrightarrow \gamma(\sigma)$ ist ein Homomorphismus der Fundamentalgruppe $\pi_1(T - S)$ in die Gruppe Γ.

Wenn Φ als Periodenabbildung einer Familie von projektiven Mannigfaltigkeiten auftritt, sind (2) und (5) notwendigerweise erfüllt.

Es geht darum, die Singularitäten der Abbildung Φ entlang der analytischen Menge S in den Griff zu bekommen. Dazu ist es zweckmässig, S soweit als möglich aus dem Problem herauszuhalten. Falls die Kodimension von S mindestens zwei beträgt, hängt T - S lokal einfach zusammen. Weil D als beschränktes Gebiet in einem $\mathbb{C}^N$ realisiert werden kann, ist es dann immer möglich, Φ holomorph auf T fortzusetzen. Also sei S ein Divisor. Laut Hironaka lässt sich T entlang S so modifizieren, dass (das modifizierte) S höchstens normale Durchschnitte ("normal crossings") als Singularitäten hat. Wenn man das Problem nun lokalisiert, erhält man $T = \Delta^n$, $S = \Delta^n - (\Delta^{n-m} \times \Delta^{*\,m})$; dabei bezeichnet Δ die Einheitskreisscheibe in $\mathbb{C}$ und Δ^* die punktierte Kreisscheibe. Ohne Einschränkung darf man annehmen, dass m = n. Damit ist man bei der folgenden Situation angelangt:

(6) $\Phi : \Delta^{*\,n} \longrightarrow \Gamma\backslash D$

ist eine holomorphe Abbildung, für die es ein kommutatives Diagramm

(7a)
$$\begin{array}{ccc} H^n & \xrightarrow{\tilde{\Phi}} & D \\ \tau\downarrow & & \downarrow \\ \Delta^{*\,n} & \xrightarrow{\Phi} & \Gamma\backslash D \end{array}$$

gibt, mit H = obere Halbebene in $\mathbb{C}$, und

(7b) $\tau(z_1,\dots,z_n) = (e^{2\pi i z_1},\dots,e^{2\pi i z_n})$.

Weiterhin existieren Elemente $\gamma_j \in \Gamma$, $1 \leq j \leq n$, mit

(8) $\tilde{\Phi}(z_1,\dots,z_{j-1},z_j+1,z_{j+1},\dots,z_n) = \gamma_j \circ \tilde{\Phi}(z_1,\dots,z_n)$.

Wegen (5) kann man die Wahl der γ_j so treffen, dass

(9) $\gamma_j\gamma_k = \gamma_k\gamma_j$, für $1 \leq j, k \leq n$.

Der hermitesch symmetrische Raum D kann mit einer $G_{\mathbb{R}}$-invarianten, hermiteschen Metrik ds_D^2 ausgestattet werden, deren Schnittkrümmungen auf holomorphen Ebenen von oben durch -4 beschränkt sind. Für jede holomorphe Abbildung F der oberen Halbebene H in D gilt dann $F^*ds_D^2 \leq ds_H^2$, wobei ds_H^2 die Poincaré Metrik auf H bezeichnet. Aus dieser Tatsache hat Borel mit einem eleganten und höchst einfachen Argument geschlossen, dass die Eigenwerte eines jeden γ_j Einheitswurzeln sind. Dieser Beweis ist in [6] und [11] wiedergegeben. Also:

(10) die Eigenwerte von γ_j, $1 \leq j \leq n$, sind Einheitswurzeln.

In der Situation (6) kann man natürlich zu einer endlichen Überlagerung $\Delta^{*n} \longrightarrow \Delta^{*n}$ übergehen; die Transformationen γ_j müssen dann durch entsprechende Potenzen $\gamma_j^{N_j}$ ersetzt werden. Es ist deshalb möglich, mit einer solchen endlichen Überlagerung alle γ_j unipotent zu machen. Der Einfachheit halber nehme ich an, dass dies geschehen ist; ich verschärfe die Aussage (10) zu

(11) γ_j ist unipotent, für $1 \leq j \leq n$.

Wegen (9) und (11) sind die γ_j in einer maximalen unipotenten Untergruppe $N \subset G_{\mathbb{R}}$ enthalten. Eine solche Gruppe N halte ich von jetzt ab fest.

Bevor das Resultat dieser Arbeit erklärt werden kann, ist es nötig, einige Tatsachen über hermitesch symmetrische Räume anzuführen, die in den klassischen Fällen von Piatetski-Schapiro [10] stammen, und die allgemein von Koranyi-Wolf [9] bewiesen worden sind. Es sei $\mathfrak{n}$ die Lie Algebra der maximalen unipotenten Untergruppe $N \subset G_{\mathbb{R}}$. Dann existieren

a) endlich dimensionale, reelle Vektorräume V, W, deren Komplexifizierungen mit $V_{\mathbb{C}}$, $W_{\mathbb{C}}$ bezeichnet werden sollen;

b) ein offener, konvexer Kegel $C \subset V$, dessen abgeschlossene Hülle $\bar{C}$ keine Gerade enthält;

c) eine hermitesche Bilinearform[3] $B : W_{\mathbb{C}} \times W_{\mathbb{C}} \longrightarrow V_{\mathbb{C}}$, so dass $B(w,w) \in \bar{C}$, für alle $w \in W_{\mathbb{C}}$, und $B(w,w) \neq 0$ falls $w \neq 0$;

d) eine holomorphe Einbettung $D \subset V_{\mathbb{C}} \oplus W_{\mathbb{C}}$,
$D = \{(v,w) \in V_{\mathbb{C}} \oplus W_{\mathbb{C}} \mid \operatorname{Im} v - B(w,w) \in C\}$;

e) eine Unteralgebra $\mathfrak{u} \subset \mathfrak{n}$ und zwei Ideale $\mathfrak{n}_1 \subset \mathfrak{n}$, $\mathfrak{n}_2 \subset \mathfrak{n}$, mit $\mathfrak{n} = \mathfrak{u} \oplus \mathfrak{n}_1 \oplus \mathfrak{n}_2$;

f) ein linearer Isomorphismus $\varphi : \mathfrak{n}_1 \oplus \mathfrak{n}_2 \xrightarrow{\sim} V \oplus W$, mit $\varphi(\mathfrak{n}_1) = V$, $\varphi(\mathfrak{n}_2) = W$;

g) eine Fortsetzung der Operation von N auf D zu einer Operation auf $V_{\mathbb{C}} \oplus W_{\mathbb{C}}$, so dass gilt:
$\exp(X + Y)(v,w) =$
$(v + \varphi(X) + 2i\, B(v,\varphi(Y)) + i\, B(\varphi(Y),\varphi(Y)),\ w + \varphi(Y))$, für $X \in \mathfrak{n}_1$, $Y \in \mathfrak{n}_2$, $v \in V_{\mathbb{C}}$, $w \in W_{\mathbb{C}}$; die Untergruppe $\exp \mathfrak{u} \subset N$ operiert auf $V_{\mathbb{C}} \oplus W_{\mathbb{C}}$ als eine unipotente, lineare Gruppe, wobei die Summanden $V_{\mathbb{C}}$ und $W_{\mathbb{C}}$ erhalten bleiben; für $u \in \exp \mathfrak{u}$, $w_1, w_2 \in W_{\mathbb{C}}$ gilt $B(uw_1, uw_2) = u\, B(w_1, w_2)$.

<u>(12) Satz</u> Es gibt Elemente $T_j \in \bar{C}$, $1 \leq j \leq n$, mit $\gamma_j = \exp \varphi^{-1}(T_j)$, und eine holomorphe Abbildung

$$\Psi : \Delta^n \longrightarrow V_{\mathbb{C}} \oplus W_{\mathbb{C}},$$
$$\Psi : (t) \longmapsto (v(t),\ w(t)),$$

[3] d.h. B ist $\mathbb{C}$-linear in der ersten Variablen, und $B(w_2, w_1)$ ist zu $B(w_1, w_2)$ komplex konjugiert, bezüglich der reellen Form $V \subset V_{\mathbb{C}}$.

so dass

$\tilde{\Phi}(z_1,\ldots,z_n) = (v\circ\tau(z_1,\ldots,z_n) + \sum_{j=1}^n z_jT_j,\ w\circ\tau(z_1,\ldots,z_n))$ (vgl. (7)).

Um die Aussage des Satzes zu verdeutlichen, mag es helfen, Φ als mehrdeutige Abbildung von Δ^{*n}, mit Werten in D, zu betrachten. Auf diese Weise erhält man die Formel

$$\Phi(t_1,\ldots,t_n) = \tag{13}$$
$$(v(t_1,\ldots,t_n) + \sum_{j=1}^n \frac{\log t_j}{2\pi i} T_j,\ w(t_1,\ldots,t_n)).$$

Ist D die Siegelsche obere Halbebene, so findet man

$W = 0$,

V = Raum der reellen, symmetrischen $g\times g$ Matrizen,

C = Kegel der positiv semidefiniten Matrizen in V.

Dadurch vereinfacht sich die Identität (13) zu

$$\Phi(t_1,\ldots,t_n) = v(t_1,\ldots,t_n) + \sum_{j=1}^n \frac{\log t_j}{2\pi i} T_j; \tag{14}$$

nun ist $v(t_1,\ldots,t_n)$ eine holomorphe Funktion auf Δ^n, mit Werten im Raum der komplexen, symmetrischen $g\times g$ Matrizen, und die T_j sind reelle, symmetrische, positiv semidefinite $g\times g$ Matrizen.

Wie noch erwähnt werden sollte, ist die Annahme, dass $\Gamma\subset G_{\mathbb{R}}$ arithmetisch definiert werden kann, für den Satz nur indirekt wichtig. Borels Beweis der Behauptung (10) hängt von dieser Voraussetzung ab; sobald aber (10) bekannt ist, wird die Voraussetzung unnötig. Der Satz (12) entspricht dem "nilpotent orbit theorem" und dem "SL_2-orbit theorem" aus [11] ; darauf werde ich nach dem Beweis des Satzes noch näher eingehen. Übrigens ist der Satz (12) auch mit der -von Borel [3] bewiesenen- Tatsache verwandt, dass sich die Abbildung (6) holomorph zu einer Abbildung des Polyzylinders Δ^n in die Borel-Baily Kompaktifizierung des Quotienten $\Gamma\backslash D$ fortsetzen lässt. Am Ende dieser Arbeit wird davon noch die Rede sein.

Der Beweis des Satzes besteht aus mehreren Hilfssätzen, die nun folgen.

(15) Hilfssatz Jede Koordinatenfunktion f der Abbildung $\tilde{\Phi}$, bezüglich der Einbettung $D \subset V_{\mathbb{C}} \oplus W_{\mathbb{C}}$, genügt der Abschätzung

$$|f(z_1,\ldots,z_n)| = O(\prod_{j=1}^{n}(\mathrm{Im}\, z_j)^{\alpha}),$$

für $\mathrm{Im}\, z_j \longrightarrow \infty$, $1 \leq j \leq n$. Die Abschätzung gilt gleichmässig auf jedem Produkt von vertikalen Streifen

$$\left\{(z_1,\ldots,z_n) \in H^n \,\middle|\, |\mathrm{Re}\, z_j| \leq \text{const.}\right\},$$

und α ist eine geeignete positive Konstante.

Beweis Es sei $\lambda \in V_{\mathbb{C}}^*$ ein lineares Funktional, reellwertig auf V, mit $\lambda(C) > 0$. Wenn man die $V_{\mathbb{C}}$-Komponente von $\tilde{\Phi}$ mit λ verknüpft, erhält man eine holomorphe Funktion $f : H^n \longrightarrow H$. Für eine solche Koordinatenfunktion f folgt die gewünschte Abschätzung aus bekannten, elementaren Argumenten. Der Dualraum V^* lässt sich von Elementen λ, mit $\lambda(C) > 0$, aufspannen. Also erhält man die Aussage des Satzes zumindest für die $V_{\mathbb{C}}$-Koordinate von $\tilde{\Phi}$. Nun bezeichne ich die $V_{\mathbb{C}}$-Koordinate mit $v(z_1,\ldots,z_n)$ und die $W_{\mathbb{C}}$-Koordinate mit $w(z_1,\ldots,z_n)$. Wenn man geeignete lineare Normen auf $V_{\mathbb{C}}$ und $W_{\mathbb{C}}$ aussucht, folgt aus

$$\mathrm{Im}\, v(z_1,\ldots,z_n) \in B(w(z_1,\ldots,z_n), w(z_1,\ldots,z_n)) + C$$

die Ungleichung

$$\begin{aligned}\|\mathrm{Im}\, v(z_1,\ldots,z_n)\| &\leq \text{const.}\|B(w(z_1,\ldots,z_n), w(z_1,\ldots,z_n))\| \\ &\leq \text{const.}\|w(z_1,\ldots,z_n)\|^2.\end{aligned}$$

Wendet man jetzt die Abschätzung von $v(z_1,\ldots,z_n)$ an, so folgt die gewünschte Aussage.

Die Komplexifizierung $N_{\mathbb{C}}$ der Gruppe N operiert ebenfalls auf

$V_{\mathbb{C}} \oplus W_{\mathbb{C}}$, und zwar nach den folgenden Regeln: für $X \in \mathfrak{n}_{1,\mathbb{C}}$, $Y \in \mathfrak{n}_{2,\mathbb{C}}$, $v \in V_{\mathbb{C}}$, $w \in W_{\mathbb{C}}$ gilt

$$\begin{aligned} &\exp(X+Y) = \\ &(v + \varphi(X) + 2i\,B(v,\varphi(\bar{Y})) + i\,B(\varphi(Y),\varphi(\bar{Y})), w + \varphi(Y)); \end{aligned} \tag{16}$$

dabei bedeutet das Überstreichen komplexe Konjugation, relativ zur reellen Form $W \subset W_{\mathbb{C}}$, und φ muss $\mathbb{C}$-linear erweitert werden. Die Untergruppe $\exp \mathfrak{n}_{\mathbb{C}} \subset N_{\mathbb{C}}$ operiert wieder linear auf $V_{\mathbb{C}}$ und $W_{\mathbb{C}}$. Offensichtlich ist diese Aktion der Gruppe $N_{\mathbb{C}}$ auf $V_{\mathbb{C}} \oplus W_{\mathbb{C}}$ holomorph.

Für $1 \leq j \leq n$ setze ich

$$T'_j = \log \gamma_j. \tag{17}$$

Diese Definition ist sinnvoll, wegen (11). Die T'_j liegen in $\mathfrak{n}$, und sie sind miteinander vertauschbar (vgl. (9)). Als Element von $N_{\mathbb{C}}$ operiert

$$\exp\left(\sum_{j=1}^{n} z_j T'_j\right)$$

auf $V_{\mathbb{C}} \oplus W_{\mathbb{C}}$. Ich betrachte die Abbildung

$$\begin{aligned} &\tilde{\Psi} : H^n \longrightarrow V_{\mathbb{C}} \oplus W_{\mathbb{C}}, \\ &\tilde{\Psi}(z_1,\ldots,z_n) = \exp\left(-\sum_{j=1}^{n} z_j T'_j\right) \circ \tilde{\Phi}(z_1,\ldots,z_n). \end{aligned} \tag{18}$$

Da $N_{\mathbb{C}}$ holomorph auf $V_{\mathbb{C}} \oplus W_{\mathbb{C}}$ operiert, ist auch $\tilde{\Psi}$ holomorph. Wegen (8), und weil die T'_j miteinander vertauschbar sind, findet man

$$\tilde{\Psi}(z_1,\ldots,z_{j-1}, z_j + 1, z_{j+1},\ldots,z_n) = \tilde{\Psi}(z_1,\ldots,z_n).$$

Es gibt also eine holomorphe Abbildung

$$\begin{aligned} &\Psi : \Delta^{*n} \longrightarrow V_{\mathbb{C}} \oplus W_{\mathbb{C}}, \\ &\text{mit } \tilde{\Psi} = \Psi \circ \tau . \end{aligned} \tag{19}$$

<u>(20) Hilfssatz</u> Ψ lässt sich holomorph auf den Polyzylinder Δ^n fortsetzen.

<u>Beweis</u> Die Koordinatenfunktionen von $\tilde{\Psi}$ kann man als Polynome in den z_j und den Koordinatenfunktionen von $\tilde{\Phi}$ ausdrücken; das folgt aus der expliziten Beschreibung der $N_{\mathbb{C}}$-Aktion auf $V_{\mathbb{C}} \oplus W_{\mathbb{C}}$, und weil $\exp \mathfrak{n}_{\mathbb{C}}$ als unipotente Matrizengruppe operiert. Ich wende den Hilfssatz (15) an und finde

$$\|\tilde{\Psi}(z_1,\ldots,z_n)\| = O(\prod_{j=1}^{n} (\mathrm{Im}\, z_j)^{\beta}),$$

für $\mathrm{Im}\, z_j \longrightarrow \infty$, $1 \leqslant j \leqslant n$, gleichmässig auf jedem Produkt von vertikalen Streifen. Für Ψ folgt daraus die Abschätzung

$$\|\Psi(t_1,\ldots,t_n)\| = O(\prod_{j=1}^{n} (\log|t_j|^{-1})^{\beta}),$$

für $t_j \longrightarrow 0$, $1 \leqslant j \leqslant n$. Folglich kann Ψ entlang $\Delta^n - \Delta^{*n}$ keine Singularitäten haben.

Nun sei $f : H \longrightarrow \mathbb{C}$ eine holomorphe Funktion, die sich als Summe

$$f(z) = \sum_{k=o}^{m} a_k(e^{2\pi i z}) z^k \tag{21}$$

ausdrücken lässt, wobei $a_o,\ldots,a_m$ holomorph auf Δ sind.

<u>(22) Hilfssatz</u> Falls $\mathrm{Im}\, f(z) > 0$, für alle z mit hinreichend grossem Imaginärteil, so gilt $a_k \equiv 0$ für alle $k \geqslant 2$, und a_1 ist konstant, reell und nicht negativ.

<u>Beweis</u> Angenommen $a_m \not\equiv 0$. Für jedes z_o mit genügend grossem Imaginärteil muss $a_m(e^{2\pi i z_o})$ ungleich Null sein, und $\mathrm{Im}\, f(z_o + x) > 0$, für alle $x \in \mathbb{R}$. Daraus ergibt sich

$$\begin{aligned} 0 &\leqslant \mathrm{Im}\left\{\lim_{k\to\infty} k^{-m} f(z_o + k)\right\} \\ &= \mathrm{Im}\left\{\lim_{k\to\infty} \sum_{j=o}^{m} k^{-(m-j)} a_j(e^{2\pi i z_o})(1 + z_o/k)^j\right\} \\ &= \mathrm{Im}\, a_m(e^{2\pi i z_o}). \end{aligned}$$

Betrachtet man die Folge $\{k^{-m} f(z_o - k)\}$, so erhält man auf gleiche Weise $(-1)^m \operatorname{Im} a_m(e^{2\pi i z_o}) \geq 0$. Diese beiden Ungleichungen sind aber nur möglich, wenn

(*) $$a_m(0) \neq 0, \text{ und } a_m = \text{const. falls } m \notin 2\mathbb{Z}.$$

Gegeben sei θ, $0 < \theta < \pi$. Damit konstruiere ich eine Folge komplexer Zahlen $\{z_k\}_{k=o}^{\infty}$, so dass $\operatorname{Im} z_k \to \infty$, und $\operatorname{Arg} z_k = \theta$. Dann gilt

$$0 \leq \operatorname{Im}\{\lim_{k\to\infty} |z_k|^{-m} f(z_k)\}$$

(**) $$= \operatorname{Im}\{\lim_{k\to\infty} \sum_{j=o}^{m} |z_k|^{-(m-j)} a_j(e^{2\pi i z_k}) e^{ij\theta}\}$$

$$= \operatorname{Im}(e^{im\theta} a_m(0)).$$

Weil θ zwischen Null und π beliebig gewählt werden kann, muss m kleiner als zwei sein. Für $m = 1$ liefert (*) die Aussage $a_1 = \text{const.}$; aus (**) folgt sodann $a_1 \in \mathbb{R}$, $a_1 \geq 0$.

Ich bezeichne die $V_{\mathbb{C}}$- und $W_{\mathbb{C}}$- Koordinaten der Abbildung $\tilde{\Phi} : H^n \to V_{\mathbb{C}} \oplus W_{\mathbb{C}}$ mit $\tilde{v}(z_1,\dots,z_n)$ und $\tilde{w}(z_1,\dots,z_n)$.

<u>(23) Hilfssatz</u> Es existieren Elemente $T_j \in \overline{C}$, und holomorphe Abbildungen $v : \Delta^n \to V_{\mathbb{C}}$, $w : \Delta^n \to W_{\mathbb{C}}$, so dass $\tilde{w} = w \circ \tau$, und

$$\tilde{v}(z_1,\dots,z_n) = v \circ \tau(z_1,\dots,z_n) + \sum_{j=1}^{n} z_j T_j.$$

<u>Beweis</u> Die Definition (18) ist gleichbedeutend mit

$$\tilde{\Phi}(z_1,\dots,z_n) = \exp(\sum_{j=1}^{n} z_j T_j') \circ \tilde{\Psi}(z_1,\dots,z_n).$$

Aus dieser Gleichung, aus Hilfssatz (20) und aus der expliziten Beschreibung der $N_{\mathbb{C}}$-Aktion auf $V_{\mathbb{C}} \oplus W_{\mathbb{C}}$ folgt: für jedes lineare Funktional $\lambda \in V_{\mathbb{C}}^*$ lässt sich die Funktion

(*) $$f(z_1,\dots,z_n) = \lambda \circ v(z_1,\dots,z_n)$$

als Polynom

$$\sum_{j_1,\ldots,j_n} a_{j_1\ldots j_n} \circ \tau(z_1,\ldots,z_n)\, z_1^{j_1}\ldots z_n^{j_n}$$

ausdrücken, wobei die $a_{j_1\ldots j_n}$ holomorphe Funktionen auf Δ^n sind. Die analoge Behauptung über $\tilde{w}$ trifft natürlich ebenso zu.

Ich wähle nun ein $\lambda \in V^*_{\mathbb{C}}$, das reelle Werte auf V annimmt, und positive Werte auf C. Die Funktion f in (*) hat dann einen positiven Imaginärteil, weil ja

$$\operatorname{Im} \tilde{v}(z_1,\ldots,z_n) \in C + B(\tilde{w}(z_1,\ldots,z_n),\tilde{w}(z_1,\ldots,z_n)) \subset C.$$

Wendet man den Hilfssatz (22) auf f an, wobei jeweils alle Koordinaten mit einer einzigen Ausnahme festzuhalten sind, so findet man

$$f(z_1,\ldots,z_n) =$$
$$g\circ\tau(z_1,\ldots,z_n) + \sum_{j=1}^{n} h_j(z_1,\ldots,z_{j-1},z_{j+1},\ldots,z_n)z_j,$$

mit einer holomorphen Funktion g auf Δ^n. Der j-te Koeffizient h_j hängt holomorph von $z_1,\ldots,z_{j-1},z_{j+1},\ldots,z_n$ ab, ist andererseits aber immer reell und nicht negativ. Folglich ist h_j konstant. Diejenigen $\lambda \in V^*$, die auf C positive Werte annehmen, spannen V^* auf. Also gilt

$$v(z_1,\ldots,z_n) = v\circ\tau(z_1,\ldots,z_n) + \sum_{j=1}^{n} z_j T_j,$$

mit v holomorph auf Δ^n und $T_j \in V$. Wegen

$$\lambda \in V^*,\ \lambda(C) > 0 \implies (\lambda, T_j) > 0$$

müssen die X_j in $\overline{C}$ liegen.

Es bleibt zu zeigen, dass $\tilde{w} = w\circ\tau$, für eine geeignete $W_{\mathbb{C}}$-wertige, holomorphe Funktion w auf Δ^n. Zunächst betrachte ich den Fall n = 1. Wie man mit den Überlegungen am Anfang dieses Beweises schliessen kann, gibt es holomorphe Funktionen $w_j : \Delta \longrightarrow W_{\mathbb{C}}$, $1 \leq j \leq m$, so dass

$$\tilde{w}(z) = \sum_{j=1}^{m} w_j(e^{2\pi i z}) z^j .$$

Angenommen, w_m sei nicht identisch Null, und $m \geq 1$. Ich wähle ein $z \in H$, mit $w_m(e^{2\pi i z}) \neq 0$. Dann findet man

$$\begin{aligned} &\lim_{k \to \infty} k^{-2m} \left\{ \operatorname{Im} \tilde{v}(z+k) - B(\tilde{w}(z+k), \tilde{w}(z+k)) \right\} \\ &= \lim_{k \to \infty} \left\{ - B(w_m(e^{2\pi i z}), w_m(e^{2\pi i z})) + k^{-1}(\ldots) \right\} \\ &= - B(w_m(e^{2\pi i z}), w_m(e^{2\pi i z})). \end{aligned}$$

Dieser Grenzwert muss in $\overline{C}$ liegen, weil alle Glieder der Folge aus C stammen. Aber auch

$$B(w_m(e^{2\pi i z}), w_m(e^{2\pi i z}))$$

liegt in $\overline{C}$, und ist ungleich Null. Widerspruch! Für beliebiges n folgt nun, dass $\tilde{w} = w \circ \tau$, wobei w eine holomorphe, $W_{\mathbb{C}}$-wertige Funktion auf der Menge

$$\bigcup_{1 \leq i < j \leq n} \left\{ (z_1, \ldots, z_n) \in \Delta^n \;\middle|\; z_i \neq 0, z_j \neq 0 \right\}$$

ist. Von dieser Menge lässt sich jede Funktion holomorph auf Δ^n fortsetzen. Das beweist den Hilfssatz.

Der Satz (12) folgt jetzt leicht. Für $1 \leq j \leq n$ ist $\gamma_j' = \exp(\varphi^{-1} T_j)$ ein Element der Gruppe N. Wie das vorhergehende Lemma zeigt, gilt

$$\tilde{\Phi}(z_1, \ldots, z_{j-1}, z_j + 1, z_{j+1}, \ldots, z_n) = \gamma_j' \circ \tilde{\Phi}(z_1, \ldots, z_n).$$

Mit γ_j anstelle von γ_j' besteht diese Identität ebenfalls (vgl. (8)), und γ_j liegt auch in N. Insbesondere hat $\gamma_j^{-1} \gamma_j'$ mindestens einen Fixpunkt auf D. Weil aber $G_{\mathbb{R}}$ mit kompakten Isotropiegruppen auf D operiert, und weil N eine kompakte Untergruppe von $G_{\mathbb{R}}$ nur in der Identität schneiden kann, müssen γ_j und γ_j' zusammenfallen. In anderen Worten,

$$\gamma_j = \exp(\varphi^{-1} T_j), \quad \text{für } 1 \leq j \leq n.$$

Damit ist der Satz bewiesen.

Wie schon erwähnt wurde, enthält der Satz (12) das "nilpotent orbit theorem" und das "SL_2-orbit theorem" aus [11] , sofern -wie in dieser Arbeit angenommen- D hermitesch symmetrisch ist. Im allgemeinen Fall sehen diese beiden Theoreme wesentlich komplizierter aus; das liegt daran, dass man sich nicht auf ein globales Koordinatensystem stützen kann. Die beiden Theoreme behaupten, grob gesprochen, dass im Fall n = 1 die Abbildung $\tilde{\Phi} : H \longrightarrow D$ asymptotisch durch eine holomorphe, äquivariante, total geodätische Einbettung der oberen Halbebene H in D approximiert werden kann. Ich will nun kurz erläutern, wie der Satz (12) eine solche Einbettung liefert. Um technische Komplikationen zu vermeiden, beschränke ich mich auf den Fall D = Siegelsche obere Halbebene. Für einen beliebigen hermitesch symmetrischen Raum kann man ganz analog argumentieren[4], wenn man die Ergebnisse von Koranyi-Wolf [9] benutzt.

Es sei also n = 1 und D = Siegelsche obere Halbebene; W, V, C haben die gleiche Bedeutung wie unterhalb der Gleichung (13). Ich betrachte den Homomorphismus

$$\rho : Gl(g,\mathbb{R}) \longrightarrow G_{\mathbb{R}} = Sp(g,\mathbb{R}) \quad ,$$

$$\rho : A \longmapsto \begin{pmatrix} A & 0 \\ 0 & {}^{t}A^{-1} \end{pmatrix} \quad ,$$

dessen Bild ich mit H bezeichne. Als Untergruppe von $G_{\mathbb{R}}$ operiert H auf D. Wie man nachrechnen kann, lässt sich diese Operation auf $V_{\mathbb{C}}$ erweitern: für $h = \rho(A)$, mit $A \in Gl(g,\mathbb{R})$, und $Z \in V_{\mathbb{C}}$ gilt $hZ = AZ\,{}^{t}A$.

4) Dabei sollte man beobachten, dass die $W_{\mathbb{C}}$-Koordinaten der Abbildung $\tilde{\Phi}$ asymptotisch konstant bleiben. In der Einbettung $D \subset V_{\mathbb{C}} \oplus W_{\mathbb{C}}$ entspricht die Faser über einem festen $w \in W_{\mathbb{C}}$ einem Tubengebiet. Man kann also zunächst $\tilde{\Phi}$ durch eine Abbildung in einen Unterraum $D' \subset D$ approximieren, so dass D' ein Tubengebiet ist. Im Falle eines Tubengebietes schliesslich kann man die Argumente für die Siegelsche obere Halbebene genau kopieren.

Wenn man nun $\tilde{\Phi}$ mit dem von $h = \rho(A)$ definierten Automorphismus des Raumes D zusammenschaltet, muss man $T = T_1$ im Satz (12) durch AT^tA ersetzen. Statt $\tilde{\Phi}$ mit h zusammenzuschalten, kann man auch -mit dem gleichen Ergebnis- die Einbettung $D \subset V_{\mathbb{C}}$ um h modifizieren. In anderen Worten, für jedes gegebene $A \in Gl(g,\mathbb{R})$ darf ich T durch AT^tA ersetzen. Als positiv semidefinite Matrix ist T einer Matrix der Form

$$(24) \qquad \begin{pmatrix} 1 & & & & & \\ & \ddots & & & 0 & \\ & & 1 & & & \\ & & & 0 & & \\ & 0 & & & \ddots & \\ & & & & & 0 \end{pmatrix} \begin{matrix} \left.\vphantom{\begin{matrix}1\\ \ddots\\ 1\end{matrix}}\right\} r \\ \left.\vphantom{\begin{matrix}0\\ \ddots\\ 0\end{matrix}}\right\} g-r \end{matrix}$$

ähnlich. Ich darf und werde deshalb annehmen, dass T schon die Form (24) hat.

In dieser Situation wird $\tilde{\Phi}$ durch die Abbildung

$$(25) \qquad \begin{aligned} H &\longrightarrow D \subset V_{\mathbb{C}} \quad , \\ z &\longmapsto \begin{pmatrix} z & & & & & \\ & \ddots & & & & \\ & & z & & & \\ & & & i & & \\ & & & & \ddots & \\ & & & & & i \end{pmatrix} \end{aligned}$$

approximiert. Das ist nun eine holomorphe, äquivariante, total geodätische Einbettung der oberen Halbebene H in D. Und zwar entspricht die äquivariante Einbettung (25) dem Homomorphismus

$$SL(2,\mathbb{R}) \longrightarrow G_{\mathbb{R}} = Sp(g,\mathbb{R}) \quad ,$$

$$\begin{pmatrix} a & b \\ c & d \end{pmatrix} \longmapsto \left(\begin{array}{cccccc|cccccc} a & & & & & 0 & b & & & & & 0 \\ & \ddots & & & & & & \ddots & & & & \\ & & a & & & & & & b & & & \\ & & & 1 & & & & & & 0 & & \\ & & & & \ddots & & & & & & \ddots & \\ 0 & & & & & 1 & 0 & & & & & 0 \\ \hline c & & & & & 0 & d & & & & & \\ & \ddots & & & & & & \ddots & & & & \\ & & c & & & & & & d & & & \\ & & & 0 & & & & & & 1 & & \\ & & & & \ddots & & & & & & \ddots & \\ 0 & & & & & 0 & & & & & & 1 \end{array}\right)$$

Es ist natürlich möglich, den Satz (12) und die beiden Theoreme aus [11] noch genauer zu vergleichen. Das will ich aber unterlassen.

Borel [3] hat bewiesen, dass sich die Abbildung (6) zu einer Abbildung des Polyzylinders Δ^n in die Borel-Baily Kompaktifizierung fortsetzen lässt. Wegen der Beschreibung der Topologie der Borel-Baily Kompaktifizierung in [1] würde die Fortsetzbarkeit der Abbildung auch aus den beiden Aussagen (26) und (27) folgen:

(26) es gibt ein Siegel Gebiet (im Sinne von [1]) $S \subset D$, welches das $\tilde{\Phi}$-Bild der Menge $\{(z_1,\ldots,z_n) \in H^n \mid |\mathrm{Re}\, z_j| \leqslant 1,\ \mathrm{Im}\, z_j \geqslant 1\}$ enthält;

und

(27) für jede Folge $(z_1^{(k)},\ldots,z_n^{(k)})$ in H^n, mit $|\mathrm{Re}\, z_j^{(k)}| \leqslant 1$ und $\mathrm{Im}\, z_j^{(k)} \longrightarrow \infty$, konvergiert $\tilde{\Phi}(z_1^{(k)},\ldots,z_n^{(k)})$ zu einem Punkt der abgeschlossenen Hülle $\bar{D}$.

Im Falle der Siegelschen oberen Halbebene kann man (26) und (27) relativ leicht vom Satz (12) ableiten. Für n = 1 und beliebiges D erhält man ebenfalls (26) und (27) aus (12), mit ein wenig mehr Anstrengung (vgl. auch Korollar (5.29) in [11]). Will man im allgemeinen Fall (26) und (27) aus dem Satz (12) herleiten, so scheint mir, dass das fast darauf hinausläuft, die Gleichheit der von Piatetski-Schapiro [10] und von Baily-Borel [1] definierten Topologien zu beweisen[5]. Die Äquivalenz dieser beiden Topologien ist aber sicherlich ein tiefergehendes Problem als die Fortsetzbarkeit der Abbildung (6) (vgl. die Arbeit [7]).

---

[5] Zusammen mit Satz (12) impliziert die Gleichheit der beiden Topologien ziemlich direkt die Fortsetzbarkeit der Abbildung $\tilde{\Phi}$.

Literatur

[1] Baily, W.L. und A. Borel: Compactification of arithmetic quotients of bounded symmetric domains. Ann. of Math. 84 (1966), 442-528.

[2] Borel, A.: Introduction aux Groupes Arithmétiques. Paris, Hermann 1969.

[3] Borel, A.: Some metric properties of arithmetic quotients of symmetric spaces and an extension theorem. J. Diff. Geom. 6 (1972), 543-560.

[4] Griffiths, P.: Periods of integrals on algebraic manifolds I, II. Amer. J. Math. 90 (1968), 568-626 und 805-865.

[5] Griffiths, P.: Periods of integrals on algebraic manifolds III. Publ. Math. I.H.E.S. 38 (1970), 125-180.

[6] Griffiths, P.: Periods of integrals on algebraic manifolds: summary of main results and discussion of open problems. Bull. Amer. Math. Soc. 76 (1970), 228-296.

[7] Kiernan, P.: On the compactifications of arithmetic quotients of symmetric spaces. Bull. Amer. Math. Soc. 80 (1974), 109-110.

[8] Kiernan, P. und S. Kobayashi: Satake compactification and extension of holomorphic mappings. Inventiones Math. 16 (1972), 237-248.

[9] Koranyi, A. und J. Wolf: Realization of hermitian symmetric spaces as generalized half planes. Ann. Math. 81 (1965), 265-288.

[10] Piatetski-Schapiro, I.I.: Géométrie des Domaines Classiques et Théorie des Fonctions Automorphes. Paris, Dunod 1966. Eine überarbeitete englische Übersetzung erschien bei Gordon and Breach, New York, 1969.

[11] Schmid, W.: Variation of Hodge structure: the singularities of the period mapping. Inventiones Math. 22 (1973), 211-319.

Singular abelian surfaces and binary quadratic forms

Tetsuji SHIODA and Naoki MITANI

By a singular abelian surface we mean a complex abelian variety of dimension 2 whose Picard number is equal to the maximum possible value 4. In this note we prove that singular abelian surfaces are in one-to-one correspondence with equivalence classes of positive definite even integral binary quadratic forms with respect to $SL_2(\mathbb{Z})$ (Theorem 3.1). As a consequence of this classification, every singular abelian surface turns out to be a product of two elliptic curves.

The result in this note depends on a general theorem on period map of abelian surfaces [7]. Both this note and [7] have been motivated by the work of Pjateckii-Šapiro and Šafarevič [4].

We wish to thank Professor Hirzebruch for stimulating conversations.

§1. "Singular" algebraic surfaces.

Let X be a non-singular complex algebraic surface. Consider the exact sequence of sheaves on X:

$$(1.1) \qquad 0 \longrightarrow \mathbb{Z} \xrightarrow{j} \mathcal{O} \longrightarrow \mathcal{O}^* \longrightarrow 0 ,$$

and the associated exact sequence of cohomology groups :

$$(1.2) \qquad H^1(X, \mathcal{O}^*) \xrightarrow{\delta} H^2(X, \mathbb{Z}) \xrightarrow{j^*} H^2(X, \mathcal{O}) .$$

Let

$$(1.3) \qquad \begin{aligned} S_X &= \mathrm{Im}(\delta) = \mathrm{Ker}(j^*) \subset H^2(X, \mathbb{Z}) , \\ \rho(X) &= \mathrm{rank}\, S_X . \end{aligned}$$

The group S_X, consisting of "algebraic cocycles", is called the <u>Néron-Severi group</u> of X, and its rank $\rho(X)$ the <u>Picard number</u> of X. It follows from (1.2) that

$$(1.4) \qquad \rho(X) \leq h^{1,1} = b_2 - 2p_g ,$$

where b_2 is the second Betti number of X and $p_g = \dim_{\mathbb{C}} H^2(X, \mathcal{O})$ is the geometric genus of X. Following a classical terminology, we call a surface X <u>singular</u> if $\rho(X) = h^{1,1}$.

For any complex surface X, $H_X = H^2(X, \mathbb{Z})/(\text{torsion})$ is a Euclidean lattice[1]. Namely it has a (non-degenerate) $\mathbb{Z}$-valued scalar product defined by the cup product:

$$(1.5) \qquad H^2(X, \mathbb{Z}) \times H^2(X, \mathbb{Z}) \longrightarrow H^4(X, \mathbb{Z}) \overset{\mathrm{id.}}{=} \mathbb{Z} .$$

(The last identification is made with respect to the natural orientation of X as a complex manifold.) The image $\overline{S}_X$ of S_X in H_X is a primitive sublattice of H_X, i.e. a sublattice with

---

1) The definitions concerning Euclidean lattices are summarized in the appendix of this note.

the torsion-free quotient $H_X/\bar{S}_X$, because the latter can be embedded in the vector space $H^2(X, \mathcal{O})$ by (1.2). Let T_X denote the orthogonal complement of $\bar{S}_X$ in H_X; it is also a primitive sublattice of H_X. We call T_X the <u>group of transcendental cocycles</u> of X. By the Hodge index theorem, $\bar{S}_X$ has the signature $(1, \rho-1)$ and T_X has the signature $(2p_g, h^{1,1}-\rho)$.

<u>Proposition</u> 1.1. Assume that X is a singular algebraic surface. Then

(i) the group T_X of transcendental cocycles of X is a positive definite Euclidean lattice of rank $2p_g$.

(ii) There is a structure of complex torus of dimension p_g on the quotient

$$H^2(X, \mathcal{O})/j^*H^2(X, \mathbb{Z}).$$

<u>Proof</u>. The first assertion is a special case of the index theorem stated above. The second one follows from the two facts:

(a) the image $j^*H^2(X, \mathbb{Z})$ has rank $2p_g$ for X singular,

(b) it generates over real numbers the vector space $H^2(X, \mathcal{O})$ (cf. [5] pp.32-33), q. e. d.

Now suppose that X is an abelian surface (i.e. abelian variety of dimension 2). In this case, $H^2(X, \mathbb{Z})$ is a (torsion-free) even Euclidean lattice of rank $b_2 = 6$, and $p_g = 1$, $h^{1,1} = 4$. For a <u>singular abelian surface</u> X, we obtain therefore the following two objects :

(1.6) $\qquad T_X$ = a positive definite even Euclidean lattice of rank 2 ,

$$(1.7) \qquad C_X = H^2(X, \mathcal{O})/j^*H^2(X, \mathbb{Z}) \quad \text{(an elliptic curve)}.$$

The Euclidean lattice T_X determines (and is determined by) an equivalence class of 2×2 positive definite even integral matrices Q with respect to $GL_2(\mathbb{Z})$. Namely, taking a basis $\{t_1, t_2\}$ of T_X, we put

$$(1.8) \qquad Q = \begin{pmatrix} t_1^2 & t_1 t_2 \\ t_1 t_2 & t_2^2 \end{pmatrix} = \begin{pmatrix} 2a & b \\ b & 2c \end{pmatrix},$$

where a, b, c are integers such that

$$a > 0, \quad c > 0, \quad b^2 - 4ac < 0 .$$

By a change of basis of T_X, Q is transformed to a matrix ${}^tMQM$ for some $M \in GL_2(\mathbb{Z})$. We call

$$(1.9) \qquad m = (a, b, c) \quad (= \text{g.c.d. of integers } a, b, c)$$

the <u>degree of primitivity</u> of T_X. The integral matrix $m^{-1}Q$ is called the <u>primitive even part</u> of Q.

§2. Period map of an abelian surface

In this section, we quote without proof some results in [7]. Let X be an abelian surface. By (1.2) and $p_g = 1$, we have a homomorphism

$$(2.1) \qquad p_X : H^2(X, \mathbf{Z}) \xrightarrow{j^*} H^2(X, \mathcal{O}) \simeq \mathbf{C} ,$$

uniquely defined up to a constant multiple. We call p_X the <u>period map</u> of X. The sublattices S_X and T_X of $H^2(X, \mathbf{Z})$, introduced in §1, can be defined in terms of the period map as

$$(2.2) \qquad S_X = \mathrm{Ker}(p_X), \quad T_X = \mathrm{Ker}(p_X)^{\perp} .$$

Let us identify the two vector spaces

$$(2.3) \qquad \mathrm{Hom}(H^2(X, \mathbf{Z}), \mathbf{C}) = H^2(X, \mathbf{C})$$

by means of the cup product. With respect to the natural scalar product in (2.3), we have

$$(2.4) \qquad p_X^2 = 0 , \quad p_X \overline{p_X} > 0 .$$

Next we represent our abelian surface X as a complex torus $\mathbf{C}^2/L$ (L a lattice in $\mathbf{C}^2$), and make the identifications (cf. [3]):

$$(2.5) \qquad \begin{cases} H_1(X, \mathbf{Z}) = L \\ H^1(X, \mathbf{Z}) = L^* = \mathrm{Hom}(L, \mathbf{Z}) \\ H^2(X, \mathbf{Z}) = \wedge^2(L^*) . \end{cases}$$

Take a basis $\{v_1, v_2, v_3, v_4\}$ of L, and the dual basis $\{u^1, u^2, u^3, u^4\}$ in L^* (i.e. $u^i(v_j) = \delta_{ij}$). Then $\{u^i \wedge u^j \mid 1 \leq i < j \leq 4\}$ forms a basis of $H^2(X, \mathbf{Z})$, which is also a $\mathbf{C}$-basis of $H^2(X, \mathbf{C})$. The period map p_X, considered as an element of $H^2(X, \mathbf{C})$ by (2.3), is given by the formula:

$$p_X = \sum_{i<j} \det(v_i \; v_j) u^i \wedge u^j \quad . \tag{2.6}$$

Recall that an abelian variety X is called <u>auto-dual</u> if X is isomorphic to the Picard variety of X.

<u>Theorem</u> 2.1. Let X, Y be abelian surfaces and assume X is auto-dual. Suppose that there exists an isometry (i.e. isomorphism preserving scalar products):

$$\psi : H^2(X, \mathbb{Z}) \xrightarrow{\sim} H^2(Y, \mathbb{Z}) \tag{2.7}$$

satisfying

$$p_Y \circ \psi = \text{const.}\ p_X \quad . \tag{2.8}$$

Then Y is isomorphic to X.

The proof can be found in [7]. We remark that, if we drop the assumption of auto-duality, the statement is <u>not</u> true.

§3. Classification of singular abelian surfaces

Theorem 3.1. Singular abelian surfaces are in one-to-one correspondence with equivalence classes of positive definite even integral binary quadratic forms with respect to $SL_2(\mathbb{Z})$.

First we prove a little weaker result.

Theorem 3.2. Let T_X denote the group of transcendental cocycles on a singular abelian surface X. Then $X \longrightarrow T_X$ is a surjective, generically two-to-one correspondence between singular abelian surfaces and positive definite even Euclidean lattices of rank 2.

(I) Construction of certain abelian surfaces. Take an arbitrary positive definite even integral matrix

$$(3.1) \qquad Q = \begin{pmatrix} 2a & b \\ b & 2c \end{pmatrix}, \qquad a,\ b,\ c \in \mathbb{Z}$$

$$a > 0, \quad c > 0, \quad \Delta = b^2 - 4ac < 0 .$$

Let T be the Euclidean lattice of rank 2 defined by Q. We shall construct in a canonical manner abelian surfaces A and A' such that

$$(3.2) \qquad T_A \simeq T , \quad T_{A'} \simeq T .$$

Putting

$$(3.3) \qquad \tau_1 = \frac{-b+\sqrt{\Delta}}{2a} , \qquad \tau_2 = \frac{b+\sqrt{\Delta}}{2} ,$$

we denote by C_ν $(\nu = 1, 2)$ the elliptic curve with the periods $1,\ \tau_\nu$:

$$(3.4) \qquad C_\nu = \mathbb{C}/(\mathbb{Z}+\mathbb{Z}\tau_\nu) \qquad (\nu = 1, 2) .$$

We consider the abelian surface

(3.5) $$A = C_1 \times C_2 = \mathbb{C}^2/L ,$$

where L is a lattice of $\mathbb{C}^2$ generated by

(3.6) $$v_1 = \begin{pmatrix} 1 \\ 0 \end{pmatrix}, \quad v_2 = \begin{pmatrix} 0 \\ 1 \end{pmatrix}, \quad v_3 = \begin{pmatrix} -\tau_1 \\ 0 \end{pmatrix}, \quad v_4 = \begin{pmatrix} 0 \\ \tau_2 \end{pmatrix}.$$

As in §2, we take the dual basis $\{u^i\}$ of $\{v_j\}$, and put $u^{ij} = u^i \wedge u^j$. Note that $u^{12} \wedge u^{34} = 1$ under the natural identification $H^4(A, \mathbb{Z}) = \mathbb{Z}$, (1.5). Computing the period map p_A of A by (2.6), we get

(3.7) $$p_A = u^{12} + cu^{34} + \tau_1 u^{23} + \tau_2 u^{14} .$$

Using (2.2) and (3.7), we can easily see that the group S_A (or T_A) of algebraic (or transcendental) cocycles on A has the following basis $\{s_i\}$ (or $\{t_k\}$) :

(3.8) $$\begin{cases} s_1 = u^{13} , \quad s_2 = u^{42} \\ s_3 = u^{23} - au^{14} - bu^{34} \\ s_4 = u^{12} - cu^{34} \end{cases}$$

(3.9) $$\begin{cases} t_1 = u^{23} + au^{14} \\ t_2 = bu^{14} + u^{12} + cu^{34} . \end{cases}$$

Since $t_1^2 = 2a$, $t_1 t_2 = b$, $t_2^2 = 2c$, the abelian surface A satisfies the condition (3.2). Replacing τ_1, τ_2 by their complex conjugates $\bar{\tau}_1$, $\bar{\tau}_2$, we obtain another abelian surface

(3.10) $$A' = C_1' \times C_2' , \quad C_\nu' = \mathbb{C}/(\mathbb{Z} + \mathbb{Z}\bar{\tau}_\nu) ,$$

and the same argument shows (3.2) for A'. Note that the period map $p_{A'}$ of A' is the complex conjugate of the period map p_A of A :

(3.11) $$p_{A'} = \overline{p_A} .$$

We remark also that (3.7) implies

$$\text{(3.12)} \qquad \mathrm{Im}(p_A) = \mathbb{Z} + c\mathbb{Z} + \tau_1\mathbb{Z} + \tau_2\mathbb{Z} = \mathbb{Z} + \tau_1\mathbb{Z},$$

and hence the elliptic curve C_A of (1.7) is isomorphic to the elliptic curve C_1; i.e.

$$\text{(3.13)} \qquad C_A \simeq C_1, \qquad C_{A'} \simeq C_1'.$$

(II) Next let X be an arbitrary singular abelian surface such that $T_X \simeq T$, T being the Euclidean lattice defined by Q (3.1). We want to prove

$$\text{(3.14)} \qquad X \simeq A \quad \text{or} \quad X \simeq A'.$$

By assumption we have an isometry $\psi_0 : T_A \simeq T_X$. In view of Theorem 1 in the appendix, ψ_0 can be extended to an isometry

$$\text{(3.15)} \qquad \psi : H^2(A, \mathbb{Z}) \xrightarrow{\sim} H^2(X, \mathbb{Z}), \qquad \psi|_{T_A} = \psi_0.$$

Let p_X and p_A be the period maps of X and A. Then two functionals p_A and $p_X \circ \psi$ on $H^2(A, \mathbb{Z})$ have the property:

$$\text{(3.16)} \qquad \begin{cases} p^2 = 0, \quad p\bar{p} > 0 \quad (\text{cf. } (2.4)) \\ p|_{T_A^{\perp}} = 0. \end{cases}$$

Using the uniqueness of such p in Proposition 2 in the appendix, we see

$$\text{(3.17)} \qquad p_X \circ \psi = \begin{cases} \text{const.}\ p_A \\ \text{or} \\ \text{const.}\ \bar{p}_A = \text{const.}\ p_{A'}. \end{cases}$$

The abelian surfaces A and A' are auto-dual, since they are products of two elliptic curves. Therefore we can apply Theorem 2.1 to (3.17) and obtain

$$X \simeq A \quad \text{or} \quad X \simeq A'.$$

This proves (3.14), and consequently Theorem 3.2.

(III) Proof of Theorem 3.1. Let us denote by A_Q the abelian surface A (3.5) constructed from a matrix Q (3.1). When Q is replaced by

$$Q' = {}^tMQM , \quad M \in SL_2(\mathbb{Z}) ,$$

the points τ_1, τ_2 in (3.3) are replaced by τ_1', τ_2' such that

$$\begin{cases} \tau_1' = M^{-1}\cdot\tau = \dfrac{\alpha\tau+\beta}{\gamma\tau+\delta} , & M^{-1} = \begin{pmatrix} \alpha & \beta \\ \gamma & \delta \end{pmatrix} \\ \tau_2' = \tau_2 + n , & n \in \mathbb{Z} . \end{cases}$$

Therefore (the isomorphism class of) the abelian surface A_Q depends only on the equivalence class of Q with respect to $SL_2(\mathbb{Z})$. We also note that the abelian surface A' (3.10) can be written as

$$A' = A_{\begin{pmatrix} 1 & 0 \\ 0 & -1 \end{pmatrix} Q \begin{pmatrix} 1 & 0 \\ 0 & -1 \end{pmatrix}} . \tag{3.18}$$

Let $\mathcal{Q}$ denote the set of all positive definite even integral 2×2 matrices Q, and let $\mathcal{Q}/SL_2(\mathbb{Z})$ or $\mathcal{Q}/GL_2(\mathbb{Z})$ be the set of equivalence classes in $\mathcal{Q}$ with respect to $SL_2(\mathbb{Z})$ or $GL_2(\mathbb{Z})$. Then we have the following commutative diagram :

$$\begin{array}{ccc} \mathcal{Q}/SL_2(\mathbb{Z}) & \xrightarrow{\ f\ } & \{\text{singular abelian surfaces}\}/\text{isom.} \\ \downarrow \text{natural} & & \downarrow g \\ \mathcal{Q}/GL_2(\mathbb{Z}) & \xleftrightarrow{\ 1:1\ } & \left\{\begin{matrix}\text{pos. def. even Euclid. lattices} \\ \text{of rank 2}\end{matrix}\right\}/\text{isom.} \end{array} \tag{3.19}$$

in which f, g are the maps induced by the maps

$$Q \longrightarrow A_Q \quad \text{and} \quad X \longrightarrow T_X .$$

It follows from (3.14) and (3.18) that f is a surjective map. In order to complete the proof of Theorem 3.1, we have only to show that f is injective. Assume

$$(3.20) \qquad A_Q \simeq A_{Q^*} \qquad \text{for } Q, Q^* \in \mathcal{Q} \ .$$

Then it is immediate from (3.19) that

$$(3.21) \qquad Q \sim Q^* \quad \text{w.r.t. } GL_2(\mathbb{Z}) \ .$$

On the other hand, considering the elliptic curve C_A (1.7) associated with A, we obtain from (3.13)

$$C_1 \simeq C_1^* = \mathbb{C}/\mathbb{Z} + \mathbb{Z}\tau_1^* \ ,$$

where τ_1^* is defined for Q^* in the same way as τ_1 for Q (3.3). Hence τ_1 and τ_1^* are equivalent points in the upper half plane under $SL_2(\mathbb{Z})$. If we denote by Q_0 (or Q_0^*) the "primitive even part" of Q (or Q^*), i.e.

$$(3.22) \qquad \begin{cases} Q_0 = \begin{pmatrix} 2a_0 & b_0 \\ b_0 & 2c_0 \end{pmatrix}, & (a_0, b_0, c_0) = 1 \\ Q = mQ_0 & \text{for some integer } m \geqslant 1, \end{cases}$$

and similarly for Q^*, this latter fact implies that

$$(3.23) \qquad Q_0 \sim Q_0^* \quad \text{w.r.t. } SL_2(\mathbb{Z}).$$

Combining (3.21) and (3.23), we conclude that Q and Q^* are equivalent with respect to $SL_2(\mathbb{Z})$. This proves Theorem 3.1.

In the course of the above proof, we have also proved the following facts :

<u>Corollary</u> 3.3. The inverse map f^{-1} of the bijective map f in (3.19) is described as follows. For a singular abelian surface X, let Q_0 be the primitive even matrix determined

(up to $SL_2(\mathbb{Z})$) by the elliptic curve C_X (1.7), and let m be the degree of primitivity of T_X, (1.8). Then $f^{-1}(X)$ is the equivalence class of mQ_0.

Corollary 3.4. Let X, Y be two singular abelian surfaces. Then

$$X \simeq Y \iff T_X \simeq T_Y\ , \quad C_X \simeq C_Y\ . \tag{3.24}$$

Corollary 3.5. Every singular abelian surface X is a product of two elliptic curves. More precisely,

$$X = C_X \times C\ , \tag{3.25}$$

with an elliptic curve C isogenous to C_X.

§4. Decomposition of a singular abelian surface

Theorem 4.1. For a complex abelian surface X, the following conditions are equivalent to each other :

(i) X is singular.

(ii) X is isogenous to a self-product $C \times C$ of an elliptic curve C with complex multiplications.

(iii) X is a product $C_1 \times C_2$, where C_1, C_2 are mutually isogenous elliptic curves with complex multiplications.

Proof. "(iii) $\Rightarrow$ (ii)" is obvious. To see "(ii) $\Rightarrow$ (i)", note first that the Picard number $\rho(X)$ does not change under isogeny. Next we have

$$\rho(C \times C) = 2 + \text{rank End}(C) = 4 \ , \quad \text{if}$$

C has complex multiplications. Finally the assertion "(i) $\Rightarrow$ (iii)" is contained in Corollary 3.5.

Remark 4.2. For a complex abelian variety X of dimension $g \geq 2$, the following are equivalent (see [3], [6]) :

(i)' The Picard number of X is equal to $h^{1,1} = g^2$.

(ii)' X is isogenous to a g-th power of an elliptic curve with complex multiplications.

Thus the above (i) $\Leftrightarrow$ (ii) is a special case of this fact, but we do not know whether the corresponding statement for (iii) holds for an abelian variety of dimension $g \geq 3$.

Remark 4.3. We understand that F. Oort has recently generalized Theorem 4.1 for an abelian surface in an arbitrary characteristic, using some lifting theorems.

Now, given a singular abelian surface A, we know that A is decomposed into a product of two elliptic curves, but in general such a decomposition is not unique (cf. [2]). Two decompositions $A \simeq C_1 \times C_2$ and $A \simeq C_1' \times C_2'$ are called distinct if $C_i \not\simeq C_j'$ for $i, j = 1, 2$. We study the number of distinct decompositions of A.

We can assume without loss of generality that A is the abelian surface constructed in § 3, (3.5):

(4.1) $$A = \mathbb{C}/(\mathbb{Z}+\mathbb{Z}\tau_1) \times \mathbb{C}/(\mathbb{Z}+\mathbb{Z}\tau_2)$$

where

(4.2) $$\tau_1 = \frac{-b+\sqrt{\Delta}}{2a}, \quad \tau_2 = \frac{b+\sqrt{\Delta}}{2}, \quad \Delta = b^2 - 4ac < 0 .$$

We put also

$$m = \text{g.c.d.}\{a, b, c\} \quad (= \text{the degree of primitivity})$$

(4.3) $$\begin{pmatrix} 2a & b \\ b & 2c \end{pmatrix} = m \begin{pmatrix} 2a_0 & b_0 \\ b_0 & 2c_0 \end{pmatrix}.$$

Take two points ω_1, ω_2 in the upper half plane and consider the abelian surface

(4.4) $$X = \mathbb{C}/(\mathbb{Z}+\mathbb{Z}\omega_1) \times \mathbb{C}/(\mathbb{Z}+\mathbb{Z}\omega_2) .$$

If X is isomorphic to A, the period maps p_X and p_A must be the same up to a constant multiple. Computing $\mathrm{Im}(p_X)$ and $\mathrm{Im}(p_A)$ explicitly (cf. (3.7), (3.12)), we have

(4.5) $$\mathbb{Z}+\mathbb{Z}\omega_1+\mathbb{Z}\omega_2+\mathbb{Z}\omega_1\omega_2 = \lambda(\mathbb{Z}+\mathbb{Z}\tau_1+\mathbb{Z}\tau_2+\mathbb{Z}\tau_1\tau_2) = \lambda(\mathbb{Z}+\mathbb{Z}\tau_1) .$$

This shows in particular that λ, τ_1, τ_2 are contained in the imaginary quadratic field

$$K = \mathbb{Q}(\tau_1) = \mathbb{Q}(\sqrt{\Delta}). \tag{4.6}$$

Therefore, in order to study the decompositions of A, we can work in this fixed field K.

Let $\mathcal{O}$ be the ring of integers in K, and let $\mathcal{O}_f$ denote the order in K with the conductor f, i.e. the unique subring of $\mathcal{O}$ of index f.[2)] By a <u>module</u> in K we mean a $\mathbb{Z}$-submodule of K of rank 2. Let $\mathcal{O}_M$ denote the ring of "multiplicators" of M:

$$\mathcal{O}_M = \{x \in K \mid xM \subset M\}. \tag{4.7}$$

Then $\mathcal{O}_M$ is an order in $\mathcal{O}$, and hence $\mathcal{O}_M = \mathcal{O}_f$ for some f. We call f the <u>conductor of</u> M. Two modules M_1, M_2 in K are called equivalent (and written $M_1 \sim M_2$) if there is an element $\lambda \in K$ such that $\lambda M_1 = M_2$. If $M_1 \sim M_2$, $\mathcal{O}_{M_1}$ is equal to $\mathcal{O}_{M_2}$. Any module in K is equivalent to a module of the form $\mathbb{Z} + \mathbb{Z}\omega$, $\omega \in K - \mathbb{Q}$. Given two modules M_1, M_2, the product M_1M_2 is defined as the submodule generated by xy ($x \in M_1$, $y \in M_2$). M_1M_2 is again a module, and its conductor is the greatest-common divisor of conductors of M_1, M_2; i.e.

$$\mathcal{O}_{M_1M_2} = \mathcal{O}_{(f_1, f_2)} \quad \text{if} \quad \mathcal{O}_{M_\nu} = \mathcal{O}_{f_\nu} \quad (\nu = 1, 2). \tag{4.8}$$

For a fixed integer f, the set of equivalence classes of modules of conductor f forms a <u>finite</u> abelian group $\mathcal{J}_f$. We denote by

$$h(\mathcal{O}_f) \quad (\text{or} \quad h_2(\mathcal{O}_f)) \tag{4.9}$$

the order of the group $\mathcal{J}_f$ (or the order of the subgroup of $\mathcal{J}_f$ consisting of elements of order 2).

---

2) For what follows, see e.g. the book of Borevič-Šafarevič [1].

Let us go back to our problem and consider

(4.10) $$M_0 = \mathbb{Z} + \mathbb{Z}\tau_1, \quad M_0' = \mathbb{Z} + \mathbb{Z}\tau_2 .$$

It is easy to see

(4.11) $$\begin{cases} \mathcal{O}_{M_0} = \mathbb{Z} + \mathbb{Z} a_0 \tau_1 , & a_0 = \frac{a}{m} \\ \mathcal{O}_{M_0'} = M_0' . \end{cases}$$

The conductor f_0 of M_0 is computed from the relation :

(4.12) $$Df_0^2 = b_0^2 - 4a_0 c_0 ,$$

in which D is the discriminant of the field K.

The following lemma is due to F. Hirzebruch.

Lemma 4.4. Let M_ν $(\nu = 1, 2)$ be a module in K of conductor f_ν, and let $X = \mathbb{C}/M_1 \times \mathbb{C}/M_2$. Then the degree of primitivity of T_X (cf. (1.9)) is equal to $f_1 f_2/(f_1, f_2)^2$.

Proposition 4.5. Let A be the abelian surface (4.1), and let M_ν be the modules with conductor f_ν $(\nu = 1, 2)$. Then

(4.13) $$A \simeq \mathbb{C}/M_1 \times \mathbb{C}/M_2 \iff \begin{cases} \text{(i)} & M_1 M_2 \sim M_0 \\ \text{(ii)} & (f_1, f_2) = f_0 \\ \text{(iii)} & f_1 f_2 = m f_0^2 . \end{cases}$$

Proof. Put $X = \mathbb{C}/M_1 \times \mathbb{C}/M_2$. If $A \simeq X$, we saw in (4.5) that

$$M_1 M_2 \sim M_0 .$$

This implies (ii) by (4.8). Then, comparing the degree of primitivity of T_X and T_A, we have (iii) by Lemma 4.4. Conversely, $M_1 M_2 \sim M_0$ implies that $C_X \simeq C_A$ (cf. (1.7)), and (ii), (iii) imply that T_X and T_A have the same degree of primitivity. Applying Corollary 3.3, we see $X \simeq A$, q.e.d.

Corollary 4.6. The number of distinct decompositions of a

singular abelian surface is finite.

This is obvious from the above proposition and the finiteness of the class number $h(\mathcal{O}_f)$.

Finally, to obtain a more explicit result, we consider the case $m = 1$, i.e. the abelian surface A corresponds to a primitive even matrix Q. In this case, (4.13) reduces to

$$(4.14) \qquad A \simeq \mathbb{C}/M_1 \times \mathbb{C}/M_2 \iff \begin{cases} M_1 M_2 \sim M_0 \\ f_1 = f_2 = f_0 . \end{cases}$$

Since equivalence classes having a fixed conductor f_0 form the group $\mathcal{I}_{f_0}$, there are exactly $h(\mathcal{O}_{f_0})$ distinct choice of M_1. Considering the change of factors M_1 and M_2, we obtain

Theorem 4.7. Let A be a singular abelian surface with primitive T_A. Then the number of distinct decompositions of A is equal to

$$\begin{cases} \frac{1}{2} h(\mathcal{O}_{f_0}) & \text{if the module class of } M_0 \text{ is not a square in } \mathcal{I}_{f_0} , \\ \frac{1}{2}(h(\mathcal{O}_{f_0}) + h_2(\mathcal{O}_{f_0})) & \text{otherwise} . \end{cases}$$

(Cf. (4.9) $\sim$ (4.12) for the notation.)

For the special case $f_0 = 1$, this result can be found in [2].

§5. Some remarks on singular K3 surfaces

Let X denote for a moment a non-singular algebraic surface such that

$$(*)\qquad \begin{cases} \text{the geometric genus } p_g = 1, \quad \text{and} \\ H^2(X, \mathbb{Z}) \text{ is a torsion-free even Euclidean lattice.} \end{cases}$$

If moreover X is <u>singular</u> in the sense of §1 (i.e. the Picard number $\rho = h^{1,1}$), then we obtain

$$(5.1)\qquad \begin{cases} T_X : \text{ a positive definite even Euclidean lattice} \\ \qquad \text{of rank } 2, \\ C_X = H^2(X, \mathcal{O})/j^*H^2(X, \mathbb{Z}), \text{ an elliptic curve}, \end{cases}$$

exactly in the same way as for the case of abelian surfaces (1.6), (1.7). Also we have

$$(5.2)\qquad C_X \simeq \mathbb{C}/(\mathbb{Z} + \mathbb{Z}\tau),$$

in which τ is an imaginary quadratic number with $\mathrm{Im}(\tau) > 0$. Let

$$(5.3)\qquad \begin{aligned} &a_0\tau^2 + b_0\tau + c_0 = 0, \quad a_0, b_0, c_0 \in \mathbb{Z} \\ &a_0 > 0, \quad (a_0, b_0, c_0) = 1 \end{aligned}$$

be the equation of τ. Let m denote the degree of primitivity of T_X. Then

$$(5.4)\qquad Q = m\begin{pmatrix} 2a_0 & b_0 \\ b_0 & 2c_0 \end{pmatrix}$$

is a positive definite even integral matrix associated with the Euclidean lattice T_X. The equivalence class q_X of Q with respect to $SL_2(\mathbb{Z})$ is uniquely defined by the surface X.

As we saw in §3 (Theorem 3.1, Corollary 3.3), the correspondence $X \longrightarrow q_X$ defines a bijection :

$$(5.5) \qquad \underbrace{\{\text{singular abelian surfaces}\}/\text{isom.}}_{\mathcal{S}_{Ab} \ \overset{\text{def.}}{=}} \xrightarrow{f^{-1}} \mathcal{Q}/SL_2(\mathbb{Z}) .$$

Now K3 surfaces satisfy the condition (*) for X. Hence we obtain a map :

$$(5.6) \qquad \underbrace{\{\text{singular K3 surfaces}\}/\text{isom.}}_{\mathcal{S}_{K3} \ \overset{\text{def.}}{=}} \xrightarrow{F} \mathcal{Q}/SL_2(\mathbb{Z}) .$$

There is a map

$$(5.7) \qquad Km : \mathcal{S}_{Ab} \longrightarrow \mathcal{S}_{K3} ,$$

associating a singular abelian surface A with the Kummer surface Km(A), which is a singular K3 surface (cf. [6]). Let $\mathcal{S}_{Km}$ denote the set of isomorphism classes of singular Kummer surfaces. Let $[2] : \mathcal{Q} \longrightarrow \mathcal{Q}$ denote the map $Q \longrightarrow 2Q$, and let $\mathcal{Q}^{[2]}$ be the image of the map [2]. Then we have the commutative diagram (cf. [4] § 6) :

$$(5.8) \qquad \begin{array}{ccc} \mathcal{S}_{Ab} & \xrightarrow{f^{-1}} & \mathcal{Q}/SL_2(\mathbb{Z}) \\ Km \downarrow & & \downarrow [2] \\ \mathcal{S}_{Km} & \xrightarrow{F'} & \mathcal{Q}^{[2]}/SL_2(\mathbb{Z}) \\ \cap & & \cap \\ \mathcal{S}_{K3} & \xrightarrow{F} & \mathcal{Q}/SL_2(\mathbb{Z}) . \end{array} \qquad , \quad F' = F \mid \mathcal{S}_{Km}$$

Since both f^{-1} (5.5) and [2] are bijective, we see that Km (and F') in (5.8) is also a bijective map. In other words, we have

Theorem 5.1. Let A_1, A_2 be two singular abelian surfaces. Suppose $Km(A_1)$ and $Km(A_2)$ are isomorphic (i.e. biholomorphic) to each other. Then A_1 and A_2 are isomorphic.

In the paper of Pjateckii-Šapiro, Šafarevič [4], it is claimed that F is bijective. But the proof given there seems incomplete. First, for the injectivity of F, one needs a theorem for K3 surfaces corresponding to Theorem 2.1 for abelian surfaces, but the "proof" in [4] has a certain gap. Secondly, the surjectivity of F is also not trivial at all. One has to construct some K3 surface out of a given matrix $Q \in \mathcal{Q}$, but, except for the case of Kummer surfaces, no canonical method of construction is known (at least to us). At any rate, the following example shows that there exists a singular K3 surface which is not Kummer.

Example. Let $\Gamma_0'(7)$ denote the discontinuous subgroup of $SL_2(\mathbb{Z})$:

$$\Gamma_0'(7) = \left\{ \begin{pmatrix} \alpha & \beta \\ \gamma & \delta \end{pmatrix} \;\middle|\; \gamma \equiv 0 \ (7),\ \left(\frac{a}{7}\right) = 1 \right\}.$$

Let $X = B_{\Gamma_0'(7)}$ be the elliptic modular surface attached to $\Gamma_0'(7)$ (cf. [5], p.42). (Roughly speaking, this surface is obtained as a non-singular compactification of $H \times \mathbb{C}/\Gamma_0'(7)\cdot\mathbb{Z}^2$, H being the upper half plane.) Since X is an elliptic surface without multiple fibres such that $p_g = 1$ and $q = 0$ (q = irregularity of X), X is a K3 surface. It is singular, because any elliptic modular surface is singular (cf. [5]).

If X were a Kummer surface, T_X would have an even degree

of primitivity (cf. (5.8)), hence in particular

$$\det(T_X) \equiv 0 \quad (2).$$

Let us show that $\det T_X$ for our X is odd. Note first that $|\det S_X| = \det T_X$, because T_X is defined as the orthogonal complement of S_X in $H^2(X, \mathbb{Z})$, which is a unimodular lattice. Let E denote the generic fibre of the elliptic surface $X \longrightarrow \mathbb{P}^1$. The group of rational points of the elliptic curve E over $\mathbb{C}(\mathbb{P}^1)$ is finite, and its order n is 1 or 3. Furthermore, there exist 4 singular fibres of $X \longrightarrow \mathbb{P}^1$, which are of type I_1, I_7, IV^*, IV^*. The number of simple components in these singular fibres is respectively 1, 7, 3 and 3. Therefore, by Corollary 1.7 of [5], we have

$$|\det S_X| = \frac{1\cdot 7\cdot 3\cdot 3}{n^2} \qquad (n = 1 \text{ or } 3)$$

$$= 7 \text{ or } 3^2\cdot 7 .$$

This proves that X is not a Kummer surface.

Appendix. Euclidean lattices

We recall in this appendix some definitions and results concerning Euclidean lattices in [4], relevant to our problem.

(I) By a <u>Euclidean lattice</u> we mean a free $\mathbb{Z}$-module of finite rank, given with a $\mathbb{Z}$-valued scalar product (i.e. non-degenerate, symmetric bilinear form). We denote the scalar product by xy, and write x^2 for xx. Let E, E' be two Euclidean lattices. By an <u>isometry</u> (or <u>isomorphism</u>) of E onto E', we mean an isomorphism of modules $\psi : E \longrightarrow E'$ such that $\psi(x)\,\psi(y) = xy$ for all $x, y \in E$. An <u>automorphism</u> of E is an isometry of E onto itself. $\mathrm{Aut}(E)$ denotes the group of automorphisms of E.

A Euclidean lattice E is called <u>even</u> if $x^2 \equiv 0\ (2)$ for all $x \in E$. It is called <u>positive-definite</u> if $x^2 > 0$ for all $x \in E$, $x \neq 0$.

Let $\{x_1, \cdots, x_n\}$ be a basis of E. The square matrix $Q = (x_i x_j)$ is a non-degenerate integral symmetric matrix. The isomorphism classes of Euclidean lattices are in one-to-one correspondence with the equivalence classes of such matrices Q with respect to the equivalence $Q \sim {}^tMQM$, $M \in GL_n(\mathbb{Z})$. We write $\det(E) = \det(Q)$, which is independent of choice of base. E is called <u>unimodular</u> if $\det(E) = \pm 1$.

A <u>sublattice</u> F of a Euclidean lattice E is a submodule of E such that the restriction of the scalar product of E to F is non-degenerate. F is called <u>primitive</u> if E/F is torsion-free.

For a sublattice F of E, the <u>orthogonal complement</u> $F^{\perp}$

of F in E is defined as

$$F^{\perp} = \{x \in E \mid xy = 0 \text{ for all } y \in F\}.$$

$F^{\perp}$ is a primitive sublattice of E such that

$$F \cap F^{\perp} = 0, \quad E/(F + F^{\perp}) \text{ is finite.}$$

Further F is primitive if and only if $(F^{\perp})^{\perp} = F$.

(II) Let E_k $(k = 1, 2, \cdots)$ denote the even unimodular Euclidean lattice of rank $2k$ with a standard basis $\{e_1, \cdots, e_k, e'_1, \cdots, e'_k\}$ such that

$$(1) \qquad e_i e'_j = \delta_{ij}, \quad e_i e_j = e'_i e'_j = 0.$$

From now on, the notation E will be exclusively used for $E = E_3$. Note that $H^2(X, \mathbb{Z})$ for a complex abelian surface X is a Euclidean lattice isometric to E.

Theorem 1. Let $E = E_3$ be as above.

(i) Given an arbitrary even lattice T of rank 2, there exists a primitive sublattice F of E isometric to T.

(ii) Suppose F_1, F_2 are primitive sublattices of rank 2 of E, isometric to each other. Then there exists an automorphism of E mapping F_1 onto F_2.

This is a special case of Theorem 1 of §6, Appendix in [4]. For the sake of completeness, we recall the proof.

Proof of (i). Take a basis $\{v_1, v_2\}$ of T. We put

$$(2) \qquad v_1^2 = 2a, \quad v_1 v_2 = b, \quad v_2^2 = 2c.$$

Put

$$(3) \qquad \begin{cases} f_1 = e_1 + a e'_1 \\ f_2 = \quad b e'_1 + e_2 + c e'_2, \end{cases} \qquad \text{(cf. (3.9))}$$

and let F be the sublattice of E generated by f_1, f_2. The map $v_\nu \longrightarrow f_\nu$ ($\nu = 1, 2$) is an isometry of T onto F. F is primitive, because e_ν has the coefficient $\delta_{\nu\mu}$ in f_μ (ν, μ = 1 or 2). This proves (i).

Before proving (ii), we define two kinds of "elementary transformations" of E. (a) Let $E = E_1 \oplus E'$ be the orthogonal decomposition with $E_1 = \mathbb{Z}e_1 + \mathbb{Z}e_1'$, and take $x_0 \in E'$. Then the map

$$(4) \qquad \begin{cases} e_1 \longrightarrow e_1 \\ e_1' \longrightarrow e_1' + x_0 - \frac{1}{2}(x_0^2)e_1 \\ x \longrightarrow x - (xx_0)e_1 \quad , \qquad x \in E' \end{cases}$$

defines an automorphism of E, denoted by $\phi_{(e_1, e_1', x_0)}$.

(b) Consider next the lattice E_2 with the standard basis $\{e_1, e_2, e_1', e_2'\}$. Let us represent an element x of E_2

$$(5) \qquad x = \alpha_1 e_1 + \alpha_1' e_1' + \alpha_2 e_2 + \alpha_2' e_2'$$

by the matrix

$$(6) \qquad M_x = \begin{pmatrix} \alpha_1 & -\alpha_2 \\ \alpha_2' & \alpha_1' \end{pmatrix} .$$

We note the following relations :

$$(7) \qquad x = \operatorname{tr}\left(M_x \begin{pmatrix} e_1 & e_2' \\ -e_2 & e_1' \end{pmatrix} \right) ,$$

$$(8) \qquad x^2 = 2 \det(M_x) .$$

We define a linear transformation ψ of E_2 by

$$(9) \qquad \begin{pmatrix} \psi(e_1) & -\psi(e_2) \\ \psi(e_2') & \psi(e_1') \end{pmatrix} = A \begin{pmatrix} e_1 & -e_2 \\ e_2' & e_1' \end{pmatrix} B , \qquad A, B \in SL_2(\mathbb{Z}).$$

Then we have

$$\psi(x) = \mathrm{tr}\left(M_x\ {}^tB\begin{pmatrix} e_1 & e_2' \\ -e_2 & e_1' \end{pmatrix}{}^tA\right)$$

$$= \mathrm{tr}\left({}^tA\, M_x\ {}^tB\begin{pmatrix} e_1 & e_2' \\ -e_2 & e_1' \end{pmatrix}\right).$$

Hence

$$\psi(x)^2 = 2\det({}^tA\, M_x\ {}^tB) = 2\det(M_x) = x^2 .$$

This shows that ψ is an automorphism of E_2. We call it the <u>elementary transformation</u> of E_2 defined by A, B.

<u>Lemma</u> 2. For an element $x \in E_2$ of (5), $x \neq 0$, let $d = \mathrm{g.c.d.}\,(\alpha_1, \alpha_1', \alpha_2, \alpha_2')$. Then there exists an elementary transformation ψ of E_2 such that

$$\psi(x) = de_1 + d'e_1' , \qquad d \mid d' .$$

<u>Proof</u>. This follows immediately from the elementary divisor theory.

<u>Proof of</u> (ii). Applying the assertion (i) to $T = F_1$, we can assume that

$$F_2 = F = \mathbf{Z}f_1 + \mathbf{Z}f_2$$

constructed in (i). It is sufficient to show that there is an automorphism ψ of $E = E_3$ such that

$$\psi(v_1) = f_1 , \qquad \psi(v_2) = f_2 . \tag{10}$$

<u>Step 1</u>. To find $\psi \in \mathrm{Aut}(E)$ with $\psi(v_1) = f_1$.

Write

$$v_1 = \sum_{i=1}^{3} (a_i e_i + a_i' e_i') , \qquad a_i, a_i' \in \mathbf{Z} .$$

Since F_1 is primitive, $\{a_1, a_2, a_3, a_1', a_2', a_3'\}$ have g.c.d. = 1. Put $d = \text{g.c.d.}\{a_1, a_1', a_3, a_3'\}$. Applying Lemma 2 to the element

$$a_1e_1 + a_1'e_1' + a_3e_3 + a_3'e_3' \quad \text{in} \quad \sum_{\nu=1,3} (\mathbb{Z}e_\nu + \mathbb{Z}e_\nu') ,$$

we can find $\varphi_1 \in \text{Aut}(E)$ such that

$$\varphi_1(v_1) = (de_1 + d'e_1') + a_2e_2 + a_2'e_2' .$$

Since $\text{g.c.d.}\{d, d', a_2, a_2\} = 1$, applying Lemma 2 again, we find $\varphi_2 \in \text{Aut}(E)$ such that

$$\varphi_2(\varphi_1(v_1)) = e_1 + \lambda e_1' .$$

Comparing the square of both sides, we have $\lambda = a$. Hence

$$(\varphi_2\varphi_1)(v_1) = e_1 + ae_1' = f_1 .$$

<u>Step 2</u>. (Assume $v_1 = f_1$.) To find $\psi \in \text{Aut}(E)$ such that

$$\psi(f_1) = f_1, \quad \psi(v_2) = f_2.$$

Write

$$v_2 = \sum_{i=1}^{3} (b_ie_i + b_i'e_i') , \quad b_i, b_i' \in \mathbb{Z} .$$

Applying Lemma 2 as before to the sublattice $\mathbb{Z}e_2 + \mathbb{Z}e_2' + \mathbb{Z}e_3 + \mathbb{Z}e_3'$, we may assume $b_3 = b_3' = 0$, without moving f_1. Since

$$v_2 - b_1f_1 = (b_1' - ab_1)e_1' + b_2e_2 + b_2'e_2'$$

is primitive in E, we have

$$\text{g.c.d.}\ (b_1' - ab_1, b_2, b_2') = 1.$$

Let $\psi_1 = \psi_{(e_3, e_3', e_1 - ae_1')} \in \text{Aut}(E)$ (cf. (4)). Then

$$\begin{cases} \psi_1(f_1) = f_1 \\ v_2' = \psi_1(v_2) = (b_1e_1 + b_1'e_1') + \{b_2e_2 + b_2'e_2' - (b_1' - ab_1)e_3\}. \end{cases}$$

By Lemma 2, applied to $\mathbb{Z}e_2 + \mathbb{Z}e_2' + \mathbb{Z}e_3 + \mathbb{Z}e_3'$, we see that the second summand $\{\cdots\}$ is mapped to $e_2 + \mu e_2'$ (some $\mu \in \mathbb{Z}$). Therefore we have $\psi_2 \in \mathrm{Aut}(E)$ such that

$$\begin{cases} \psi_2(f_1) = f_1 \\ v_2'' = \psi_2(v_2') = b_1 e_1 + b_1' e_1' + e_2 + \mu e_2' . \end{cases}$$

Finally the elementary transformation $\psi_3 = \psi_{(e_2', e_2, -b_1 f_1)}$ maps v_2'' to an element of the form :

$$v_2''' = \beta e_1' + e_2 + \gamma e_2' .$$

From (2), we get $\beta = b$, $\gamma = c$. Hence, putting

$$\psi = \psi_3 \psi_2 \psi_1 \in \mathrm{Aut}(E) ,$$

we have $\psi(f_1) = f_1$, $\psi(v_2) = f_2$. This completes the proof of (10), hence of Theorem 1.

(III) Let $E_{\mathbb{C}}^*$ denote the vector space of complex-valued linear functionals on E :

$$(11) \qquad E_{\mathbb{C}}^* = \mathrm{Hom}(E, \mathbb{C}) .$$

By means of the scalar product of E, $E_{\mathbb{C}}^*$ is canonically identified with $E_{\mathbb{C}} = E \otimes \mathbb{C}$, and hence has a natural scalar product.

Proposition 2. Let T be a positive definite primitive sublattice of rank 2 of E. Then there exists a linear functional p on E satisfying

$$\begin{cases} \text{(i)} & p^2 = 0 , \quad p\bar{p} > 0 \\ \text{(ii)} & p \big|_{T^\perp} = 0 . \end{cases}$$

Such a functional p is unique up to a constant multiple and the complex conjugation $p \longrightarrow \bar{p}$.

Proof. Take a basis $\{t_1, t_2\}$ of T, and put

$$\begin{cases} t_1^2 = 2a , \quad t_1 t_2 = b , \quad t_2^2 = 2c \\ \Delta = b^2 - 4ac < 0 . \end{cases}$$

Then p satisfies the condition (ii) if and only if p can be written (as an element of $E_{\mathbb{C}}$) as a linear combination of t_1, t_2:

$$p = \lambda_1 t_1 + \lambda_2 t_2 , \qquad \lambda_1, \lambda_2 \in \mathbb{C} .$$

Such p satisfies (i) if and only if

$$\begin{cases} p^2 = 2(a\lambda_1^2 + b\lambda_1\lambda_2 + c\lambda_2^2) = 0 \\ p\bar{p} = 2(a|\lambda_1|^2 + b\mathcal{R}(\lambda_1\bar{\lambda}_2) + c|\lambda_2|^2) > 0 \end{cases}$$

i.e.

$$\begin{cases} \lambda_1\lambda_2 \neq 0 \quad \text{and} \\ \lambda_1 = \dfrac{-b \pm \sqrt{\Delta}}{2a} \lambda_2 . \end{cases}$$

This shows the existence and the uniqueness of p, , q.e.d.

References

1. Z. I. Borevič and I. R. Šafarevič, Number theory, Academic Press, 1966.

2. T. Hayashida and M. Nishi, Existence of curves of genus two on a product of two elliptic curves, J. Math. Soc. Japan, 17 (1965), 1-16.

3. D. Mumford, Abelian varieties, Tata-Oxford Univ. Press, 1970.

4. I. I. Pjateckii-Šapiro and I. R. Šafarevič, Torelli theorem for algebraic K3 surfaces, Izv. Akad. Nauk SSSR, Ser. Mat. 35 (1971), 530-572.

5. T. Shioda, On elliptic modular surfaces, J. Math. Soc. Japan, 24 (1972), 20-59.

6. ——— , Algebraic cycles on certain K3 surfaces in characteristic p, Proc. Intern. Conf. on Manifolds and Related Topics in Topology, Tokyo, 1973.

7. ——— , Period map of abelian surfaces (to appear).

INTRODUCTION TO CLASSIFICATION THEORY
OF ALGEBRAIC VARIETIES AND COMPACT COMPLEX SPACES

Kenji Ueno*

Introduction

The first definite result on classification theory for compact complex manifolds of dimension > 2 was obtained by Kawai [13], I. He succeeded to show that the general fibres of the algebraic reduction of a three - dimensional compact complex manifold of algebraic dimension two are elliptic curves. After that, Iitaka [10] has introduced the notion of "Kodaira dimension" for algebraic varieties and complex varieties and has proved the fundamental theorem on the fibration determined by the m-canonical mappings. In [11], Iitaka has shown that the Kodaira dimension plays an important role in classification theory and has discussed several problems and conjectures concerning the classification of compact complex manifolds. Ueno [32] has studied Albanese mappings of algebraic manifolds and the canonical bundle formula for an elliptic threefold and has enriched the classification theory.

The present note is intended to give an outline of our classification theory. More systematic treatments can be found in [33]. The present note is based on the lectures which I gave at the Mannheim meeting. Several topics which I did not mention there have been added. I hope that this note will serve as an introduction to the lecture note [33].

---

* This work was supported by SFB (Theoretische Mathematik), Mathematisches Institut der Universität Bonn.

In Chapter I , we shall study the Kodaira dimension of complex varieties. First we shall recall the classification theory of curves and surfaces. Then the Kodaira dimension will be defined (Definition 1.3.3) and its fundamental properties will be given (see 1.3). In 1.4 , we shall provide the fundamental theorem of classification theory. A sketch of the proof of the fundamental theorem will be given in 1.5. In 1.6 we shall discuss complex varieties of hyperbolic type. Finally in 1.7 we shall define the D - dimension of a complex variety due to Iitaka. In Chapter II, we shall use the D - dimension to calculate the Kodaira dimension of a certain fibre space (see 2.6).

In Chapter II , we shall study algebraic manifolds of parabolic type using the Albanese mapping. In 2.1 , the Albanese torus of a complex manifold will be defined and its fundamental properties will be given. The fibre space associated with the Albanese mapping will also be defined. To study this fibre space it is natural to study the structure of the image variety of the Albanese mapping. For that purpose, in 2.2 , the results on subvarieties of a complex torus due to Ueno [32] will be given. Using these results, in 2.3 , we shall discuss algebraic manifolds of parabolic type. Conjecture K_n concerning the structure of the fibre space associated with the Albanese mapping of an algebraic manifold of parabolic type will be proposed. In 2.4 , we shall show that Conjecture C_n concerning the Kodaira dimension of a fibre space is important to prove Conjecture K_n. A few affirmative answers to Conjecture C_n due to Nakamura and Ueno [25] , Ueno [31] , 32 , will be given. Important examples which support Conjecture K_n are Kummer manifolds (Definition 2.5.1). In 2.5 , the main results on Kummer manifolds will be given. In 2.6 , we shall give the canonical bundle formula for a certain elliptic fibre space. Moreover, we shall show that the canonical bundle formula is deeply related to Conjecture K_n. Finally, in 2.7 , we shall provide the example due to Nakamura [24] which shows that the Kodaira dimension and several bimeromorphic invariants introduced in Definition 1.3.1 and Definition 2.1.4 are not invariant under small deformations.

In Chapter III, we shall study the fibre space introduced by the algebraic reduction of a complex variety (Definition 3.2.1) and the fibre space associated with the Albanese mapping of a complex manifold of algebraic dimension zero. These two fibre spaces have similar properties (see Theorem 3.2.8 , Theorem 3.3.6

and Theorem 3.3.7). In 3.1 , the algebraic dimension of a complex variety will be defined. In 3.2 , we shall study the algebraic reduction of a complex variety. The algebraic reduction was first considered by Kodaira [18], I, in the case of surfaces. Kawai [13] has generalized the result in the case of complex manifolds of dimension three. Hironaka [8] has given the new idea to study algebraic reductions, which we shall follow. In 3.3 , we shall study complex manifolds of algebraic dimension zero. This theory was started also by Kodaira [18]. I. Kawai [13], II, has studied three dimensional Kähler manifolds of algebraic dimension zero. Since we have obtained good informations on the Albanese mapping (see Corollary 2.2.4), we can remove the assumption that the complex manifold is Kähler. We emphasize here, though we consider a completely non - algebraic object, we need a deep fact on algebraic objects, that is, the moduli spaces of curves and algebraic surfaces of general type exist as algebraic spaces.

Table of Contents

Introduction.

Notations and Conventions.

Chapter I. Kodaira dimensions of complex varieties.

1.1 Classification of curves.
1.2 Classification of surfaces.
1.3 Kodaira dimension.
1.4 Fundamental theorem of classification theory.
1.5 Proof of the fundamental theorem.
1.6 Classification and complex varieties of hyperbolic type.
1.7 D - dimension.

Chapter II. Albanese mappings and algebraic manifolds of parabolic type.

2.1 Albanese mapping.
2.2 Subvarieties of a complex torus.
2.3 Algebraic manifolds of parabolic type.
2.4 Conjecture C_n.
2.5 Kummer manifolds.
2.6 Canonical bundle formula.
2.7 The Kodaira dimension is not a deformation invariant.

Chapter III. Algebraic reductions of complex varieties and complex manifolds of algebraic dimension zero.

3.1 Algebraic dimension.
3.2 Algebraic reduction.
3.3 Complex manifolds of algebraic dimension zero.

Notations and Conventions

$a(V)$	the algebraic dimension of a complex variety V(Definition 3.1.2).
$\alpha: M \to A(M)$	the Albanese mapping of a complex manifold M (Definition 2.1.1).
$\mathrm{Aut}(V)$	the group of all analytic automorphisms of a complex variety V.
$b_i(V)$	the i - th Betti number of a complex variety V.
$\mathbb{C}(V)$	the field consisting of all meromorphic functions on a complex variety V.
$[D]$	the line bundle associated with a Cartier divisor D.
$g_k(V)$	see Definition 2.1.4.
$K_M = K(M)$	the canonical line bundle (a canonical divisor) of a complex manifold M.
mK_M means $K_M^{\otimes m}$	
$\kappa(V)$	the Kodaira dimension (Definition 1.3.3).
$p_g(V)$	the geometric genus (Definition 1.3.1 and Definition 1.3.2).
$P_m(V)$	the m - genus (Definition 1.3.1 and Definition 1.3.2).
$q(V)$ $r(V)$ $t(M)$	see Definition 2.1.4.

Unless otherwise explicitly stated to the contrary, the following conventions will be in force throughout this note.

1) All algebraic varieties are defined over $\mathbb{C}$ and complete. A non - singular algebraic variety is called an algebraic manifold.

2) An irreducible reduced complex space is called a complex variety. All complex manifolds and complex varieties are assumed to be compact.

3) By GAGA (Serre [30]), all algebraic varieties are considered as complex varieties.

4) A subvariety of a complex variety is assumed to be irreducible.

5) By a fibre space $f:V \to W$ of complex varieties we mean that f is surjective and all the fibres of f are connected.

6) By a line bundle we mean a complex line bundle. If a line bundle is analytically trivial, we often say that it is trivial.

7) As the definition of meromorphic mappings we use the one due to Remmert [28] (see also Ueno [33], §2).

Chapter I

Kodaira dimensions of complex varieties

In this chapter we shall provide an outline of the theory of classification of algebraic and analytic varieties due to Iitaka. In his paper [10] , Iitaka has introduced the notion of Kodaira dimensions of algebraic and analytic varieties and has shown the fundamental theorem on classification theory (see Theorem 1.4.1 , below). Before we shall consider varieties of arbitrary dimension, we shall recall classification theory of curves and surfaces.

(1.1) Classification of Curves

For a non - singular curve C , the genus g(C) of the curve C is defined by

$$g(C) = \dim_{\mathbb{C}} H^{0}(C, \Omega^{1}_{C}).$$

The genus is a birational invariant (that is, if two non - singular curves C_1 and C_2 have the same function field, then $g(C_1) = g(C_2)$). Let K_C be the canonical line bundle of the curve C. It is well- known that $3K_C$ is very ample if $g(C) \geqq 2$. If $g(C) = 1$ (resp. $g(C) = 0$) , then K_C is trivial (resp. $- K_S$ is very ample). Hence we can classify isomorphism classes of non - singular curves into the following three classes.

$\kappa(C)$	$g(C)$	structure	canonical bundle K_C	universal covering
1	$\geqq 2$		ample	$D = \{z \mid z < 1\}$
0	1	elliptic curve	trivial	$\mathbb{C}$
$-\infty$	0	projective line	negative	$\mathbb{P}^1$

In the above table, $\kappa(C)$ is called the Kodaira dimension of a curve C. The precise definition will be given in 1.3. Here we only remark that the Kodaira dimension classifies curves into three big classes.

This is a quite rough classification of curves. A fine classification of curves is theory of moduli. Let $\underline{M}_g$ be the set of all isomorphism classes of non - singular curves of genus g. On $\underline{M}_g$ we can introduce the structure of a quasi-projective variety (see Baily [2], Mumford [23]). The quasi-projective variety $\underline{M}_g$ is called the moduli space of curves of genus g and has the following properties:

1) there is a one-to-one correspondence between isomorphism classes of curves of genus g and points of $\underline{M}_g$;
2) let $f: \underline{C} \to \Delta$ be a complex analytic family of non - singular curves of genus g ; then the mapping

$$\begin{array}{ccc} \Delta & \to & \underline{M}_g , \\ \Psi & & \Psi \\ x & \to & [C_x] \end{array}$$

where $C_x = f^{-1}(x)$ and $[C_x]$ is the isomorphism class of the curve C_x , is holomorphic.

(1.2) Classification of Surfaces

In this section, a surface is always assumed to be non - singular. Rough classification of surfaces is given as follows. First we shall define several bimeromorphic invariants of surfaces.

DEFINITION 1.2.1. The m - genus $P_m(S)$, the geometric genus $p_g(S)$ and the irregularity $q(S)$ of a surface S are defined as follows.

$$P_m(S) = \dim_{\mathbb{C}} H^0(S , \underline{O}(mK_S)) , \quad m = 1 , 2 , \ldots..$$

$$p_g(S) = P_1(S) ,$$

$$q(S) = \dim_{\mathbb{C}} H^1(S , \underline{O}_S) .$$

These invariants are bimeromorphic invariants of surfaces.

To state classification of surfaces, we need the important notion of an exceptional curve of the first kind.

DEFINITION 1.2.2. A curve C in a surface S is called an exceptional curve of the first kind if C is isomorphic to $\mathbb{P}^1$ and $C^2 = -1$ (that is, the degree of the normal bundle $N_{C/S}$ of C in S is -1) .

THEOREM 1.2.3. Let S and C be as in Definition 1.2.2 then there exists a surface $\hat{S}$ and a proper surjective morphism $f: S \to \hat{S}$ which satisfies the following properties:

1) $f(C)$ is a point $\hat{p}$;
2) $\hat{S}$ is non - singular at p ;
3) f induces an isomorphism between $S - C$ and $\hat{S} - \hat{p}$.

DEFINITION 1.2.4. A surface S is called a relatively minimal model if the surface S does not contain any exceptional curve of the first kind.

It is easy to show that any surface is bimeromorphically equivalent to a relatively minimal model (use the fact that $H^2(S, \mathbb{Z})$ is of finite rank for any surface S and that, using the same notations as those in Theorem 1.2.3 , we have $H^2(S, \mathbb{Z}) \simeq H^2(\hat{S}, \mathbb{Z}) \oplus \mathbb{Z})$. Therefore, in what follows we shall always assume that a surface is a relatively minimal model.

(1.2.7) The following is the classification table of surfaces due to Kodaira [19] .

κ	p_g	P_{12}	q	b_1	structure
2		> 0			algebraic surface of general type
1					elliptic surface of general type
	1	1	2	4	complex torus
	1	1	2	3	elliptic surface with a trivial canonical bundle
0	0	1	1	2	hyperelliptic surface
	0	1	1	1	elliptic surface belonging to class VII with mK trivial for a positive integer m.
	1	1	0	0	K 3 surface
	0	1	0	0	Enriques surface
			0	0	rational surface
$-\infty$	0	0	$\geqq 1$	$2q$	ruled surface of genus q
			1	1	surface of class VII

Now we must define several surfaces. Note that we always assume that a surface is a relatively minimal model.

1) A surface S is called an <u>algebraic surface of general type</u> if

$$\overline{\lim_{m \to +\infty}} \frac{P_m(S)}{m^2}$$

is a positive number. An algebraic surface of general type is characterized by

$$P_2(S) > 0 \ , \ K_S^2 = c_1^2 > 0 \ .$$

2) A surface S is called an <u>elliptic surface</u> if there exist a curve C and a surjective morphism $f: S \to C$ such that a general fibre of f is an elliptic curve. An elliptic surface S is called an <u>elliptic surface of general type</u> if

$$\overline{\lim_{m \to +\infty}} \frac{P_m(S)}{m}$$

is a positive number.

3) An algebraic surface S is called a <u>hyperelliptic surface</u> if S is a non - trivial elliptic bundle over an elliptic curve. A hyperelliptic surface has a finite unramified covering which is a product of two elliptic curves.

4) A surface S is called a <u>K3 - surface</u> if K_S is trivial and $q(S) = 0$. Any K3 - surface is a deformation of a non - singular quartic surface in $\mathbb{P}^3$. Therefore S is simply connected.

5) A surface S is called an <u>Enriques surface</u> if $q(S) = 0$, $p_g(S) = 0$, and $2K_S$ is trivial. The two sheeted unramified covering of an Enriques surface is a K3 - surface.

6) An algebraic surface S is called a <u>rational surface</u> if S is birationally equivalent to $\mathbb{P}^2$.

7) An algebraic surface S is called a <u>ruled surface</u> of genus g if S is birationally equivalent to $\mathbb{P}^1 \times C$ where C is a non - singular curve of genus g.

8) A surface S belonging to class VII is characterized by $b_1(S) = 0$ (see Kodaira [19] , II , Theorem 26). All known examples of surfaces of class VII are certain elliptic surfaces, Hopf surfaces or Inoue surfaces (see Kodaira [19] , II , III , and Inoue [12]).

As for a fine classification, especially the problem of moduli, Popp [26] has shown the following:

THEOREM 1.2.8. Moduli spaces of algebraic surfaces of general type exist as algebraic spaces.

See also the paper of Popp in this volume.

(1.3) Kodaira Dimension

First we shall define certain bemeromorphic invariants of varieties.

DEFINITION 1.3.1. Let M be a complex manifold. The m - genus $P_m(M)$, $m \geq 1$, and the geometric genus $p_g(M)$ of the manifold M are defined by

$$P_m(M) = \dim_{\mathbb{C}} H^0(M, \underline{O}(mK_M)),$$

$$p_g(M) = P_1(M),.$$

It can be shown that if two complex manifolds M and M′ are bimeromorphically equivalent, then $P_m(M) = P_m(M')$ (for the proof see, for example, Ueno [33], §6). Using this fact, we can define the above invariants for singular varieties.

DEFINITION 1,3.2. Let V be a singular variety and let V^* be a non - singular model of V. The m - genus $P_m(V)$ and the geometric genus $p_g(V)$ of the variety are defined by

$$P_m(V) = P_m(V^*), \quad m = 1, 2, \ldots..$$

$$p_g(V) = p_g(V^*),$$

Now we shall define the Kodaira dimension $\kappa(V)$ of a variety V. First we assume that V is non - singular. We set $\underline{\underline{N}}(V) = \{m > 0 \mid P_m(V) \geq 1\}$. Suppose that $\underline{\underline{N}}(V)$ is not empty. For any positive integer $m \in \underline{\underline{N}}(V)$, let $\{\phi_0, \phi_1, \ldots, \phi_N\}$ be a basis of $H^0(V, \underline{O}(mK_V))$. Using this basis, we define a meromorphic mapping (we call it a pluricanonical mapping)

$$\begin{array}{cccc} \Phi_{mK}: & V & \longrightarrow & \mathbb{P}^N \\ & \Psi & & \Psi \\ & z & \to & (\phi_0(z) : \phi_1(z) : \dots : \phi_N(z)). \end{array}$$

DEFINITION 1.3.3. The Kodaira dimension $\kappa(V)$ of V is defined by

$$\kappa(V) = \begin{cases} \max_{m \in \underline{\underline{N}}(V)} \dim \Phi_{mK}(V) & \text{if } \underline{\underline{N}}(V) \neq \emptyset , \\ -\infty & \text{if } \underline{\underline{N}}(V) = \emptyset \end{cases}$$

By definition, it is easy to show that if V and V^* are bimeromorphically equivalent, then

$$\kappa(V) = \kappa(V^*) .$$

Hence we can define the Kodaira dimension $\kappa(V)$ of a singular variety as follows.

DEFINITION 1.3.4. Let V be a singular variety and let V^* be a non - singular model of the variety V. The Kodaira dimension $\kappa(V)$ of V is defined by

$$\kappa(V) = \kappa(V^*).$$

By definition, we have

$$\kappa(V) \leqq \dim V.$$

Moreover, we have

$$\kappa(V) \leqq a(V) ,$$

where a(V) is the algebraic dimension of the variety V (see Definition 3.1.2 , below).

LEMMA 1.3.5. Let M be a complex compact manifold. Suppose that $\kappa(M) \geqq 0$. Then we have

$$\kappa(M) = \text{tr.} \deg_{\mathbb{C}}\Big(\bigoplus_{m \geqq 0} H^0(M , \underline{O}(mK_M))\Big) - 1.$$

PROOF Let $\psi_0, \psi_1, \ldots, \psi_k$ be algebraically independent elements of the ring $R = \bigoplus_{m \geq 0} H^0(M, \underline{O}(mK_M))$ where $\psi_i \in H^0(M, O(m_i K_M))$. Let m be the least common multiple of m_i's. We set $m = m_i n_i$. Then $\psi_0^{n_0}, \psi_1^{n_1}, \ldots, \psi_k^{n_k}$ are algebraically independent and $\psi^n \in H^0(M, \underline{O}(mK_M))$. Hence, we have

$$k \leq \dim \Phi_{mK}(M).$$

Therefore, we obtain

$$\text{tr.}\deg_{\mathbb{C}} R \leq \kappa(M) + 1.$$

On the other hand, if $k = \dim \Phi_{mK}(M)$, there exist $(k + 1)$ elements $\phi_0, \phi_1, \ldots, \phi_k$ of $H^0(M, \underline{O}(mK_M))$ which are algebraically independent. Hence we have

$$\kappa(M) \leq \text{tr.}\deg_{\mathbb{C}} R - 1.$$

Q.E.D.

EXAMPLES 1.3.6 1) $\kappa(M \times V) = \kappa(M) + \kappa(V)$

2) An algebraic variety V is called a ruled variety if V is birational equivalent to $\mathbb{P}^1 \times W$ where W is an algebraic variety. If V is ruled, then $\kappa(V) = -\infty$.

3) If the canonical bundle of a complex manifold M is analytically trivial, (for example, M is a complex torus), then $\kappa(M) = 0$.

4) Let D be a bounded domain in $\mathbb{C}^n$ and let Γ be a discrete group operating on D properly discontinuously and freely. Suppose that the quotient manifold $V = D/\Gamma$ is compact. Kodaira [17] has shown that K_V is ample. Hence $\kappa(V) = \dim V$.

5) Let $V(a_1, a_2, \ldots, a_m) \subset \mathbb{P}^{m+n}$ be a non-singular complete intersection of type $(a_1, a_2, \ldots, a_m)$. The canonical bundle of the manifold $V(a_1, a_2, \ldots, a_m)$ has the form

$$K(V(a_1, a_2, \ldots, a_m)) = \left[\{(a_1 + a_2 + \ldots + a_m) - (m + n + 1)\}H\right]\Big|_V,$$

where H is a hyperplane divisor of $\mathbb{P}^{m+n}$. Therefore we have

$$\kappa(V(a_1, a_2, \ldots, a_m)) = \begin{cases} -\infty, & a_1 + a_2 + \ldots + a_m < m + n + 1, \\ 0, & a_1 + a_2 + \ldots + a_m = m + n + 1, \\ n, & a_1 + a_2 + \ldots + a_m > m + n + 1. \end{cases}$$

The following theorems show the important properites of Kodaira dimensions. The proof can be found in Iitaka [10] and Ueno [33], §5 , §6.

THEOREM 1.3.5. 1) Let $f: V \to W$ be a surjective morphism. Suppose that $\dim V = \dim W$. Then we have

$$\kappa(V) \geqq \kappa(W)$$

2) Let $f: V \to W$ be a finite unramified covering of a variety W. Then we have the equality

$$\kappa(V) = \kappa(W).$$

THEOREM 1.3.6. Let $f: V \to W$ be a surjective morphism of complex varieties with connected fibres. Then there exists an open dense subset U of W such that, for any point $w \in U$, the fibre $V_w = f^{-1}(w)$ is non - singular and the inequality

$$\kappa(V) \leqq \kappa(V_w) + \dim W$$

holds.

COROLLARY 1.3.7. Let $f: V \to W$ be as above. Suppose that there exists an open set U of W such that for any point $w \in U$ the fibre V_w is non - singular and $\kappa(V_w) = -\infty$. Then we have

$$\kappa(V) = -\infty.$$

(1.4) Fundamental Theorem of Classification Theory

Now we shall state the fundamental theorem of classification theory due to Iitaka [10] .

THEOREM 1.4.1. (Fundamental Theorem). Let V be a complex (resp. algebraic) variety of positive Kodaira dimension. Then there exist a complex (resp. projective) manifold V^*, a projective manifold W^* and a surjective morphism $f: V^* \to W^*$ which satisfy the following conditions:

1) V^* is bimeromorphically (resp. birationally) equivalent to V ;

2) $\dim W^* = \kappa(V)$;

3) There exists a dense subset U of W^* (in the complex topology) such that, for any point $w \in U$, $V^*_w = f^{-1}(w)$ is irreducible and non - singular;

4) $\kappa(V^*_w) = 0$, for $w \in U$;

5) if $f^\#: V^\# \to W^\#$ satisfies the above conditions 1)~ 4), then there exist bimeromorphic (rep. birational) mappings $g: V^* \to V^\#$ and $h: W^* \to W^\#$ such that the following diagram is commutative;

$$\begin{array}{ccc} V^* & \xrightarrow{g} & V^\# \\ f\downarrow & \circlearrowright & \downarrow f^\# \\ W^* & \xrightarrow[h]{} & W^\# \end{array}$$

Moreover, if V is a complex (resp. algebraic) manifold, then the fibre space $f: V^* \to W^*$ is bimeromorphically (resp. birationally) equivalent to a pluricanonical mapping $\Phi_{mK}: V \to W_m = \Phi_{mK}(V) \subset \mathbb{P}^N$ for a sufficiently large m, $m \in \underline{\underline{N}}(V)$.

<u>REMARK 1.4.2.</u> It is not known whether we can choose the above dense set U as a Zariski open set. However, if $\kappa(V) \leqq \dim V - 2$, then by virtue of the deformation invariance of Kodaira dimensions of surfaces and curves (see Iitaka [9]), this is the case. In general, Kodaira dimensions and plurigenera are <u>not</u> invariant under (small) deformations (see 2.7., below). But if the following problem is affirmative, then U can be chosen as a dense open set.

<u>PROBLEM 1.4.3.</u> Are Kodaira dimensions upper semi - continuous under small deformations? That is, for a complex analytic family $\underline{M} = \{M_t\}_{t \in \Delta}$ and a point $t_0 \in \Delta$, does there exist a small neighbourhood U of t_0 in Δ such that the inequality

$$\kappa(M_{t_0}) \geqq \kappa(M_t) , \quad t \in U$$

holds?

It is not known whether Kodaira dimensions are invariant under algebraic (or Kähler) deformations.

Another important property of Kodaira dimensions is the following:

THEOREM 1.4.4. Let V be a complex manifold and let d be the largest common divisor of integers in $\underline{\underline{N}}(V)$.

1) There exist positive numbers α , β and a positive integer m_0 such that we have the inequalities

$$(1.4.5) \qquad \alpha m^{\kappa(V)} \leqq P_{md}(V) \leqq \beta m^{\kappa(V)} ,$$

for any integer $m > m_0$.

2) For any positive integer p there exist a positive number γ and a positive integer m_1 , such that

$$P_{md}(V) - P_{(m-p)d}(V) \leqq \gamma m^{\kappa(V)-1},$$

for any integer $m > m_1$.

For the proof, see Iitaka [10] and Ueno [33] , §8.

REMARK 1.4.6. Iitaka has used the above inequalities 1.4.5 to define the Kodaira dimension (see Iitaka [10]).

(1.5) Proof of The Fundamental Theorem

Now we shall provide an outline of the proof of the fundamental theorem. We can assume that the variety V is non - singular. We set

$$W_m = \Phi_{mK}(V) \subset \mathbb{P}^N \text{ for } m \in \underline{\underline{N}}(V).$$

LEMMA 1.5.1. There exists a positive integer m_0 such that $\mathbb{C}(W_m)$ is algebraically closed in $\mathbb{C}(V)$ for any positive integer $m \geqq m_0$, $m \in \underline{\underline{N}}(V)$.

Now we fix a positive integer $m \geqq m_0$. Let $\tilde{V}$ be a non - singular model of the graph of the meromorphic mapping Φ_{mK} and let $\pi : \tilde{V} \to V$ be the natural morphism. We set $f = \Phi_{mK} \circ \pi$ and $W = W_m$. It is easy to show that $f = \Phi_{mK_{\tilde{V}}}$. Therefore, we can assume that $\tilde{V} = V$, that is, $f = \Phi_{mK} : V \to W$ is a morphism. Under this assumption, any divisor in the pluricanonical system $|mK_V|$ can be written in the form

$$E_\lambda = f^* H_\lambda + F$$

where H_λ , $\lambda = (\lambda_0 : \lambda_1 : \dots : \lambda_N)$, is a hyperplane divisor of W and F is the fixed component of $|mK_V|$. Since f is a proper morphism, by the Grauert proper mapping theorem, $f_*(\underline{L}^{\otimes n})$, $\underline{L} = \underline{O}(mK_V)$, is a coherent sheaf. By GAGA (see Serre [30]), there exists an algebraic coherent sheaf $\underline{F}_n$ such that the associated analytic sheaf $\underline{F}_n^{an}$ of $\underline{F}_n$ is isomorphic to $\underline{L}^{\otimes n}$. By W_λ^s and W^s we mean $\mathbb{C}$ - schemes of finite type associated with the projective variety W and the affine variety $W_\lambda = W - H_\lambda$, respectively. Since W_λ^s is an affine $\mathbb{C}$ - scheme of finite type, $H^o(W_\lambda^s , \underline{F}^n)$ is spanned by its global sections ψ_1 , ψ_2 , ... , ψ_M as an $H^o(W_\lambda^s , \underline{O}_{W_\lambda^s})$ module. We can consider ψ_i as an element of $H^o(W_\lambda , f^*(\underline{L}^{\otimes n})) = H^o(f^{-1}(W_\lambda) , \underline{L}^{\otimes n})$. By $\underline{F}_n(eH_\lambda)$, where e is a positive integer, we mean the sheaf of germs of rational sections in the algebraic coherent sheaf $\underline{F}_n$ which have a pole of order at most e on H_λ . Then we have the inclusion

$$H^o(W_\lambda^s , \underline{F}_n) \subset \bigcup_{e=1}^{\infty} H^o(W^s , \underline{F}_n(eH_\lambda)).$$

Hence there exists a positive integer e such that

$$\psi_i \in H^o(W^s , \underline{F}_n(enH_\lambda)) , i = 0, 1, ... , M.$$

By GAGA we have

$$\begin{aligned} H^o(W^s , \underline{F}_n(eH_\lambda)) &= H^o(W , f_*(\underline{L}^{\otimes n})(enH_\lambda)) \\ &= H^o(V , \underline{L}^{\otimes n}(enf^*H_\lambda)) \subset H^o(V , \underline{L}^{\otimes n}(enE_\lambda)). \end{aligned}$$

Therefore, ψ_i , i = 0, 1, ... , M, can be considered as elements of $H^o(V , \underline{L}^{\otimes n}(enE_\lambda))$. There exists an element $\mu \in H^o(V , \underline{O}([enE_\lambda]))$ such that the divisor enE_λ is defined by the equation $\mu = 0$. Then $\mu\psi_i$ is an element of $H^o(V , \underline{L}^{\otimes n} \otimes \underline{O}([enE_\lambda])) = H^o(V , L^{\otimes (eH)n})$. Let ψ_{M+1}, ... , ψ_N be all monomials of degree n in ϕ_0, ϕ_1, ... , ϕ_N, where $\{\phi_0, \phi_1, ... , \phi_N\}$ is a basis of $H^o(V , \underline{L})$. Then $\mu\psi_i \in H^o(V , \underline{L}^{\otimes (e+1)n})$, i = M + 1, ... , N, and we can define a meromorphic mapping

$$\begin{array}{ccc} h^{(n)}: V & \longrightarrow & \mathbb{P}^N \\ \Downarrow & & \Downarrow \\ z & \mapsto & (\psi_0(z) : \psi_1(z) : .... : \psi_M(z) : \psi_{M+1}(z) : .... : \psi_N(z)). \end{array}$$

We set $V_N = h^{(n)}(V)$. Since $\mu\psi_i \in H^o(V , \underline{L}^{\otimes (e+1)n})$, i = 0, 1, ..., N,

we infer that

$$\mathbb{C}(W) \subset \mathbb{C}(V_n) \subset \mathbb{C}(W_{(e+1)nm}).$$

By lemma 1.5.1, we conclude that $\mathbb{C}(W) = \mathbb{C}(V_n)$. Moreover, there exists a natural birational mapping $g^{(n)} : V_n \to W$ such that the following diagram is commutative.

$$\begin{array}{ccc} V & \xrightarrow{h^{(n)}} & V_n \\ & f \searrow \quad \swarrow g^{(n)} & \\ & W & \end{array}$$

Since $f_*(\underline{L}^{\otimes n})$ is coherent, there exists nowhere dense algebraic subset S_n of W such that $f_*(\underline{L}^{\otimes n})\big|_{W-S_n}$ is locally free and that $\underline{L}^{\otimes n}$ is flat over $W - S_n$. By the Grauert proper mapping theorem, we have the canonical isomorphism

$$f_*(\underline{L}^{\otimes n})_w \xrightarrow{\sim} H^0(V_w , \underline{L}_w^{\otimes n}) ,$$

for any point $w \in W - S_n$. Moreover, there exists a nowhere dense algebraic subset T of W such that the fibre $V_w = f^{-1}(w)$ is non - singular for any point $w \in W - T$. Since, on W_λ^s , the algebraic coherent sheaf $\underline{F}_n$ is spanned by global sections ψ_i , $i = 0 , 1 , 2 , \ldots , M$ as an $\underline{O}_{W^s}$ - module, by GAGA, $H^0(V_w , \underline{L}_w^{\otimes n})$ is spanned by $\psi_{i,w} = \psi_i\big|_{V_w}$. On the other hand, there exist nowhere dense analytic subsets A_n and B_n of W and V_n , respectively, such that $g^{(n)}$ induces an isomorphism between $W - A_n$ and $V_n - B_n$. Therefore, the meromorphic mapping $h_w^{(n)} : V_w \to V_{n,w}$ which is the restriction of $h^{(n)}$ to V_w where $w \in W_\lambda - (S_n \cup A_n \cup T)$ and $V_{n,w} = g^{(n)-1}(w)$, is bimeromorphically equivalent to a pluricanonical mapping $\Phi_{mnK(V_w)}$. Since $V_{n,w}$ is a point for any point w $W_\lambda - (S_n \cup A_n \cup T)$, we must have

$$\dim_{\mathbb{C}} H^0(V_w , \underline{L}_w^{\otimes n}) = 1.$$

Let $\pi^* : W^* \to W$, $g^* : V^* \to V \times_W W^*$ be a resolution of singularities of W and $V \times_W W^*$, respectively. We set $f^* = p_2 \circ g^*$ where $p_2 : V \times_W W^*$ is the projection to the second factor. There exist nowhere dense algebraic subsets S^* of W^* and S of W such that π^* induces an isomorphism between $W^* - S^*$ and $W - S$. We set

$$U = W^* - (S^* \cup \pi^{*-1}(\bigcup_{n=1}^{\infty}(S_n \cup A_n) \cup T \cup H_\lambda)).$$

Then it is not difficult to show that W^*, V^*, $f^* : V^* \to W^*$ and U have the desired properties.

(1.6) Classification and Complex Varieties of Hyperbolic Type

DEFINITION 1.6.1. A complex variety V is called a variety of hyperbolic type (parabolic type, elliptic type, respectively) if $\kappa(V) = \dim V$ ($\kappa(V) = 0$, $\kappa(V) = -\infty$, respectively).

The fundamental theorem on classification theory shows that classification theory is reduced to

1) the study of varieties of hyperbolic, parabolic and elliptic type;

2) the study of fibre spaces whose general fibres are of parabolic type.

In the next chapter we shall show that the study of algebraic varieties of parabolic type is deeply related to the fibre spaces associated with Albanese mappings (see 2.1. and 2.4). Here we shall discuss briefly problems on complex varieties of hyperbolic type and elliptic type. Algebraic varieties of parabolic type and fibre spaces whose general fibres are of parabolic type will be discussed in the next chapter.

First we shall consider a complex manifold V of hyperbolic type. Since $a(V) \geqq \kappa(V) = \dim V$, V is a Moishezon manifold. If V is a curve, V is of hyperbolic type if and only if the genus of V is bigger than one. In this case, the canonical bundle K_V is ample and $3K_V$ is very ample. Moreover, if V is not hyperelliptic, then K_V is very ample. A surface S of hyperbolic type is usually called a surface of genral type (see 1.3). If S is a surface of general type which is free from exceptional curves of the first kind, the the pluricanonical system $|mK_S|$ is free from base points and fixed components for $m \geqq 4$ and Φ_{mK} is a birational morphism for $m \geqq 5$ (see Kodaira [20] and Bombieri [4]). It is natural to ask whether these properties hold for a complex manifold of hyperbolic type of dimension $n \geqq 3$. After the Mannheim meeting, the author found an algebraic manifold V of hyperbolic type of dimension $n \geqq 3$ such that, for any positive integer m and for any bimeromorphically equivalent model V^* of V , the pluricanonical system $|mK_{V^*}|$ has always fixed components (see Ueno [33] , §16 and Ueno [34]). But it is interesting to study the following problem.

PROBLEM 1.6.2. Does there exist a positive integer m which only depends on the dimension of a complex manifold V of hyperbolic type such that, for any integer $m \geqq m_0$, the pluricanonical mapping Φ_{mK} is bimeromorphic?

As for complex varieties of elliptic type, there are very few results. The problem to find the criterion of rationality of algebraic threefolds is an old and difficult problem. Recently, it **was shown** that there are algebraic threefolds which are unirational but not rational. A non - singular cubic threefold is one of such examples (see the paper of Murre in this volume). The following problem seems interesting.

PROBLEM 1.6.3. Does there exist an algebraic manifold V of elliptic type such that V is not unirational and $g_k(V) = 0$, $k = 1, 2, \ldots, \dim V$ (See Definition 2.1.4.) ?

A complex manifold of algebraic dimension zero is of parabolic type or of elliptic type. Such a manifold will be studied in Chapter III.

(1.7) D - Dimension

To define the Kodaira dimension of a complex variety, we use the canonical line bundle. Instead of the canonical bundle, if we use a Cartier divisor on a complex variety, we obtain a notion of the D - dimension.

First we shall assume that a variety V is normal. Let D be a Cartier divisor on V. We set

$$\underline{\underline{N}}(D, V) = \{m > 0 \mid \dim_{\mathbb{C}} H^0(V, \underline{O}_V(mD)) \geqq 1\}.$$

Suppose that $\underline{\underline{N}}(D, V)$ is not empty. For any integer $m \in \underline{\underline{N}}(V)$, let $\{\phi_0, \phi_1, \ldots, \phi_N\}$ be a basis of $H^0(V, \underline{O}_V(mD))$. Using this basis, we define a meromorphic mapping

$$\begin{array}{rcl} \Phi_{mD} : V & \longrightarrow & \mathbf{P}^N \\ \cup & & \cup \\ z & \rightarrow & (\phi_0(z) : \phi_1(z) : \ldots : \phi_N(z)). \end{array}$$

DEFINITION 1.7.1. The D - dimension $\kappa(D, V)$ of a normal complex variety is defined by

$$\kappa(D, V) = \begin{cases} \max\limits_{m \in \underline{\underline{N}}(D, V)} \dim \Phi_{mD}(V) & \text{if } \underline{\underline{N}}(D, V) \neq \emptyset, \\ -\infty & \text{if } \underline{\underline{N}}(D, V) = \emptyset. \end{cases}$$

If V is not normal and if D is a Cartier divisor on V , the D - dimenison $\kappa(D , V)$ of V is defined by

$$\kappa(D , V) = \kappa(\iota^*D , V^*) ,$$

where $\iota : V^* \to V$ is the normalization of V.

It is easy to generalize results on Kodaira dimensions to those on D - dimensions. We only mention the following theorem which we shall use later.

<u>THEOREM 1.7.2.</u> Let $f: V \to W$ be a surjective morphism of complex varieties and let D be a Cartier divisor on W. We have the equality

$$\kappa(f^*D , V) = \kappa(D , W).$$

The proof can be found in Ueno [33] , §5.

Chapter II

Albanese Mappings and Algebraic Manifolds of Parabolic Type

(2.1) Albanese Mapping

The study of Albanese mappings gives an important tool to study algebraic varieties of parabolic type.

DEFINITION 2.1.1. $(A(M), \alpha)$ is called an Albanese torus of a complex manifold M if it satisfies the following properties.
1) $A(M)$ is a complex torus and $\alpha: M \to A(M)$ is a holomorphic mapping (we call it an Albanese mapping).
2) For any complex torus T and a holomorphic mapping $g: M \to T$, there exist a complex Lie group homomorphism $h: A(M) \to T$ and an element a of T such that

$$g(z) = h(\alpha(z)) + a , \quad z \in M.$$

The above property 2) is often called the universal property of the Albanese mapping. By virtue of the property 2) , if an Albanese torus $(A(M), \alpha)$ of a complex manifold M exists, then $(A(M), \alpha)$ is unique up to translations by elemnts of $A(M)$. Therefore, we can say "the" Albanese torus of M.

THEOREM 2.1.2. For any complex manifold M, there exists the Albanese torus $(A(M), \alpha)$.

An outline of the proof is as follows. Let $\{\omega_1 , \omega_2 , \dots , \omega_r\}$ be a basis of $H^0(M , d\mathcal{O}_M)$ and let $\{\gamma_1 , \gamma_2 , \dots , \gamma_b\}$ be a basis of the free part of $H_1(M , \mathbb{Z})$. We shall consider a lattice Δ in $\mathbb{C}^r$ generated by the r vectors

$$\begin{pmatrix} \int_{\gamma_i} \omega_1 \\ \int_{\gamma_i} \omega_2 \\ \vdots \\ \int_{\gamma_i} \omega_r \end{pmatrix} \qquad i = 1 , 2 , \dots , r.$$

We let $\bar{\Delta}$ be the smallest closed Lie subgroup of $\mathbb{C}^r$ containing Δ such that its connected component containing the origin is a vector subspace of $\mathbb{C}^r$. We set $\mathbb{C}^r/_{\bar{\Delta}} = A(M)$. We define a holomorphic mapping $\alpha: M \to A(M)$ via

$$\alpha(x) = \begin{pmatrix} \int_{x}^{x}\omega_1 \\ \int_{x}^{x}\omega_2 \\ \vdots \\ \int_{x}^{x}\omega \end{pmatrix} \in \mathbb{C}^r/_{\bar{\Delta}}$$

It is not difficult to show that $(A(M), \alpha)$ is the Albanese torus of M. For the detailed discussion, see Blanchard [3] and Ueno [33] .

REMARK 2.1.3. 1) If M is Kähler or algebraic, then $r = \frac{1}{2}b_1$, and $\Delta = \bar{\Delta}$.

2) If M_1 and M_2 are bimeromorphically equivalent, then $A(M_1)$ and $A(M_2)$ are isomorphic.

Now we shall introduce some bimeromorphic invariants.

DEFINITION 2.1.4. For a complex manifold M, we shall define

$$q(M) = \dim_{\mathbb{C}} H^1(M , \underline{O}_M)$$

$$g_k(M) = \dim_{\mathbb{C}} H^0(M , \Omega_M^k) , \quad k = 1 , 2 , \dots , \dim M ,$$

$$r(M) = \dim_{\mathbb{C}} H^0(M , d\underline{O}_M) ,$$

$$t(M) = \dim A(M).$$

Let M be a singular variety and let M^* be a non - singular model of M. We define

$$q(M) = q(M^*)$$

$$g_k(M) = g_k(M^*) ,$$

$$r(M) = r(M^*) ,$$

$$t(M) = t(M^*).$$

These are well defined (see Ueno [33] , §9).

<u>REMARK 2.1.5.</u> If M is a Kähler manifold, then we have

$$t(M) = r(M) = g_1(M) = \frac{1}{2} b_1(M).$$

This is also true if there exists a surjective morphism $g: \tilde{M} \to M$ where $\tilde{M}$ is Kähler and $\dim \tilde{M} = \dim M$ (see Ueno [32], §9).
For example, if M is a Moishezon
manifold, then this is the case (see Theorem 3.1.3., below).

<u>EXAMPLE 2.1.5.</u> Let A be an abelian surface and let D be an ample divisor. We set $V = A \times \mathbb{P}^1$. Since a divisor $E = A \times p + D \times \mathbb{P}^1$ (where p is a point of $\mathbb{P}^1$) is ample, the mapping $\Phi_{mE}: V \to \mathbb{P}^N$ associated wiθh the complete linear system $|mE|$ (see 1.7) is embedding for a sufficiently large m. We fixed a sufficiently large m and we let S be a non - singular surface obtained by a generic hyperplane section of $\Phi_{mE}(V)$ in $\mathbb{P}^N$. By Lefschetz´ theorem, we have isomorphisms

$$H_1(A, \mathbb{Z}) \tilde{\to} H_1(S, \mathbb{Z}),$$

$$H^0(A, \Omega_A^1) \tilde{\to} H^0(S, \Omega_S^1).$$

Hence $A(S)$ is isomorphic to A and the Albanese mapping $\alpha: S \to A(S) = A$ is nothing other than the morphism induced by the projection of V onto A. By the adjunction formula, the canonical bundle K_S has the form

$$K_S = K_V|_S \otimes [S]|_S = [(m-2) A \times p + mE].$$

As m is sufficiently large, K_S is ample. Hence $\kappa(S) = 2$. Therefore $\alpha: S \to A$ is a ramified covering.

This example shows that fibres of the Albanese mapping $\alpha: M \to \alpha(M)$ of a complex manifold M are <u>not</u> necessarily connected. Let

$$\begin{array}{ccc} M & \xrightarrow{\alpha} & \alpha(M) \subset A(M) \\ & {}_{\beta}\searrow \quad \nearrow_{\gamma} & \\ & W & \end{array}$$

be the Stein factorization of the morphism $\alpha: M \to \alpha(M)$ (see Cartan [5]).
The morphism $\beta: M \to W$ has connected fibres and $\gamma: W \to \alpha(M)$ is a finite morphism. The fibre space $\beta: M \to W$ is called the <u>fibre space associated with the Albanese mapping</u> $\alpha: M \to A(M)$. We note that

we have

$$\kappa(W) \geqq \kappa(\alpha(M)$$

by Theorem 1.3.5., 1).

(2.2) Subvarieties of a Complex Torus

To study the fibre space associated with the Albanese mapping $\alpha: M \to \alpha(M)$, it is necessary to study the structure of the image variety $\alpha(M)$ of α. More generally, we shall study subvarieties of a complex torus. In what follows, we let B be an n - dimensional subvariety of a complex torus T.

LEMMA 2.2.1.

$$g_k(B) \geqq \binom{n}{k}, \; k = 1 , 2 , \dots , n$$

$$P_m(B) \geqq 1, \quad m = 1 , 2 , \dots ,$$

$$\kappa(B) \geqq 0.$$

THEOREM 2.2.2. The following conditions are equivalent.

1) $g_1(B) = n$
2) $p_g(B) = 1$,
3) $P_m(B) = 1$ for a positive integer m.
4) $\kappa(B) = 0$
5) B is a translation of a complex subtorus T_1 of T by an element $a \in T$.

COROLLARY 2.2.3. If the subvariety B generates A (that is, the morphism

$$\overbrace{B \times \dots \times B}^{m} \longrightarrow A$$
$$\begin{array}{ccc} \cup & & \cup \\ (z_1, \dots , z_m) & \to & z_1 + \dots + z_m \end{array}$$

is surjective for a suitable positive integer m), then $\kappa(B) > 0$. By the universal property of Albanese mappings, it is easy to show that the image variety $\alpha(M)$ generates the Albanese variety $A(M)$ of a complex manifold M. Therefore, we obtain the following:

COROLLARY 2.2.4. For the Albanese mapping $\alpha: M \to A(M)$ of a complex manifold we have

$$\kappa(\alpha(M)) \geqq 0.$$

Moreover, $\kappa(\alpha(M)) = 0$ if and only if the Albanese mapping α is surjective.

THEOREM 2.2.5. Suppose that B is algebraic and that $\kappa(B) > 0$. Then there exist a finite unramified covering $\pi: \tilde{B} \to B$ of B, a projective variety W and a complex subtorus A_1 of T such that $\tilde{B}$ is isomorphic to $A_1 \times W$ and that $\kappa(W) = \kappa(B) = \dim W$.

For the more detailed discussions in this section, we refer the reader to Ueno [33], § 10 and Ueno [32], I, § 3.

(2.3) Algebraic Manifolds of Parabolic Type

Now we shall consider algebraic manifolds of parabolic type. The following important theorem is due to Matsushima [21].

THEOREM 2.3.1. Let V be a projective manifold. Suppose that mK_V is analytically trivial for a positive integer m. Then there exists a finite unramified covering $f: \tilde{V} \to V$ of V, an abelian variety A and a projective manifold with a trivial canonical bundle such that $\tilde{V} = A \times W$ and $q(\tilde{V}) = \dim A$.

If V is an (algebraic) surface of parabolic type and if V does not contain exceptional curves of the first kind, then $12K_V$ is analytically trivial. But if $\dim V \geqq 3$, the following example shows that this is not the case.

EXAMPLE 2.3.2. Let T be an n - dimensional abelian variety and let ι be the involution of T defined by

$$\begin{array}{ccc} \iota : T & \to & T \\ \cup\!\!\!\!\!\!\in & & \cup\!\!\!\!\!\!\in \\ z & \to & -z. \end{array}$$

The quotient space $T/\langle\iota\rangle$ has 2^{2n} isolated singular points corresponding to the fixed points of ι. Let $K^{(n)}$ be a non-singular model of $T/\langle\iota\rangle$ obtained by the canonical resolution of its singularities (see Ueno [33], §16, Example 16.11). $K^{(n)}$ is called a classical Kummer manifold. $K^{(1)}$ is the projective line $\mathbb{P}^1$ and $K^{(2)}$ is a K3 surface. It is easy to show that $\kappa(K^{(n)}) = 0$ for $n \geq 2$. Let A be an abelian variety. Then we can show that, for any bimeromorphically equivalent model V^* of the algebraic manifold $K^{(n)} \times A$, mK_{V^*} is <u>not</u> analytically trivial for any positive integer m. For the proof see Ueno [33], §16, Proposition 16.17.

The above example shows that Theorem 2.3.1 is insufficient to study algebraic manifolds of parabolic type. On the other hand, all known examples of algebraic manifolds of parabolic type support the following:

<u>CONJECTURE K_n</u> Let V be an n - dimensional algebraic manifold of parabolic type. Then the Albanese mapping $\alpha: V \to A(V)$ is surjective and $\alpha: V \to A(V)$ is birationally equivalent to a fibre bundle over A(V) in the etale topology whose fibre is an algebraic manifold of parabolic type.

If Conjecture K_n is true, we have $q(V) \leq \dim V$ as a corollary.

<u>(2.4) Conjecture C_n.</u>

Since we have already studied the structure of the image varieties of Albanese mappings, to study Conjecture K_n, it is natural to consider the following conjecture due to Iitaka [11].

<u>CONJECTURE C_n</u> Let $f: V \to W$ be a surjective morphism of algebraic varieties with connected fibres. Then for a general point $w \in W$, we have

$$\kappa(V) \geq \kappa(W) + \kappa(V_w) ,$$

where $V_w = f^{-1}(w)$.

If Conjecture C_n is true, then from Corollary 1.3.7 and Corollary 2.2.4, we infer that the Albanese mapping $\alpha: V \to A(V)$ is surjective for algebraic manifolds V of parabolic type. Therefore, if we can prove the following conjecture which is a special case of Conjecture K_n and if we can find a canonical bundle formula for fibre spaces whose general fibres are algebraic manifolds of parabolic type (see 2.6 , below), then Conjecture C_n implies Conjecture K_n.

CONJECTURE B_n Let V be an n - dimensional algebraic manifold of parabolic type. Suppose that $\mathring{q}(V) = \dim V$. Then the Albanese mapping $\alpha: V \to A(V)$ is a birational morphism.

Up to now all conjectures above are solved only in special cases. As for Conjecture C_n , we have the following affirmative answers.

THEOREM 2.4.1. Let $f: V \to W$ be a fibre bundle over a complex manifold W whose fibre and structure group are an algebraic manifold F and the automorphism group Aut (F) of F, respectively. Then we have an equality

$$\kappa(V) = \kappa(W) + \kappa(F).$$

For the proof, we refer the reader to Nakamura and Ueno [25] , and Ueno [33] , §15.

THEOREM 2.4.2. Conjecture C_n is true in the following cases.

1) General fibres of the fibre space $f: V \to W$ are elliptic curves and for any w of W there exist an open neighbourhood U of w in W and a meromorphic section $o: U \to f^{-1}(U)$.

2) W is a curve and $f: V \to W$ is a fibre space of principally polarized abelian surfaces (for the definition of such a fibre space, see Ueno [31] , I).

This theorem is a corollary of **the canonical bundle formula for such fibre spaces (see Theorem 2.6.1, below).**

On the other hand, the following example due to Iitaka shows that Conjecture C_n is not necessarily true if $f: V \to W$ is a fibre space of complex manifolds.

EXAMPLE 2.4.3. Let a, b, c be three roots of the equation

$$x^3 + 3x + 1 = 0 ,$$

such that a is real. We let α_1 , α_2 , β_1 , β_2 , γ_1 , γ_2 be six roots of the equation

$$z^6 + 3z^2 + 1 = 0 ,$$

such that

$$\alpha_i^2 = a \ , \ \beta_i^2 = b \ , \ \gamma_i^2 = c \ , \ i = 1 \ , \ 2.$$

We set

$$\Omega = \begin{pmatrix} 1 & \alpha_1 & \alpha_1^2 & \alpha_1^3 & \alpha_1^4 & \alpha_1^5 \\ 1 & \beta_1 & \beta_1^2 & \beta_1^3 & \beta_1^4 & \beta_1^5 \\ 1 & \beta_2 & \beta_2^2 & \beta_2^3 & \beta_2^4 & \beta_2^5 \end{pmatrix}.$$

There exists a three - dimensional complex torus T with period matrix Ω. Left multiplication of a matrix

$$\begin{pmatrix} \alpha_1 & 0 & 0 \\ 0 & \beta_1 & 0 \\ 0 & 0 & \beta_2 \end{pmatrix}$$

to $\mathbb{C}^3$ induces an analytic automorphism g of the complex torus T. Let E be an elliptic curve with period matrix $(1. \ \omega)$, $\mathrm{Im}(\omega) > 0$. We let G be a free abelian group of analytic automorphisms of $\mathbb{C} \times T$ generated by two automorphisms

$$g_1 : (z \ , \ p) \to (z + 1 \ , \ p) \ ,$$

$$g_2 : (z \ , \ p) \to (z + \omega \ , \ g(p)).$$

The group G acts on $\mathbb{C} \times T$ freely and properly discontinuously. The quotient manifold $M = \mathbb{C} \times T/_G$ is a fibre bundle over the elliptic curve E whose fibre and structure group are the complex torus T and Aut (T) , respectively. We can prove that

$$\kappa(M) = - \infty,$$

(see Nakamura and Ueno [25] and Ueno [33] , §15).

(2.5) Kummer Manifolds

Other important examples which support Conjecture K_n are Kummer manifolds. All results in this section can be found in Ueno [32], I, §7 and Ueno [33], §16.

DEFINITION 2.5.1. An algebraic manifold V is called a Kummer manifold if V is a non - singular model of a quotient space A/G of an abelian variety A by a finite group G of analytic automorphisms of A. An algebraic manifold V is called a generalized Kummer manifold if there exists a generically surjective rational mapping $f: A \to V$ of an abelian variety A onto V.

THEOREM 2.5.2. For a generalized Kummer manifold V, we have

$$\kappa(V) \leqq 0.$$

PROPOSITION 2.5.3. Let V be a generalized Kummer manifold.

1) $q(V) \leqq \dim V$.

2) The Albanese mapping $\alpha: V \to A(V)$ is a surjective morphism with connected fibres.

3) If $q(V) = \dim V$, the Albanese mapping $\alpha: V \to A(V)$ is a birational morphism. That is, Conjecture B_n is true.

THEOREM 2.5.4. For a Kummer manifold, Conjecture K_n is true.

THEOREM 2.5.1. Let V be a generalized Kummer manifold. Suppose that $q(V) = \dim V - 1$. Then we have

1) $\kappa(V) = 0$ if and only if V is birationally equivalent to a fibre bundle in the etale topology over the Albanese torus A(V) of V ;

2) $\kappa(V) = -\infty$ if and only if general fibres of the Albanese mapping $\alpha: V \to A(V)$ is $\mathbb{P}^1$.

(2.6) Canonical Bundle Formula

Not only to study Conjecture K_n but also to study fibre spaces whose general fibres are of parabolic type, it is important to study the canonical bundles of such fibre spaces. In the case of surfaces, Kodaira has given the canonical bundle formula for an elliptic surface. Ueno [32] has generalized the formula as follows:

Let $f: V \to W$ be a fibre space over a complex manifold W whose general fibres are elliptic curves. We can assume that an analytic subset $S = f(A)$ where $A = \{z \in V \mid f \text{ is not of maximal rank at } z\}$, is a divisor with normal crossings. Let S_i be an irreducible component of S.

Let D be a small disk in W such that D intersects S_i at the origin of D which is a general point of S_i. We restrict the fibre space $f: V \to W$ on D and obtain a fibre space $f_D: V_D \to D$.

Let M_D be the monodry matrix of the fibre space $f_D: V_D \to D$ around the origin. We say that the fibre space $f: V \to W$ has a <u>singular fibre of type</u> Kod $(*)$ over S_i if the monodry M is $SL(2, \mathbb{Z})$ - conjugate to the following matrix of type $(*)$.

type	I_o^*	I_b	I_b^*	II	II^*
matrix	$\begin{pmatrix} -1 & 0 \\ 0 & -1 \end{pmatrix}$	$\begin{pmatrix} 1 & b \\ 0 & 1 \end{pmatrix}$	$\begin{pmatrix} -1 & -b \\ 0 & -1 \end{pmatrix}$	$\begin{pmatrix} 1 & 1 \\ -1 & 0 \end{pmatrix}$	$\begin{pmatrix} 0 & -1 \\ 1 & 1 \end{pmatrix}$
		$b > 0$	$b > 0$		

III	III^*	IV	IV^*
$\begin{pmatrix} 0 & 1 \\ -1 & 0 \end{pmatrix}$	$\begin{pmatrix} 0 & -1 \\ 1 & 0 \end{pmatrix}$	$\begin{pmatrix} 0 & 1 \\ -1 & -1 \end{pmatrix}$	$\begin{pmatrix} -1 & -1 \\ 1 & 0 \end{pmatrix}$

<u>THEOREM 2.6.1.</u> Let $f: V \to W$ be as above. Suppose that for any point $x \in W$, there exist an open neighbourhhod U of x in W and a meromophic section $o: U \to f^{-1}(U)$. Then the twelfth canonical divisor $12K(V)$ has the form

$$f^*(12K(W) + F) + G,$$

where G is an effective divisor on V which does not come from a divisor on W and F is a divisor on W written in the form

$$F = \sum_b b\, S_{I_b} + \sum_b (6 + b)\, S_{I_b^*} + 2S_{II} + 10S_{II^*} + 3S_{III} + 9S_{III^*}$$

$$+ 4S_{IV} + 8S_{IV^*} ,$$

such that $S_{(*)} = \Sigma S_j$, where S_j is one of the components of S over which f has a singular fibre of type Kod (*).

The proof can be found in Ueno [32] , I, §4 ~ §6. We remark that the canonical bundle formula for a fibre space of principally polarized abelian surfaces is obtained by Ueno [31] , III.

Now we shall show that the above canonical bundle formula is deeply related to Conjecture K_n. Let V be an n - dimensional algebraic manifold of parabolic type with $q(V) = n - 1$. If the Albanese mapping $\alpha: V \to A(V)$ is surjective with connected fibres and if Conjecture C_n is true, then general fibres are elliptic curves. Suppose, moreover, that the fibre space $\alpha: V \to A(V)$ has locally meromorphic section S (in the complex topology) at any point of A(W). We set $S = \alpha(\{z \in V \mid \alpha \text{ is not of maximal rank at } z\})$. If S is not a divisor with normal crossings, by a finite succession of monoidal transformations with non - singular centres, we obtain an algebraic manifold W and a birational morphism $g: W \to A(V)$ such that the strict transform $\tilde{S}$ of S is a divisor with normal crossings. Let V^* be a non - singular model of $V \times_{A(V)} W$ obtained by a finite succession of monoidal transformations with non - singular centres contained in the singular loci. We have a surjective morphism $f: V^* \to W$. Moreover, by our construction we obtain

$$\tilde{S} = f(\{z \in V^* \mid f \text{ is not of maximal rank at } z\}).$$

Applying Theorem 2.6.1 and Theorem 1.7.2 to the fibre space $f: V^* \to W$, we obtain

$$\kappa(V) = \kappa(V^*) \geqq \kappa(f^*(12K(W) + F , V^*) = \kappa(12k(W) + F , W) = \kappa(F , W) ,$$

because the canonical divisor K(W) consists of exceptional varieties appearing in the monoidal transformations. Let F^* be a divisor on the Albanese Torus A(V) such that $F = g^* F^* + D$ where D is an effective divisor which appears in the monoidal transformations. We have

$$\kappa(F, W) = \kappa(g^* F^*, W) = \kappa(F^*, A(V)).$$

From the theory of the theta functions (see, for example, Weil [35], Théorèm 1, p. 114 and Proposition 7, p. 121), we infer that, if F^* is an effective divisor, then $\kappa(F^*, A(V)) > 0$. On the other hand, by our assumption, we have the inequality

$$0 = \kappa(V) \geqq \kappa(F^*, A(V)).$$

Since F^* is an effective divisor or the zero divisor, it follows that $F^* = 0$. By the theory of elliptic fibre spaces (see Kodaira [18], II, Kawai [14] and Ueno [32], I), this implies that V is birationally equivalent to a fibre bundle in the etale topology over A(V) whose fibre and structure group are an elliptic curve E and the automorphism group Aut(E), respectively. Hence Conjecture K_n is true in our case. Note that the above argument is used to prove Theorem 2.5.1. (See Ueno [32], I, p. 331 - p. 333).

(2.7) The Kodaira Dimension is not a Deformation Invariant.

In this section, we shall provide an example of a one parameter complex family of three - dimensional complex manifolds due to Nakamura [24]. This example shows that the Kodaira dimension, the m - genera and the invariants g_1, q, r, t defined in 2.1 are not invariant under small deformations.

Let A be a 2 × 2 unimodular matrix with tr $A \geqq 3$. The matrix A has real eigenvalues α, α^{-1} where $\alpha > 1$. There is a real 2 × 2 matrix P such that

$$\begin{pmatrix} \alpha & 0 \\ 0 & \alpha^{-1} \end{pmatrix} = P A P^{-1}.$$

We set

$$(P, \tau P) = \begin{pmatrix} \omega_{21} & \omega_{22} & \omega_{23} & \omega_{24} \\ \omega_{31} & \omega_{32} & \omega_{33} & \omega_{34} \end{pmatrix}, \quad \text{Im}(\tau) > 0.$$

Consider the group Δ of analytic automorphisms of $C^* \times C^2$

generated by

$$g: (z_1, z_2, z_3) \to (\alpha z_1, \alpha z_2, \alpha^{-1} z_3),$$

$$g_j: (z_1, z_2, z_3) \to (z_1, z_2 + \omega_{2j}, z_3 + \omega_{3j}), \ j = 1, 2, 3, 4.$$

It is easy to show that the group Γ operates on $\mathbb{C}^* \times \mathbb{C}^2$ freely and properly discontinuously. The quotient manifold $M = \mathbb{C}^* \times \mathbb{C}^2/_\Delta$ is compact. Since holomorphic 1 - forms

$$\phi_1 = \frac{dz_1}{z_1}, \ \phi_2 = \frac{dz_2}{z_1}, \ \phi_3 = z_1 dz_3$$

are Γ - invariant, they induce nowhere vanishing **holomorphic 1 - forms** on M. Therefore, M is a complex parallelizable manifold. The manifold M has the following numerical invariants.

$$\kappa(M) = 0, \ P_m(M) = 1, \ m = 1, 2, \ldots .$$

$$q(M) = 3, \ r(M) = t(M) = 1, \ g_1(M) = 3.$$

Let G be the infinite cyclic group of analytic automorphisms of $\mathbb{C}^*$ generated by **the automorphism**

$$z_1 \to \alpha z_1.$$

Then the quotient manifold $E = \mathbb{C}^*/G$ is an elliptic curve with period matrix $(2\pi i, \beta)$ where $\beta = \log\alpha > 0$. The Albanese torus $A(M)$ of M is isomorphic to the elliptic curve E and the Albanese mapping is given by

$$\begin{array}{cccc} \alpha & : & M & \longrightarrow & E \\ & & \cup & & \cup \\ & & [z_1, z_2, z_3] & \to & [z_1] . \end{array}$$

For a complex number s, we set

$$W_s = \{(\zeta_1, \zeta_2, \zeta_3) \in \mathbb{C}^3 \mid \zeta_1 - s\bar{\zeta}_2 \neq 0\} .$$

Let Δ_s be a group of analytic automorphisms of W_s generated by automorphisms

$$g : (\zeta_1 , \zeta_2 , \zeta_3) \to (\alpha\zeta_1 , \alpha\zeta_2 , \alpha^{-1}\zeta_3)$$

$$g_j : (\zeta_1 , \zeta_2 , \zeta_3) \to (\zeta_1 + s\bar{\omega}_{2j} , \zeta_2 + \omega_{2j} , \zeta_3 + \omega_{3j}) , \ j = 1,2,3,4.$$

The quotient $M_s = W_s/\Delta_s$ is a compact complex manifold and $M_0 = M$. The collection $\{M_s\}_{|s|<\varepsilon}$ is a one parameter complex family. Nakamura [24] has shown the following:

<u>PROPOSITION 2.7.1.</u> If $s \neq 0$, we have

$$\kappa(M_s) = -\infty , \ P_m(M_s) = 0 , \ m = 1, 2, \ldots\ldots ,$$

$$q(M_s) = 2 , \ r(M_s) = t(M_s) = 0 , \ g_1(M_s) = 2.$$

The Kuranishi family of the manifold M is constructed in Nakamura [24] . The Kuranishi space has several branches at the point which corresponds to the manifold M.

Chapter III

Algebraic Reductions of Complex Varieties and Complex Manifolds of Algebraic Dimension Zero,

(3.1) Algebraic Dimension

Let $\mathbb{C}(V)$ be the field of all meromorphic functions on a complex variety V (we call it the meromorphic function field of V).

THEOREM 3.1.1. $\mathbb{C}(M)$ is a finitely generated extension over $\mathbb{C}$ (that is, $\mathbb{C}(M)$ is an algebraic function field) satisfying the inequality

$$\text{tr. deg.}_{\mathbb{C}}\ \mathbb{C}(V) \leqq \dim V.$$

The proof is found in Remmert [27] .

DEFINITION 3.1.2. The algebraic dimension $a(V)$ of a complex variety V is defined by

$$a(V) = \text{tr. deg.}_{\mathbb{C}}\ \mathbb{C}(V) \leqq \dim V.$$

A complex variety V is called a Moishezon variety if $a(V) = \dim V$. Moishezon varieties have been studied by Moishezon [22] . Artin [1] has shown that Moishezon varieties are algebraic spaces. The following theorem can be found in Moishezon [22] .

THEOREM 3.1.3. 1) Let M be a Moishezon variety. Then by a finite succession of monoidal transformations with non - singular centres we obtain a projective manifold.
2) A Moishezon manifold M is a projective manifold if and only if M carries a Kähler metric.

THEOREM 3.1.4. 1) Let $f: V \to W$ be a fibre space of complex varieties. Then there exists a nowhere dense analytic subset W_1 of W such that for any point $w \in W_1$, the fibre $V_w = f^{-1}(w)$ is irreducible and we have

$$a(V) \leq a(W) + \operatorname{codim} W.$$

The proof is found in Ueno [33] , §7.

COROLLARY 3.1.5. 1) Subvarieties of a Moishezon variety are Moishezon varieties.
2) If V is a Moishezon variety and $f: V \to W$ is a surjective morphism, then W is a Moishezon manifold.

(3.2) Algebraic Reduction

Let V be a complex variety. By a resolution of singularities due to Hironaka [6] , there exists a projective manifold W such that the rational function field $\mathbb{C}(W)$ is isomorphic to the meromorphic function field $\mathbb{C}(V)$ of the variety V. Let $\mathbb{C}[W] = \mathbb{C}[\zeta_0 , \zeta_1 , \ldots , \zeta_N]$ be the homogeneous coordinate ring of W. Since ζ_i/ζ_0 , $i = 1 , \ldots , N$ are rational functions on W, there exist meromorphic functions ϕ_i , ..., N of $\mathbb{C}(V)$ which correspond to ζ_i/ζ_0 by the isomorphism between $\mathbb{C}(W)$ and $\mathbb{C}(V)$. We can define a meromorphic mapping

$$\begin{array}{cccc} \psi: V & \longrightarrow & W \subset \mathbb{P}^N \\ \psi & & \psi \\ z \to (1 : \phi_1(z) : \phi_2(z) : \ldots : \phi_N(z)). \end{array}$$

Let $f: V^* \to G$ be a resolution of singularities of the graph G of the meromorphic mapping. Then $\psi^* = p_W \circ f: V^* \to W$ is a surjective morphism where $p_W: G \to W$ is the natural projection. The surjective morphism $\psi^*: V^* \to W$ induces an isomorphism between $\mathbb{C}(W)$ and $\mathbb{C}(V) = \mathbb{C}(V^*)$.

DEFINITION 3.2.1. The surjective morphism $\psi^*: V^* \to W$ is called an algebraic reduction of **the complex variety V.**

REMARK 3.2.2. The following example shows that, in general, we cannot take V as V^* in the algebraic reduction $\psi^*: V^* \to W$ of a complex manifold V. But this is the case if V is a surface or $\mathbb{C}(V)$ is an algebraic function field of one varieable of genus $g \geq 1$.

EXAMPLE 3.2.3. Let a and b be algebraically independent complex numbers such that

$$1 < |a| < |b| .$$

Let G be the infinite cyclic group of analytic automorphisms of $\mathbb{C}^3 - \{0\}$ generated by the automorphism

$$g: (z_1 , z_2 , z_3) \to (az_1 , az_2 , bz_3).$$

The group G operates on $\mathbb{C}^3 - \{0\}$ freely and properly discontinuously. The quotient manifold $V = \mathbb{C}^3 - \{0\}/G$ is compact and diffeomorphic to $S^1 \times S^5$. It is easy to see that $\mathbb{C}(V) = \mathbb{C}(\frac{z_1}{z_2})$. Let V^* be a complex manifold obtained by a monoidal transformation with centre T where T is defined by the equation

$$z_1 = z_2 = 0.$$

Then V^* is a non - singular model of the graph of a meromorphic mapping

$$\begin{array}{cccc} \psi: & V & \longrightarrow & \mathbb{P}^1 \\ & \ni & & \ni \\ & [z_1 , z_2 , z_3] & \to & (z_1 : z_2) . \end{array}$$

We set $\psi^* = \psi \circ f$ where $f : V^* \to V$ is a natural morphism. $\psi^* : V^* \to \mathbb{P}^1$ is an algebraic reduction of V.

PROPOSITION 3.2.4. Any fibre of an algebraic reduction is connected.

This is an easy consequence of Zariski's connectedness theorem.

Now we shall study the structure of general fibres of algebraic reductions. We begin with the following theorem.

THEOREM 3.2.5. Let $\psi^* : V^* \to W$ be an algebraic reduction of a complex variety V. For any divisor D on V^*, there exists a dense subset U of W such that, for any point $w \in U$, the fibre $V_w^* = \psi^{*-1}(w)$ is non - singular and we have

$$\kappa(D_w , V_w^*) \leqq 0 ,$$

where D_w is the restriction of the divisor D to the fibre V_w^*.

For the proof we use a similar method as was given in the proof of the fundamental theorem (see 1.5). For the detailed discussion, see Ueno [33] , §12.

COROLLARY 3.2.6. For an algebraic reduction $\psi^*: V^* \to W$ of a complex variety V , there exists a dense set U of W such that, for any point $w \in U$, the fibre $V_w = \psi^{*-1}(w)$ is non - singular and $\kappa(V_w^*) \leqq 0$.

By a similar argument as in the proof of Theorem 3.2.7 , we can prove the following:

PROPOSITION 3.2.7. Let $f: V \to W$ be a fibre space of complex manifolds. Suppose that W is a Moishezon manifold and that there exist a divisor D on V and an open set U in W such that, for any $w \in U$, the fibre $V_w = f^{-1}(w)$ is non - singular and $\kappa(D_w , V_w) = m > 0$, where D_w is the restriction of the divisor D to V_w. Then we have

$$a(V) \gneqq \dim W + m.$$

It is not known whether we can choose the dense set U in Corollary 3.2.6 as a Zariski open set or not. But if $a(V) \leqq \dim V - 2$, this is the case and we have the following theorem.

THEOREM 3.2.8. Let $\psi^*: V^* \to W$ be an algebraic reduction of a complex variety V. If $a(V) \leqq \dim V - 2$, then there exists a Zariski open set U of W which satisfies the following properties.

1) If $a(V) = \dim V - 1$, then the fibre $V_w^* = \psi^{*-1}(w)$ is an elliptic curve for any point $w \in U$.

2) If $a(V) = \dim V - 2$, then $\kappa(V_w^*) \leqq 0$ for any point $w \in U$. Moreover, V_w^* is not $\mathbb{P}^2$ for any point $w \in U$.

The proof depends on the fact that the m - genera are deformation invariants for curves and surfaces (see Iitaka [9]).

REMARK 3.2.9. By classifiction of surfaces (see 1.2), a general fibre of the algebraic reduction in Theorem 3.2.8., 2) is one of the following surfaces.

(1) K3 surface. (2) complex torus. (3) hyperelliptic surface. (4) Enriques surface. (5) elliptic surface with a trivial canonical bundle. (6) surface of class VII. (7) rational surface. (8) ruled surface.

It is not known whether a hyperelliptic surface, an Enrigues surface and a rational surface appear as a general fibre of an algebraic reduction.

(3.3) Complex Manifolds of Algebraic Dimension Zero

In what follows, M is always assumed to be a complex manifold of algebraic dimension zero. We consider the Albanese mapping $\alpha: M \to A(M)$ of M.

LEMMA 3.3.1. The Albanese mapping $\alpha: M \to A(M)$ is surjective. Hence $t(M) \leqq \dim M$. Moreover, $a(A(M)) = 0$.

Since we have $a(M) \geqq \kappa(\alpha(M))$, this is a consequence of Corollary 2.2.4.

COROLLARY 3.3.2. $t(M) \neq 1$.

It is easy to show that, if a complex torus A contains a non - zero divisor, then $a(A(M)) \geqq 1$. Therefore, we have the following:

LEMMA 3.3.3. We set

$$S = \{z \in M \quad \alpha \text{ is not of maximal rank at } z\}.$$

Then $\alpha(S)$ is an analytic set of codimension at least two. Using this lemma, we can prove the following important fact:

LEMMA 3.3.4. Any fibre of the Albanese mapping $\alpha: M \to A(M)$ is connected.

COROLLARY 3.3.5. If $t(M) = \dim M$, then the Albanese mapping is a modification.

By Lemma 3.3.4., for a complex manifold M of algebraic dimension zero, the Albanese mapping $\alpha: M \to A(M)$ gives a structure of a fibre space. First we shall study the case where $t(M) = \dim M - 1$.

THEOREM 3.3.6. If $t(M) = \dim M - 1$, there exists an analytic subset T of A(M) of codimension at least two such that $\alpha' = \alpha|_{M'}: M' = M - \alpha^{-1}(T) \to A' = A(M) - T$ is an analytic fibre bundle whose fibre is a non - singular elliptic curve or $\mathbb{P}^1$.

OUTLINE OF THE PROOF By Lemma 3.3.3., there exists an anylytic subset

T of A(M) of codimension at least two such that α is of maximal rank at any point $x \in M' = M - \alpha^{-1}(T)$. Hence, for any point $x \in A' = A(M) - T$, the fibre $M_x = \alpha^{-1}(x)$ is a non - singular curve of genus g. Suppose that $g \geq 1$. Then there exists a holomorphic mapping $f: A' \to \underline{M}_g$ of A' into the moduli space of curves of genus g. Since $\underline{M}_g$ is quasi projective, if $\dim f(A') \geq 1$, then A' has a non - constant meromorphic function induced by the morphism. Then, by Hartogs' theorem, this meromorphic function can be extended to a meromorphic function on A since T is of codimension at least two. Therefore, f(A') must be a point. This implies that, for any point $x \in A'$, the fibre M_x is isomorphic to a fixed curve C and $\alpha': M' \to A'$ is a fibre bundle over A' whose fibre and structure group are the curve C and Aut(S), respectively. Suppose, moreover, that $g \geq 2$. Then Aut(S) is a finite group and the fibre bundle $\alpha': M' \to A'$ can be extended to a fibre bundle $\hat{\alpha}: \hat{M} \to A$ over A. The important fact which we should prove is that M and $\hat{M}$ are bimeromorphically equivalent. Using theory of hyperbolic analysis (see Kobayashi [15]), we can prove this fact. For the detailed discussion, we refer the reader to Ueno [33] , §13. Since Aut(S) is finite, there is an Aut(S) — invariant non - constant meromorphic function on C. This meromorphic function can be considered as a meromorphic function on $\hat{M}$. Hence $a(\hat{M}) \geq 1$. Since M and $\hat{M}$ are bimeromorphically equivalent, this is a contradiction. This proves the theorem.

In the above proof, the fact that the moduli space $\underline{M}_g$ has many non - constant meromorphic functions is essential. This is one of the reasons why we are interested in constructing moduli spaces as algebraic spaces.

To generalize the above argument to the case where $t(M) = \dim M - 2$, we need the theory of moduli of surfaces and the results on hyperbolic analysis. Since the moduli space of algebraic surfaces of general type is an algebraic space (see Theorem 1.2.8.), generalizing the argument given in Kobayashi and Ochiai [16] , we can prove the following theorem.

<u>THEOREM 3.3.7.</u> Suppose that $t(M) = \dim M - 2$. Then there exists an analytic subset T of A(M) of codimension at least two such that $\kappa(M_w) = 0$ for any $w \in A(M) - T$ where $M_w = \alpha^{-1}(w)$. Moreover, M_w is <u>not</u> a ruled surface of genus $g \geq 2$, for any $w \in A(M) - T$.

For the proof, see Ueno [33] , §13.

<u>REMARK 3.3.8.</u> By classification of surfaces, a general fibre

appearing in Theorem 3.3.7 is one of the surfaces given in Remark 3.2.9. The author does not know whether a hyperelliptic surface and an Enriques surface appear as a general fibre.

EXAMPLE 3.2.10. Let T be a two dimensional torus of algebraic dimension zero with period matrix $\Omega = (\alpha_1, \alpha_2, \alpha_3, \alpha_4)$ (for example,

$$\Omega = \begin{pmatrix} 1 & 0 & \sqrt{-2} & \sqrt{-5} \\ 0 & 1 & \sqrt{-3} & \sqrt{-7} \end{pmatrix}).$$

We construct a four - dimensional complex manifold M such that $a(M) = 0$, $t(M) = 2$, $A(M) = T$ and the Albanese mapping of M has the structure of an analytic fibre bundle over T whose fibre is an abelian surface or an elliptic surface with a trivial canonical bundle.

1) Let A be an abelian surface. We choose four points $b_1, b_2, b_3, b_4 \in A$ in general position. Let G be the group of analytic automorphisms of $\mathbb{C}^2 \times A$ generated by

$$g_i : \mathbb{C}^2 \times A \to \mathbb{C}^2 \times A, \quad i = 1, 2, 3, 4,$$

$$(\zeta, z) \to (\zeta + \alpha_i, -z + b_i).$$

The group G acts on $\mathbb{C}^2 \times A$ properly discontinuously and freely. The quotient manifold $M = \mathbb{C}^2 \times A/G$ has the structure of a fibre bundle over T induced by the natural projection $p: \mathbb{C}^2 \times A \to \mathbb{C}^2$. The fibre of this bundle is the abelian variety A. By our construction, it is easy to see that A(M) is isomorphic to T. Since the meromorphic function field $\mathbb{C}(M)$ of M is isomorphic to the invariant subfield of $\mathbb{C}(A)$ by the group of analytic automorphisms of A generated by the automorphisms

$$z \to -z_1 + b_i, \quad i = 1, 2, 3, 4,$$

$\mathbb{C}(M) = \mathbb{C}$ if b_i, $i = 1, 2, 3, 4$, are in general position.

2) An elliptic surface with a trivial canonical bundle is represented by the quotient manifold $S = \mathbb{C}^2/H$ where H is a group of analytic automorphisms of $\mathbb{C}^2$ gnerated by automorphisms

$$g_j : (z_1, z_2) \to (z_1 + \alpha_j, z_2 + \bar{\alpha}_j z_1 + \beta_j), \quad j = 1, 2, 3, 4,$$

such that

$$\alpha_1 = \alpha_2 = 0 ,$$

$$\bar{\alpha}_3\alpha_4 - \bar{\alpha}_4\alpha_3 = m\beta_2 \neq 0 ,$$

where m is a positive integer and $\{\alpha_3 , \alpha_4\}$, $\{\beta_1 , \beta_2\}$ are fundamental periods of elliptic curves. Here we set

$$\alpha_3 = 1 , \alpha_4 = b\sqrt{-1} , \beta_1 = (2b^2 + 2b\sqrt{-1})/(1 + b^2)$$

$$\beta_2 = 2b\sqrt{-1} , \beta_3 = \beta_4 = 0 ,$$

where b is a transcendental number.

Let G be the group of analytic automorphisms of $\mathbb{C}^2 \times S$ generated by

$$g_1 : (\zeta, [z_1 , z_2]) \to (\zeta + \alpha_1 , [z_1 + \alpha , z_2]),$$

$$g_j : (\zeta, [z_1 , z_2]) \to (\zeta + \alpha_j , [-z_1 , z_2]) , j = 2, 3, 4,$$

where

$$\alpha = (2b^2 + 2b\sqrt{-1})/(1 + b^2)$$

and $[z_1 , z_2]$ is a point of S corresponding to a point $(z_1 , z_2) \in \mathbb{C}^2$. The quotient manifold $M = \mathbb{C}^2 \times S/G$ has the structure of a fibre bundle over T induced by the natural projection $p: \mathbb{C}^2 \times S \to \mathbb{C}^2$. The fibre of this bundle is the surface S. It is not difficult to see that the manifold M has the desired properties.

References

[1] Artin, M. Algebraization of formal moduli, II. Existence of modification, Ann. of Math., 91 (1970), 88 - 135.

[2] Baily, W.L. On the moduli of Jacobian varieties, Ann. of Math., 71 (1960), 303 - 314.

[3] Blanchard, A. Sur les variétés analytiques complexes, Ann. Ecole Norm. Sup., 73 (1956), 157 - 202.

[4] Bombieri, E. Canonical models of surfaces of general type, Publ. Math. IHES., 42 (1973), 171 - 219.

[5] Cartan, H. Quotient of complex analytic spaces. International colloquium on fuction theory, Tata Inst. Bombay (1960), 1 - 15.

[6] Hironaka, H. Resolution of singularity of an algebraic variety of characteristic zero I, II, Ann. of Math. (1964), 109 - 326.

[7] ——————. Bimeromorphic smoothing of a complex - analytic space.Preprint, University of Warwick (1970).

[8] ——————.Review of S. Kawai's paper, Math Review, 32, No 11 (1966), 87 - 88.

[9] Iitaka, S. Deformations of compact complex surfaces II, J.Math. Soc. Japan, 22 (1970), 247 - 261.

[10] ——————. On D - dimensions of algebraic varieties, J. Math. Soc. Japan, 23 (1971), 356 - 373.

[11] ——————. Genera and classification of algebraic varieties I (in Japanese), Sugaku, 24 (1972), 14 - 27.

[12] Inoue, M. On surfaces of class VII_0, to appear in Invent. Math. (see also Proc. Japan Academy, 48 (1972), 445 - 446).

[13] Kawai, S. On compact complex analytic manifold of complex dimension 3, I, II, J. Math. Soc. Japan, 17 (1965), 438 - 442, ibid., 21 (1969), 604 - 616.

[14] ——————. Elliptic fibre spaces over compact surfaces, Comment, Math. Univ. St. Paul, 15 (1967), 119 - 138.

[15] Kobayashi, S. Hyperbolic manifolds and holomorphic mappings. Marcel Dekker, INC., New York (1970).

[16] Kobayashi, S. and T. Ochiai. Mapping into compact complex manifolds with negative first chern class, J. Math. Soc. Japan, 23 (1971), 137 - 148.

[17] Kodaira, K. On Kähler varieties of restricted type (an intrinsic characterization of algebraic varieties), Ann. of Math., 60 (1954), 28 - 48.

[18] ———. On compact analytic surfaces, I, II, III, Ann. of Math., 71 (1960),111-152,ibid.,77(1963),563-626,ibid.,78(1963),1-40.

[19] ———. On the structure of compact compelx analytic surfaces I, II, III, IV, Amer. J. Math., 86(1964), 751 - 798, ibid.,88 (1966), 682 - 721, ibid. , 90 (1968), 1048 - 1066.

[20] ———. Pluricanonical systems on algebraic surfaces of general type, J. Math. Soc. Japan, 20 (1968), 170 - 192.

[21] Matsushima, Y. On Hodge manifold with zero first chern class, J. Diff. Geometry, 3 (1969), 477 - 480.

[22] Moishezon, B.G. On n - dimensional compact complex manifold with n algebraically independent meromorphic functions I, II, III, Amer, Math. Soc. Translation, 63 (1967), 51 - 177 (English translation).

[23] Mumford, D. Geometric invariant theory . Springer Verlag, Berlin, Hiedelberg, New York (1965).

[24] Nakamura, I. On classification of parallelizable manifolds and small deformations, to appear.

[25] Nakamura, I. and K. Ueno. On addition formula for Kodaira dimensions of analytic fibre bundles whose fibres are Moishezon manifolds , J. Math. Soc. Japan, 25 (1973), 363 - 391.

[26] Popp, H. On moduli of algebraic varieties II, to appear in Compositio Math., 28.

[27] Remmert, R. Meromorphe Funktionen in kompakten komplexen Räumen, Math. Ann., 132 (1956), 277 - 288.

[28] ———. Holomorphe und meromorphe Abbildungen komplexer Räume, Math. Ann. 133 (1957), 328 - 370.

[29] Šafarevič, I.R. et al. Algebraic surfaces, Proc. Steklov Inst. Moscow (1965).

[30] Serre, J.P. Géometrie algebrique et géometrie analytique, Ann. Inst. Fourier, 6 (1956), 1 - 42.

[31] Ueno, K. On fibre spaces of normally polarized abelian varieties of dimension 2, I, II, III, J. Fac. Sci. Univ. of Tokyo, Sec. IA, 18 (1971), 37 - 95, ibid., 19 (1972), 163 - 199, in preparation.

[32] ———. Classification of algebraic varieties I, II. Compositiv Math., 27 (1973), 277 - 342, in preparation.

[33] ———. Classification of algebraic varieties and compact complex spaces. Lecture Note, to appear.

[34] ———. On the pluricanonical systems of algebraic manifolds of dimension 3, to appear.

[35] Weil, A. Variétés kählériennes, Hermann, Paris (1958).

Adressen der Autoren

W. Barth: Department of Mathematics, University of Leiden, Leiden, The Netherlands

W.D. Geyer: Mathematisches Institut, Universität Erlangen, 852 Erlangen, W-Germany

H. Grauert: Mathematisches Institut, Universität Göttingen, 34 Göttingen, W-Germany

F. Hirzebruch: Mathematisches Institut, Universität Bonn, 53 Bonn, W-Germany

J. Lipman: Department of Mathematics, Purdue University, Lafayette/Indiana, USA

N. Mitani: Faculty of Science, University of Tokyo, Tokyo, Japan

B. Moishezon: Department of Mathematics, University of Tel Aviv, Tel Aviv, Israel

J.P. Murre: Department of Mathematics, University of Leiden, Leiden, The Netherlands

Y. Namikawa: Mathematisches Institut, Universität Mannheim, 68 Mannheim, W-Germany

F. Oort: Department of Mathematics, University of Amsterdam, Amsterdam, The Netherlands

H. Popp: Mathematisches Institut, Universität Mannheim, 68 Mannheim, W-Germany

W. Schmid: Department of Mathematics, Columbia University, New York, USA

T. Shioda: Faculty of Science, University of Tokyo, Tokyo, Japan

K. Ueno: Faculty of Science, University of Tokyo, Tokyo, Japan

A. Van de Ven: Department of Mathematics, University of Leiden, Leiden, The Netherlands

Vol. 310: B. Iversen, Generic Local Structure of the Morphisms in Commutative Algebra. IV, 108 pages. 1973. DM 16,-

Vol. 311: Conference on Commutative Algebra. Edited by J. W. Brewer and E. A. Rutter. VII, 251 pages. 1973. DM 22,-

Vol. 312: Symposium on Ordinary Differential Equations. Edited by W. A. Harris, Jr. and Y. Sibuya. VIII, 204 pages. 1973. DM 22,-

Vol. 313: K. Jörgens and J. Weidmann, Spectral Properties of Hamiltonian Operators. III, 140 pages. 1973. DM 16,-

Vol. 314: M. Deuring, Lectures on the Theory of Algebraic Functions of One Variable. VI, 151 pages. 1973. DM 16,-

Vol. 315: K. Bichteler, Integration Theory (with Special Attention to Vector Measures). VI, 357 pages. 1973. DM 26,-

Vol. 316: Symposium on Non-Well-Posed Problems and Logarithmic Convexity. Edited by R. J. Knops. V, 176 pages. 1973. DM 18,-

Vol. 317: Séminaire Bourbaki – vol. 1971/72. Exposés 400–417. IV, 361 pages. 1973. DM 26,-

Vol. 318: Recent Advances in Topological Dynamics. Edited by A. Beck, VIII, 285 pages. 1973. DM 24,-

Vol. 319: Conference on Group Theory. Edited by R. W. Gatterdam and K. W. Weston. V, 188 pages. 1973. DM 18,-

Vol. 320: Modular Functions of One Variable I. Edited by W. Kuyk. V, 195 pages. 1973. DM 18,-

Vol. 321: Séminaire de Probabilités VII. Edité par P. A. Meyer. VI, 322 pages. 1973. DM 26,-

Vol. 322: Nonlinear Problems in the Physical Sciences and Biology. Edited by I. Stakgold, D. D. Joseph and D. H. Sattinger. VIII, 357 pages. 1973. DM 26,-

Vol. 323: J. L. Lions, Perturbations Singulières dans les Problèmes aux Limites et en Contrôle Optimal. XII, 645 pages. 1973. DM 42,-

Vol. 324: K. Kreith, Oscillation Theory. VI, 109 pages. 1973. DM 16,-

Vol. 325: Ch.-Ch. Chou, La Transformation de Fourier Complexe et L'Equation de Convolution. IX, 137 pages. 1973. DM 16,-

Vol. 326: A. Robert, Elliptic Curves. VIII, 264 pages. 1973. DM 22,-

Vol. 327: E. Matlis, 1-Dimensional Cohen-Macaulay Rings. XII, 157 pages. 1973. DM 18,-

Vol. 328: J. R. Büchi and D. Siefkes, The Monadic Second Order Theory of All Countable Ordinals. VI, 217 pages. 1973. DM 20,-

Vol. 329: W. Trebels, Multipliers for (C, α)-Bounded Fourier Expansions in Banach Spaces and Approximation Theory. VII, 103 pages. 1973. DM 16,-

Vol. 330: Proceedings of the Second Japan-USSR Symposium on Probability Theory. Edited by G. Maruyama and Yu. V. Prokhorov. VI, 550 pages. 1973. DM 36,-

Vol. 331: Summer School on Topological Vector Spaces. Edited by L. Waelbroeck. VI, 226 pages. 1973. DM 20,-

Vol. 332: Séminaire Pierre Lelong (Analyse) Année 1971-1972. V, 131 pages. 1973. DM 16,-

Vol. 333: Numerische, insbesondere approximationstheoretische Behandlung von Funktionalgleichungen. Herausgegeben von R. Ansorge und W. Törnig. VI, 296 Seiten. 1973. DM 24,-

Vol. 334: F. Schweiger, The Metrical Theory of Jacobi-Perron Algorithm. V, 111 pages. 1973. DM 16,-

Vol. 335: H. Huck, R. Roitzsch, U. Simon, W. Vortisch, R. Walden, B. Wegner und W. Wendland, Beweismethoden der Differentialgeometrie im Großen. IX, 159 Seiten. 1973. DM 18,-

Vol. 336: L'Analyse Harmonique dans le Domaine Complexe. Edité par E. J. Akutowicz. VIII, 169 pages. 1973. DM 18,-

Vol. 337: Cambridge Summer School in Mathematical Logic. Edited by A. R. D. Mathias and H. Rogers. IX, 660 pages. 1973. DM 42,-

Vol: 338: J. Lindenstrauss and L. Tzafriri, Classical Banach Spaces. IX, 243 pages. 1973. DM 22,-

Vol. 339: G. Kempf, F. Knudsen, D. Mumford and B. Saint-Donat, Toroidal Embeddings I. VIII, 209 pages. 1973. DM 20,-

Vol. 340: Groupes de Monodromie en Géométrie Algébrique. (SGA 7 II). Par P. Deligne et N. Katz. X, 438 pages. 1973. DM 40,-

Vol. 341: Algebraic K-Theory I, Higher K-Theories. Edited by H. Bass. XV, 335 pages. 1973. DM 26,-

Vol. 342: Algebraic K-Theory II, "Classical" Algebraic K-Theory, and Connections with Arithmetic. Edited by H. Bass. XV, 527 pages. 1973. DM 36,-

Vol. 343: Algebraic K-Theory III, Hermitian K-Theory and Geometric Applications. Edited by H. Bass. XV, 572 pages. 1973. DM 38,-

Vol. 344: A. S. Troelstra (Editor), Metamathematical Investigation of Intuitionistic Arithmetic and Analysis. XVII, 485 pages. 1973. DM 34,-

Vol. 345: Proceedings of a Conference on Operator Theory. Edited by P. A. Fillmore. VI, 228 pages. 1973. DM 20,-

Vol. 346: Fučík et al., Spectral Analysis of Nonlinear Operators. II, 287 pages. 1973. DM 26,-

Vol. 347: J. M. Boardman and R. M. Vogt, Homotopy Invariant Algebraic Structures on Topological Spaces. X, 257 pages. 1973. DM 22,-

Vol. 348: A. M. Mathai and R. K. Saxena, Generalized Hypergeometric Functions with Applications in Statistics and Physical Sciences. VII, 314 pages. 1973. DM 26,-

Vol. 349: Modular Functions of One Variable II. Edited by W. Kuyk and P. Deligne. V, 598 pages. 1973. DM 38,-

Vol. 350: Modular Functions of One Variable III. Edited by W. Kuyk and J.-P. Serre. V, 350 pages. 1973. DM 26,-

Vol. 351: H. Tachikawa, Quasi-Frobenius Rings and Generalizations. XI, 172 pages. 1973. DM 18,-

Vol. 352: J. D. Fay, Theta Functions on Riemann Surfaces. V, 137 pages. 1973. DM 16,-

Vol. 353: Proceedings of the Conference on Orders, Group Rings and Related Topics. Organized by J. S. Hsia, M. L. Madan and T. G. Ralley. X, 224 pages. 1973. DM 20,-

Vol. 354: K. J. Devlin, Aspects of Constructibility. XII, 240 pages. 1973. DM 22,-

Vol. 355: M. Sion, A Theory of Semigroup Valued Measures. V, 140 pages. 1973. DM 16,-

Vol. 356: W. L. J. van der Kallen, Infinitesimally Central-Extensions of Chevalley Groups. VII, 147 pages. 1973. DM 16,-

Vol. 357: W. Borho, P. Gabriel und R. Rentschler, Primideale in Einhüllenden auflösbarer Lie-Algebren. V, 182 Seiten. 1973. DM 18,-

Vol. 358: F. L. Williams, Tensor Products of Principal Series Representations. VI, 132 pages. 1973. DM 16,-

Vol. 359: U. Stammbach, Homology in Group Theory. VIII, 183 pages. 1973. DM 18,-

Vol. 360: W. J. Padgett and R. L. Taylor, Laws of Large Numbers for Normed Linear Spaces and Certain Fréchet Spaces. VI, 111 pages. 1973. DM 16,-

Vol. 361: J. W. Schutz, Foundations of Special Relativity: Kinematic Axioms for Minkowski Space Time. XX, 314 pages. 1973. DM 26,-

Vol. 362: Proceedings of the Conference on Numerical Solution of Ordinary Differential Equations. Edited by D. Bettis. VIII, 490 pages. 1974. DM 34,-

Vol. 363: Conference on the Numerical Solution of Differential Equations. Edited by G. A. Watson. IX, 221 pages. 1974. DM 20,-

Vol. 364: Proceedings on Infinite Dimensional Holomorphy. Edited by T. L. Hayden and T. J. Suffridge. VII, 212 pages. 1974. DM 20,-

Vol. 365: R. P. Gilbert, Constructive Methods for Elliptic Equations. VII, 397 pages. 1974. DM 26,-

Vol. 366: R. Steinberg, Conjugacy Classes in Algebraic Groups (Notes by V. V. Deodhar). VI, 159 pages. 1974. DM 18,-

Vol. 367: K. Langmann und W. Lütkebohmert, Cousinverteilungen und Fortsetzungssätze. VI, 151 Seiten. 1974. DM 16,-

Vol. 368: R. J. Milgram, Unstable Homotopy from the Stable Point of View. V, 109 pages. 1974. DM 16,-

Vol. 369: Victoria Symposium on Nonstandard Analysis. Edited by A. Hurd and P. Loeb. XVIII, 339 pages. 1974. DM 26,-

Vol. 370: B. Mazur and W. Messing, Universal Extensions and One Dimensional Crystalline Cohomology. VII, 134 pages. 1974. DM 16,-